Introduction to Stochastic Control Theory

Karl J. Åström

Department of Automatic Control
Lund Institute of Technology, Lund, Sweden

Dover Publications, Inc.
Mineola, New York

Copyright

Copyright © 1970 by Karl J. Åström
All rights reserved.

Bibliographical Note

This Dover edition, first published in 2006, is an unabridged republication of the work first published by Academic Press, Inc., New York, in 1970 as Volume 70 in the Mathematics in Science and Engineering series.

Library of Congress Cataloging-in-Publication Data

Åström, Karl J. (Karl Johan), 1934–
  Introduction to stochastic control theory / Karl J. Åström.
    p. cm.
  Originally published: New York : Academic Press, 1970, in series: Mathematics in science and engineering series, v. 70.
  Includes index.
  ISBN-13: 978-0-486-44531-1
  ISBN-10: 0-486-44531-3
  1. Stochastic control theory. I. Title.

QA402.3.A18 2006
629.8'312—dc22

2005052028

Manufactured in the United States by Courier Corporation
44531303   2015
www.doverpublications.com

To Birgit

Introduction to Stochastic Control Theory

TABLE OF CONTENTS

Preface	ix
Acknowledgments	xi

Chapter 1 STOCHASTIC CONTROL

1.	Introduction	1
2.	Theory of Feedback Control	1
3.	How to Characterize Disturbances	5
4.	Stochastic Control Theory	6
5.	Outline of the Contents of the Book	9
6.	Bibliography and Comments	11

Chapter 2 STOCHASTIC PROCESSES

1.	Introduction	13
2.	The Concept of a Stochastic Process	13
3.	Some Special Stochastic Processes	16
4.	The Covariance Function	23
5.	The Concept of Spectral Density	26
6.	Analysis of Stochastic Processes	33
7.	Bibliography and Comments	42

Chapter 3 STOCHASTIC STATE MODELS

1.	Introduction	44
2.	Discrete Time Systems	45
3.	Solution of Stochastic Difference Equations	47
4.	Continuous Time Systems	51
5.	Stochastic Integrals	57
6.	Linear Stochastic Differential Equations	63
7.	Nonlinear Stochastic Differential Equations	71
8.	Stochastic Calculus—The Ito Differentiation Rule	74
9.	Modeling of Physical Processes by Stochastic Differential Equations	78
10.	Sampling a Stochastic Differential Equation	82
11.	Bibliography and Comments	86

Chapter 4 ANALYSIS OF DYNAMICAL SYSTEMS WHOSE INPUTS ARE STOCHASTIC PROCESSES

1.	Introduction	91

2.	Discrete Time Systems	92
3.	Spectral Factorization of Discrete Time Processes	98
4.	Analysis of Continuous Time Systems Whose Input Signals Are Stochastic Processes	104
5.	Spectral Factorization of Continuous Time Processes	107
6.	Bibliography and Comments	112

Chapter 5 PARAMETRIC OPTIMIZATION

1.	Introduction	115
2.	Evaluation of Loss Functions for Discrete Time Systems	116
3.	Evaluation of Loss Functions for Continuous Time Systems	128
4.	Reconstruction of State Variables for Discrete Time Systems	142
5.	Reconstruction of State Variables for Continuous Time Systems	150
6.	Bibliography and Comments	156

Chapter 6 MINIMAL VARIANCE CONTROL STRATEGIES

1.	Introduction	159
2.	A Simple Example	160
3.	Optimal Prediction of Discrete Time Stationary Processes	162
4.	Minimal Variance Control Strategies	172
5.	Sensitivity of the Optimal System	181
6.	An Industrial Application	188
7.	Bibliography and Comments	209

Chapter 7 PREDICTION AND FILTERING THEORY

1.	Introduction	210
2.	Formulation of Prediction and Estimation Problems	211
3.	Preliminaries	218
4.	State Estimation for Discrete Time Systems	225
5.	Duality	238
6.	State Estimation for Continuous Time Processes	241
7.	Bibliography and Comments	252

Chapter 8 LINEAR STOCHASTIC CONTROL THEORY

1.	Introduction	256
2.	Formulation	257
3.	Preliminaries	259
4.	Complete State Information	264
5.	Incomplete State Information 1	269
6.	Incomplete State Information 2	278
7.	Continuous Time Problems	286
8.	Bibliography and Comments	292

Index	295

PREFACE

The object of this book is to present stochastic control theory — analysis, parametric optimization and optimal stochastic control. The treatment is limited to linear systems with quadratic criteria. It covers discrete time as well as continuous time systems.

The first three chapters provide motivation and background material on stochastic processes. Chapter 4 is devoted to analysis of dynamical systems whose inputs are stochastic processes. A simple version of the problem of optimal control of stochastic systems is discussed in Chapter 6; this chapter also contains an example of an industrial application of this theory. Filtering and prediction theory are covered in Chapter 7, and the general stochastic control problem for linear systems with quadratic criteria is treated in Chapter 8.

In each chapter we shall first discuss the discrete time version of a problem. We shall then turn to the continuous time version of the same problem. The continuous time problems are more difficult both analytically and conceptually. Chapter 6 is an exception because it deals only with discrete time systems.

There are several different uses for this volume:

- A short applications-oriented course covering Chapter 1, a survey of Chapter 2, Sections 1, 2 and 3 of Chapters 3 and 4, Sections 1 and 4 of Chapter 5, Chapter 6, and a survey of Chapters 7 and 8.

- An introductory course in discrete time stochastic control covering the sections mentioned above, and also Section 2 of Chapter 5, Sections 1-5 of Chapter 7, and Sections 1-6 of Chapter 8.

- A course in stochastic control including both discrete time processes as well as continuous time processes covering the whole volume.

The prerequisites for using this book are a course in analysis, one in probability theory (preferably but not necessarily covering the elements of stochastic processes), and a course in dynamical systems which includes frequency response as well as the state space approach for continuous time and discrete time systems. A reader who is acquainted with the deterministic theory of optimal control for linear systems with quadratic criteria will get a much richer understanding of the problems discussed, although this knowledge is not absolutely required in order to read this book.

This work is an expansion of the notes from lectures delivered to various industrial and academic audiences between 1962–1969. A preliminary version was given in seminars in 1963 at the IBM Research Laboratories in San Jose, California and Yorktown Heights, New York. An expanded version was presented during 1964 and 1965 at the IBM Nordic Laboratory, the Royal Institute of Technology, and the Research Institute of National Defense, all located in Stockholm, Sweden. Part of this material has been used in graduate courses at the Lund Institute of Technology, Sweden, since 1965. The complete manuscript was presented as a graduate course in stochastic control at the Lund Institute of Technology during the 1968–1969 academic year.

ACKNOWLEDGMENTS

I would like to acknowledge my gratitude to several persons and institutions. First, I would like to thank Professor Richard Bellman who persuaded me to write this book. He has also provided constructive criticism and support, as well as a stimulating intellectual environment for the final version of the manuscript. Professor U. Grenander's seminars on stochastic processes at the University of Stockholm have been an everlasting source of inspiration. I have profited from stimulating discussions with Professor L.E. Zachrisson at the Royal Institute of Technology, Mr. T. Bohlin of the IBM Nordic Laboratory, Mr. S. Jahnberg of the Research Institute of National Defense, all located in Sweden, and Mr. R.W. Koepcke of IBM Research, San Jose California. I am also very grateful to IBM and Billerud AB who provided ample opportunities to put some of the theoretical results into practice and to the Swedish Board of Technical Development who has supported my research in the field since 1966. This book has also gained considerably from criticism by several of my students. I am particularly grateful to P. Hagander, K. Mårtensson, J. Wieslander, and B. Wittenmark. The final manuscript, as well as several preliminary versions, has been expertly typed by Miss L. Jönsson. The figures have been prepared by Mrs. B. Tell.

Introduction to Stochastic Control Theory

CHAPTER 1

STOCHASTIC CONTROL

1. INTRODUCTION

This introductory chapter will try to put stochastic control theory into a proper context. The development of control theory is briefly discussed in Section 2. Particular emphasis is given to a discussion of deterministic control theory. The main limitation of this theory is that it does not provide a proper distinction between open loop systems and closed loop systems. This is mainly due to the fact that disturbances are largely neglected in the framework of deterministic control theory. The difficulties of characterizing disturbances are discussed in Section 3. An outline of the development of stochastic control theory and the most important results are given in Section 4. Section 5 is devoted to a presentation of the contents of the different chapters of the book.

2. THEORY OF FEEDBACK CONTROL

Control theory was originally developed in order to obtain tools for analysis and synthesis of control systems. The early development was concerned with centrifugal governors, simple regulating devices for industrial processes, electronic amplifiers, and fire control systems. As the theory developed, it turned out that the tools could be applied to a large variety of different systems, technical as well as nontechnical. Results from various branches of applied mathematics have been exploited throughout the development of control theory. The control problems have also given rise to new results in applied mathematics.

In the early development there was a strong emphasis on stability theory based on results like the Routh-Hurwitz theorem. This theorem is a good example of interaction between theory and practice. The stability problem was actually suggested to Hurwitz by Stodola who had found the problem in connection with practical design of regulators for steam turbines.

The analysis of feedback amplifiers used tools from the theory of analytical functions and resulted, among other things, in the celebrated Nyquist criterion.

During the postwar development, control engineers were faced with several problems which required very stringent performance. Many of the control processes which were investigated were also very complex. This led to a new formulation of the synthesis problem as an optimization problem, and made it possible to use the tools of calculus of variations as well as to improve these tools. The result of this development has been the theory of optimal control of deterministic processes. This theory in combination with digital computers has proven to be a very successful design tool. When using the theory of optimal control, it frequently happens that the problem of stability will be of less interest because it is true, under fairly general conditions, that the optimal systems are stable.

The theory of optimal control of deterministic processes has the following characteristic features:

- There is no difference between a control program (an open loop system) and a feedback control (a closed loop system).
- The optimal feedback is simply a function which maps the state space into the space of control variables. Hence there are no dynamics in the optimal feedback.
- The information available to compute the actual value of the control signal is never introduced explicitly when formulating and solving the problem.

We can illustrate these properties by a simple example.

EXAMPLE 2.1

Consider the system

$$\frac{dx}{dt} = u \tag{2.1}$$

with initial conditions

$$x(0) = 1 \tag{2.2}$$

Suppose that it is desirable to control the system in such a way that the performance of the system judged by the criterion

$$J = \int_0^\infty [x^2(t) + u^2(t)] \, dt \tag{2.3}$$

is as small as possible. It is easy to see that the minimal value of the criterion (2.3) is $J = 1$ and that this value is assumed for the control program

$$u(t) = -e^{-t} \qquad (2.4)$$

as well as for the control strategy

$$u(t) = -x(t) \qquad (2.5)$$

Equation (2.4) represents an open loop control because the value of the control signal is determined from a priori data only, irrespective of how the process develops. Equation (2.5) represents a feedback law because the value of the control signal at time t depends on the state of the process at time t.

The example thus illustrates that the open loop system (2.4) and the closed loop system (2.5) are equivalent in the sense that they will give the same value to the loss function (2.3). The stability properties are, however, widely different. The system (2.1) with the feedback control (2.5) is asymptotically stable while the system (2.1) with the control program (2.4) only is stable. In practice, the feedback control (2.5) and the open loop control (2.4) will thus be widely different. This can be seen, e.g., by introducing disturbances or by assuming that the controls are calculated from a model whose coefficients are slightly in error.

Several of the features of deterministic control theory mentioned above are highly undesirable in a theory which is intended to be applicable to feedback control. When the deterministic theory of optimal control was introduced, the old-timers of the field particularly reacted to the fact that the theory showed no difference between open loop and closed loop systems and to the fact that there were no dynamics in the feedback loop. For example, it was not possible to get a strategy which corresponded to the well-known PI-regulator which was widely used in industry. This is one reason for the widely publicized discussion about the gap between theory and practice in control. The limitations of the deterministic control theory were clearly understood by many workers in the field from the very start, and this understanding is now widespread. The heart of the matter is that no realistic models for disturbances are used in deterministic control theory. If a so-called disturbance is introduced, it is always postulated that the disturbance is a function which is known a priori. When this is the case and the system is governed by a differential equation with unique solutions, it is clear that the knowledge of initial conditions is equivalent to the knowledge of the state of the system at an arbitrary instant of time. This explains why there are no differences in performance between an open loop system and a closed loop system, and why the assumption of a given initial condition implicitly involves that the actual value of the state is known at all times. Also when the state of the system is known, the optimal feedback will always be a function which maps the state space into the space of control variables. As will be seen later, the dynamics of the feedback arise when the state is not known but must be reconstructed from measurements of output signals.

The importance of taking disturbances into account has been known by practitioners of the field from the beginning of the development of control theory. Many of the classical methods for synthesis were also capable of dealing with disturbances in an heuristic manner. Compare the following quotation from A. C. Hall*:

> I well remember an instance in which M. I. T. and Sperry were co-operating on a control for an air-borne radar, one of the first such systems to be developed. Two of us had worked all day in the Garden City Laboratories on Sunday, December 7, 1941, and consequently did not learn of the attack on Pearl Harbor until late in the evening. It had been a discouraging day for us because while we had designed a fine experimental system for test, we had missed completely the importance of noise with the result that the system's performance was characterized by large amounts of jitter and was entirely unsatisfactory. In attempting to find an answer to the problem we were led to make use of frequency-response techniques. Within three months we had a modified control system that was stable, had a satisfactory transient response, and an order of magnitude less jitter. For me this experience was responsible for establishing a high level of confidence in the frequency-response techniques.

Exercises

1. Consider the problem of Example 2.1. Show that the control signal (2.4) and the control law (2.5) are optimal. Hint: First prove the identity

$$\int_0^T [x^2(t) + \dot{x}^2(t)]\, dt = x^2(0) - x^2(T) + \int_0^T [x(t) + \dot{x}(t)]^2\, dt .$$

2. Consider the problem of Example 2.1. Assume that the optimal control signal and the optimal control law are determined from the model

$$\frac{dx}{dt} = au$$

where a has a value close to 1 when the system is actually governed by Eq. (2.1). Determine the value of the criterion (2.3) for the systems obtained with open loop control and with closed loop control.

3. Compare the performance of the open loop control (2.4) and the closed loop control (2.5) when the system of Example 2.1 is actually governed by the equation

---

* A. C. Hall, in *Frequency Response* (R. Oldenburger, ed.), p. 4. Macmillan, New York, 1956.

How to Characterize Disturbances

$$\frac{dx}{dt} = u + v$$

where v is an unknown disturbance. In particular let v be an unknown constant.

3. HOW TO CHARACTERIZE DISTURBANCES

Having realized the necessity of introducing more realistic models of disturbances, we are faced with the problem of finding suitable ways to characterize them. A characteristic feature of practical disturbances is the impossibility of predicting their future values precisely. A moment's reflection indicates that it is not easy to devise mathematical models which have this property. It is not possible, for example, to model a disturbance by an analytical function because, if the values of an analytical function are known in an arbitrarily short interval, the values of the function for other arguments can be determined by analytic continuation.

Since analytic functions do not work, we could try to use statistical concepts to model disturbances. As can be seen from the early literature on statistical time series, this is not easy. For example, if we try to model a disturbance as

$$x(t) = \sum_{i=1}^{n} a_i(t)\xi_i \tag{3.1}$$

where $a_1(t), a_2(t), \ldots, a_n(t)$ are known functions and ξ_i is a random variable, we find that if the linear equations

$$\begin{aligned} x(t_1) &= a_1(t_1)\xi_1 + a_2(t_1)\xi_2 + \cdots + a_n(t_1)\xi_n \\ x(t_2) &= a_1(t_2)\xi_1 + a_2(t_2)\xi_2 + \cdots + a_n(t_2)\xi_n \\ &\vdots \\ x(t_n) &= a_1(t_n)\xi_1 + a_2(t_n)\xi_2 + \cdots + a_n(t_n)\xi_n \end{aligned} \tag{3.2}$$

have a solution then the particular realizations of the stochastic variables $\xi_1, \xi_2, \ldots, \xi_n$ can be determined exactly from observations of $x(t_1)$, $x(t_2)$, $\ldots, x(t_n)$ and the future values of x can then be determined exactly. The disturbance described by (3.1) is therefore called a *completely deterministic stochastic process* or a *singular stochastic process*.

A more successful attempt is to model a disturbance as a sequence of random variables. A simple example is given by the *autoregressive process* $\{x(t)\}$ given by

$$x(t+1) = ax(t) + e(t), \quad t = t_0, t_0 + 1, \ldots \tag{3.3}$$

where $x(t_0) = 1$, $|a| < 1$ and $\{e(t), t = t_0, t_0 + 1, \ldots\}$ is a sequence of independent normal $(0, \sigma)$ stochastic variables. It is also assumed that $e(t)$ is independent of $x(t)$ for all t. Assume, for example, that we want to predict the value of $x(t + 1)$ based on observations of $x(t)$. It seems reasonable to predict $x(t + 1)$ by $ax(t)$. The prediction error is then equal to $e(t)$, that is, a stochastic variable with zero mean and variance σ^2.

It turns out that one answer to the problem of modeling disturbances is to describe them as stochastic processes. The theory of stochastic processes has actually partly grown out of attempts to model the fluctuations observed in physical systems. The theory matured very quickly due to contributions from such intellectual giants as Cramér, Khintchine, Kolmogorov, and Wiener.

Problems of prediction are of central importance in the theory of stochastic processes. As will be seen in the following, they are also closely related to the problems of control.

Exercises

1. Consider a disturbance which is characterized by

$$x(t) = a \cos t$$

   where a is a stochastic variable. Give a procedure for predicting future values of x exactly.
2. Consider a disturbance characterized by (3.3). Show that the predictor $\hat{x}(t + 1) = ax(t)$ is optimal in the sense that it minimizes the least squares prediction error defined by $E[x(t + 1) - \hat{x}(t + 1)]^2$.

4. STOCHASTIC CONTROL THEORY

This section will discuss the main problems and results of stochastic control theory. It will also give a brief account of the development of the theory.

Stochastic control theory deals with dynamical systems, described by difference or differential equations, and subject to disturbances which are characterized as stochastic processes. The theory aims at answering problems of analysis and synthesis.

- Analysis—What are the statistical properties of the system variables?
- Parametric Optimization—Suppose that we are given a system and a regulator with a given structure but with unknown parameters. How are the parameters to be adjusted in order to optimize the system with respect to a given criterion?
- Stochastic Optimal Control—Given a system and a criterion, find the control law which minimizes the criterion.

The tools required to solve all these problems are fairly recent developments. Stochastic control theory was used at M. I. T. during the Second World War to synthesize fire control systems. An interesting example, design of a tracking radar, using parametric optimization is described in a book by James, Nichols, and Phillips.*

The filtering and prediction theory developed by Wiener and Kolmogorov is one of the cornerstones in stochastic control theory. This theory makes it possible to extract a signal from observations of signal and disturbances. It plays a very important role in the solution of the stochastic optimal control problem. The Wiener-Kolmogorov theory has, however, not been applied extensively. One reason for this is that it requires the solution of an integral equation (the Wiener-Hopf equation). In realistic problems the Wiener-Hopf equation seldom has analytical solutions, and it is not easy to solve the equation numerically.

The use of digital computers both for analysis and synthesis has profoundly influenced the development of the theory. A significant contribution to the filtering problem was given by Kalman and Bucy. Their results made it possible to solve prediction and filtering problems recursively. This is ideally suited for digital computers. The results of Kalman and Bucy also generalize to nonstationary processes. Using the Kalman-Bucy theory, the predictor is given as the output of a linear dynamical system driven by the observations. To determine the coefficients of the dynamical system, it is necessary to solve an initial value problem for a Riccati equation. The Riccati equation is similar to the one encountered in the theory of optimal control of linear deterministic systems with quadratic criteria. The prediction problem and the linear quadratic control problem are in fact mathematical duals. This result is of great interest both from theoretical and practical points of view. If one of the problems is solved we can easily get the solution of the other by invoking the duality. Also the same computer programs can be used to solve both the filtering and the deterministic control problem.

The solution of the stochastic optimal control problem relies heavily upon the concepts and techniques of dynamic programming. For linear systems with quadratic criteria, the solution is given by the so-called *separation theorem*. This result implies that the optimal strategy can be thought of as composed of two parts. See Fig. 1.1. One part is an optimal filter which computes an estimate of the state in terms of the conditional mean given the observed output signals. The other part is a linear feedback from the estimated state to the control signal.

It turns out that the linear feedback is the same as would be obtained if there were no disturbances and if the state of the system could be measured exactly. The linear feedback can be determined by solving a

* See Bibliography and Comments.

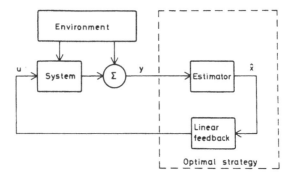

Fig. 1.1. Block diagram which illustrates the separation theorem. The control variable is denoted by u, the output by y, and the state by x.

deterministic control problem. The conditional mean of the state is obtained as the output of a Kalman filter which is essentially a mathematical model of the system driven by the observations. The filter depends on the disturbances and on the system dynamics, but it is independent of the criterion.

The separation theorem thus provides a connection between filtering theory and the theory of optimal stochastic control. A version of the separation theorem was first published by Joseph and Tou.* A related result has been known in econometrics under the name of *certainty equivalence principle*.

The optimal strategy obtained when solving the stochastic optimal control problem for linear systems with quadratic criteria thus consists of a linear dynamical system possibly with time varying parameters. This class of strategies includes those which have been used in practice for many years, usually arrived at by ad hoc methods. Since there are no difficulties in dealing with many inputs and many outputs, the linear stochastic conrol theory is a very important design tool. The result of the theory is a "closed form" solution in the sense that the parameters of the optimal strategy are given in terms of solutions of initial value problems for Riccati equations. Efficient numerical algorithms for solving such equations are available. Occasionally the problems might be numerically ill conditioned.

The linear stochastic control theory is one of the simplest structures available that includes several of the features that are desirable in a theory of feedback control systems. For example:

- The theory shows directly that there are considerable differences between open loop and closed loop systems.

* See Bibliography and Comments.

- The performance of the system depends critically on the information available at the time the value of the control signal should be determined. For example, a delay of the measured signal will lead directly to a deterioration of the performance.
- The optimal feedback consists of a linear dynamical system.

5. OUTLINE OF THE CONTENTS OF THE BOOK

Having discussed some features of stochastic control theory heuristically, it is now meaningful to outline the contents of the book. The purpose of the book is to present theory for analysis, parametric optimization, and optimal control of stochastic control processes. The results presented are limited to linear systems, however, both discrete time systems and continuous time systems are covered. The discrete time case is conceptually and analytically much simpler than the continuous time case.

It should be emphasized that in practical applications where digital computers are used to implement the control strategies, the discrete time case is sufficient. The results for the discrete time case are complete in the sense that all results are shown in full detail.

Chapter 2 gives a survey of some results and concepts from the theory of stochastic processes. Specific stochastic processes, such as stationary processes, Markov processes, processes of second order, and processes with independent increments, are discussed. Covariance functions and spectral densities are introduced. Special attention is given to the concept of white noise. This is the first case where the differences between continuous time and discrete time stochastic processes show up. For discrete time processes, white noise is simply a sequence of independent, equally distributed, random variables. The continuous time white noise is a process which is considerably more involved. It has, for example, infinite variance.

A machinery which will enable us to perform analysis of continuous time processes, e.g., differentiation and integration, is also given. The key to this is the definition of convergence. It turns out that the result will depend on the chosen topology. The practical significance of the different convergence concepts are discussed. For simplicity, the analysis is carried out for convergence in the mean square only. Chapter 2 is kept quite brief. The newcomer to the field of stochastic processes should be prepared to spend some effort in reading the references. Readers who are already familiar with the theory of stochastic processes could scan this chapter quickly.

So-called stochastic state models are developed in Chapter 3. This provides another good illustration of the differences between discrete time and continuous time processes. Discrete time processes are handled in 8 pages while continuous time processes require 45 pages. The purpose of the chapter is to develop the concept of state for stochastic processes. For deterministic systems the state is, roughly speaking, the minimal amount of information about the past history of a system which is required in order to predict its future motion. For a stochastic system it turns out that it is not possible to predict the future motion exactly. Instead the state will be the minimal amount of information which is required in order to predict the future *probability distribution of the state*. It turns out that state models for discrete time systems can be characterized as difference equations driven by discrete time white noise, i.e., a sequence of independent, equally distributed random variables. The analysis of the linear stochastic difference equation is discussed in detail. For continuous time systems, the situation is considerably more involved. To obtain a state model for continuous time processes, we are led to the concept of stochastic differential equations. These equations are explained intuitively and the machinery required to handle the equations is reviewed. Several key theorems are, however, not proven. An alternative to this analysis would be a formal analysis based on manipulation of delta functions, but this was considered less satisfactory because the formal manipulations can give incorrect results.

While Chapters 2 and 3 can be considered as introductory, the main theme is taken up in Chapter 4. This chapter presents the key theorems which are required in order to analyze dynamical systems whose inputs are stochastic processes. Discrete time systems characterized by input-output relations, such as weighting functions and transfer functions, with second order stochastic processes as inputs are discussed. The results show that it is possible to generate stochastic processes with covariance functions belonging to a large class simply by sending white noise through a linear system. These so-called representation theorems will make it possible to considerably simplify the results, since a large class of problems can be reduced to the analysis of linear systems with white noise inputs. The analogous results for continuous time systems are also given. Chapter 4 thus provides us with the tools required to analyse control systems subject to stochastic disturbances.

Evaluation of quadratic functions of state variables for linear systems is the subject of Chapter 5. By using results from the theory of analytical functions, we derive recursive formulas for the evaluation of quadratic loss functions. The results also exhibit interesting relations between stability analysis and the evaluation of quadratic loss functions. As an illustration of parametric optimization of time dependent processes, we also discuss the problem of reconstructing the state variables of a dynamical system using a mathematical model. The optimal gain for a reconstructor with a given structure is discussed. Later in Chapter 7, it will be shown that the chosen structure is actually optimal. The result will in fact be the Kalman filter reconstructing the state.

Chapter 6 is devoted to a particularly simple class of stochastic control problems, namely linear systems with one input and one output where the criterion is to minimize the mean square deviations of the output in steady state. This particular problem gives a good insight into the structure of the optimal solutions because the separation theorem can be proven with very little mathematical sophistication. The solution clearly shows the relationships between optimal filtering and optimal control. As a side result, we also get a new algorithm for solving the filtering problem for processes with rational spectral densities. An industrial application of the theory is also given in this chapter.

Prediction and filtering theory are covered in Chapter 7. The problem formulation and the general properties of the solutions are discussed. The necessary properties of Gaussian processes are developed and Kalman's recursive formulas are derived. Geometric interpretations of the results, as well as analysis of the properties of the prediction errors, are given. The duality between optimal prediction and control is also proven.

The general linear quadratic control problem is discussed in Chapter 8. For discrete time systems the separation theorem is proved, using two different approaches. One direct method uses dynamic programming and requires the state estimator explicitly. An indirect proof which exploits an identity from the calculus of variations is also given. The second proof will make it possible to give a physical interpretation to the different terms of the minimal value of the expected loss. It also clearly shows the importance of specifying the information which is available for calculating the control signal at each particular time. The continuous time version of the separation theorem is proven using the indirect method only.

6. BIBLIOGRAPHY AND COMMENTS

The book

Bellman, R. and Kalaba, R. (eds.), *Mathematical Trends in Control Theory*, Dover New York, 1963.

contains reproduction of many early papers and gives a good survey of the initial efforts in control theory.

A survey of classical as well as modern control theory is found in

Aström, K. J., *Reglerteori*, Almqvist and Wiksell, Uppsala, 1968 (in Swedish).

Introductory treatments of deterministic control theory are found in

Athans, M. and Falb, P., *Optimal Control*, McGraw-Hill, New York, 1966.
Bellman, R., *Introduction to the Mathematical Theory of Control Processes*, Vol. 1, Academic Press, New York, 1967.

Bryson, A. E. Jr. and Ho, Yu-Chi., *Applied Optimal Control*, Blaisdell Waltham, Massachusetts, 1969.

More sophisticated analysis is presented in

Markus, L. and Lee, E. B., *Foundations of the Theory of Optimal Control*, Wiley, New York, 1967.

Pontryagin, L. S., Boltyanskii, V. G., Gamkrelidze, R. V. and Mishchenko, E. F., *The Mathematical Theory of Optimal Processes*, Wiley, New York, 1962.

Stochastic control problems are discussed in

Bellman, R., *Dynamic Programming*, Princeton Univ. Press, Princeton, New Jersey, 1957.

Bellman, R., *Adaptive Control Processes: A Guided Tour*, Princeton Univ. Press, Princeton, New Jersey, 1961.

Aoki, M., *Optimization of Stochastic Systems*, Academic Press, New York, 1967.

An early example of application of stochastic control theory is found in

James, H. M., Nichols, N. B., and Phillips, R. S., *Theory of Servomechanisms*, McGraw-Hill, New York, 1947.

The paper

Yule, G. U., "On a Method of Investigating Periodicities in Disturbed Series with Special Reference to Walter's Numbers," *Phil. Trans. Roy. Soc.*, **A226**, 267-298 (1927).

and the books

Wax, N. (ed.), *Collected Papers on Noise and Stochastic Processes*, Dover, New York, 1954.

Wold, H., *Stationary Time Series*, Almqvist and Wiksell, Uppsala, 1938.

give a good feeling for the initial difficulties in modeling disturbances.

Wiener's filtering theory is found in

Wiener, N., *The Extrapolation, Interpolation, and Smoothing of Stationary Time Series with Engineering Applications*, Wiley, New York, 1949.

The Kalman-Bucy theory is found in

Kalman, R. E., "A New Approach to Linear Filtering and Prediction Problems," *ASME J. Basic Eng.* **82**, 34-45 (1960).

Kalman, R. E. and Bucy, R. S., "New Results in Linear Filtering and Prediction Theory," *ASME J. Basic Eng.* **83**, 95-107 (1961).

The separation theorem was published in

Joseph, P. D. and Tou, J. T., "On Linear Control Theory," *Trans. AIEE* (Applications and Industry), **80**, 193-196 (1961).

The certainty equivalence principle appears in

Simon, H. A., "Dynamic Programming under Uncertainty with a Quadratic Criterion Function," *Econometrica*, **24**, 74-81 (1956).

Theil, H., "A Note on Certainty Equivalence in Dynamic Planning," *Econometrica*, **25**, 346-349 (1957).

CHAPTER 2

STOCHASTIC PROCESSES

1. INTRODUCTION

This chapter will present the elements of the theory of stochastic processes which are needed in the future chapters. The presentation is kept very brief and standard textbooks are referred to for more detailed presentations. The newcomer to the field should observe that no one can be competent in a field as complicated as the theory of stochastic processes with a single reading of a short chapter of this type. The reason for including it, in spite of this, is that the results are needed for easy reference in other chapters of the book. The concept of a stochastic process is covered in Section 2. A few examples of specific stochastic processes such as normal processes, Markov processes, second order processes, and processes with independent increments are discussed in Section 3. The properties of the covariance function are discussed in Section 4. In Section 5 we introduce the notion of spectral density. Particular attention is given to the concept of white noise for discrete time and continuous time processes. The machinery required to analyze continuous time stochastic processes is developed in Section 6.

2. THE CONCEPT OF A STOCHASTIC PROCESS

A stochastic process (random process, or random function) can be defined as a family of random variables $\{x(t), t \in T\}$. The variables are indexed with the parameter t which belongs to the set T, the *index set*, or the parameter set. In our applications we shall often interpret t as time and we will consider two different index sets. When $T = \{\ldots, -1, 0, 1, \ldots\}$

or $T = \{0, 1, 2, \ldots\}$ the stochastic process is referred to as a *discrete parameter process* or a *discrete time process*. When $T = \{t; 0 \leqslant t < \infty\}$ or $T = \{t; -\infty < t < \infty\}$, the process is referred to as a *continuous parameter process* or a *continuous time process*. We will further assume that the random variables $x(t)$ take values on the real line or in an n-dimensional Euclidean space. A stochastic process $\{x(t), t \in T\}$ is a function of two arguments $\{x(t, \omega), t \in T, \omega \in \Omega\}$ where Ω is called the *sample* space. For fixed $t \in T$, $x(t, \cdot)$ is thus a random variable and for fixed ω, $x(\cdot, \omega)$ is a function of time which is called a *realization* of the process, a *sample function*, a *trajectory*, or a *path*. The sample functions can be regarded as elements of the space $\mathscr{X}$ which is called the *sample function space*.

The essential difficulty of the theory of stochastic processes is to assign a probability measure on subsets of the sample space Ω (or the sample function space $\mathscr{X}$). It is in fact not possible to assign a probability measure to all subsets of Ω, only to a Borel field B of subsets.

The assignment of a probability measure P on a Borel field B of subsets of Ω requires measure theory. For ordinary random variables whose sample function spaces are Euclidean spaces, probability measures can be assigned by ordinary distribution functions. Kolmogorov has shown that it is possible to assign a probability measure on a Borel field of subsets of the infinite dimensional sample space of a stochastic process by a similar procedure. Let $\{x(t), t \in T\}$ be a random process. Assume that it is possible to assign a probability distribution to the multidimensional random variable

$$x(t_1), x(t_2), \ldots, x(t_k)$$

for any k and arbitrary $t_i \in T$ with a distribution function

$$\begin{aligned} &F(\xi_1, \xi_2, \ldots, \xi_k; t_1, t_2, \ldots, t_k) \\ &= P\{x(t_1) \leqslant \xi_1, x(t_2) \leqslant \xi_2, \ldots, x(t_k) \leqslant \xi_k\} \end{aligned} \qquad (2.1)$$

which satisfies the conditions of *symmetry* and *consistency*. The distribution (2.1) is called the *finite-dimensional distribution* of the process.

The symmetry condition implies that F is symmetric in all pairs (ξ_i, t_i). The consistency condition is expressed by

$$\begin{aligned} &F(\xi_1, \xi_2, \ldots, \xi_{k-1}; t_1, t_2, \ldots, t_{k-1}) \\ &= \lim_{\xi_k \to \infty} F(\xi_1, \xi_2, \ldots, \xi_k, t_1, t_2, \ldots, t_k) \end{aligned} \qquad (2.2)$$

It follows from Kolmogorov's theorem that a probability measure can be assigned to a Borel field of subsets of Ω and that there exists a stochastic process $\{x(t), t \in T\}$ such that the joint distribution of the values of x at times $t_1, t_2, \ldots, t_k$ has the distribution function $F(\xi_1, \xi_2, \ldots, \xi_k; t_1, t_2, \ldots, t_k)$. The probability measure is denoted by P. The probability assigned by Kolmogorov's theorem is thus uniquely determined by the finite-di-

The Concept of a Stochastic Process

mensional distributions.

The stochastic process $\{x(t, \omega), t \in T\}$ is thus a function which maps a point $\omega \in \Omega$ into the sample function space $\mathscr{X}$. The measure P defined on subsets of Ω and the function $x(t, \omega)$ will induce a measure on $\mathscr{X}$ in the following way. Consider sets $A' \in \mathscr{X}$ such that the sets $\{\omega; x(\cdot, \omega) \in A'\} \in \Omega$ are measurable. We can then define $P'\{x \in A'\} = P\{\omega; x(\cdot, \omega) \in A'\}$.

Another way to describe a random process is to express x as a functional of a known random process. Consider for example the discrete time case with $T = \{\ldots, -1, 0, 1, \ldots\}$. Let the stochastic process $\{e(t), t \in T\}$ simply be a sequence of independent normal $(0, 1)$ random variables. Introduce the stochastic process $\{x(t), t \in T\}$ defined by

$$x(t) = e(t) + c_1 e(t-1) + \cdots + c_n e(t-n), \quad t \in T \quad (2.3)$$

This process is called a *moving average process* of order n. If the polynominal

$$z^n + a_1 z^{n-1} + \cdots + a_n = 0$$

has all roots inside the unit circle we can also introduce the process $\{x(t), t \in T\}$ defined by

$$x(t) + a_1 x(t-1) + \cdots + a_n x(t-n) = e(t), \quad t \in T \quad (2.4)$$

which is called an *autoregressive process* of order n.

We have previously remarked that it is in general not possible to assign a probability measure to all subsets of Ω but only to sets which belong to a Borel field of subsets of Ω, that is, sets which are obtained from enumerable intersections and unions of intervals. With respect to the applications, this is no serious limitation for discrete parameter processes. The restriction is, however, a serious one for continuous time processes because the set

$$\{\omega; x(t, \omega) \leqslant c \quad \text{for all } t \in (a, b)\}$$

is, for example, not a Borel set.

It can also be shown that the measure of this set is not uniquely determined by the finite-dimensional distributions of the process. Hence, for continuous parameter processes specified by finite-dimensional distributions only, it is in general not possible to give the probabilities that all sample functions are bounded, continuous, differentiable, etc.

The mean value of a process is defined by

$$m(t) = Ex(t) = \int_\Omega x(t, \omega) \, P(d\omega) = \int_{-\infty}^{\infty} \xi \, dF(\xi, t) \quad (2.5)$$

It is thus a function of time. Higher moments are defined similarly. The covariance of $x(s)$ and $x(t)$ is, for example, given by

$$\text{cov}[x(s), x(t)] = E[x(s) - m(s)][x(t) - m(t)]$$
$$= \int_\Omega [x(s, \omega) - m(s)][x(t, \omega) - m(t)] P(d\omega)$$
$$= \iint_{-\infty}^{\infty} [\xi_1 - m(s)][\xi_2 - m(t)] d^2 F(\xi_1 \xi_2; s, t) \quad (2.6)$$

Exercises

1. Let Ω be the segment $[0, 1]$ on the real line and the measure P be a uniform distribution. Let the index set T also be the interval $[0, 1]$; consider the stochastic processes $\{x(t), t \in T\}$ and $\{y(t), t \in T\}$ defined by
$$x(t, \omega) = 0 \quad \text{for all } t \text{ and } \omega$$
$$y(t, \omega) = \begin{cases} 1 & \text{for } t = \omega \\ 0 & \text{otherwise} \end{cases}$$
Show that the stochastic processes have the same finite-dimensional distributions and that
$$P\{\omega; x(t, \omega) < 0.5 \text{ for all } t\} = 1$$
$$P\{\omega; y(t, \omega) < 0.5 \text{ for all } t\} = 0$$

2. Consider the moving average process of order one
$$x(t) = e(t) + ce(t - 1)$$
where $\{e(t), t = \ldots, -1, 0, 1, \ldots\}$ is a sequence of independent normal $(0, 1)$ random variables. Determine the covariance of $x(t)$ and $x(s)$.

3. Consider the autoregressive process
$$x(t) + ax(t - 1) = e(t)$$
where $|a| < 1$ and $\{e(t), t = \ldots, -1, 0, 1, \ldots\}$ is a sequence of independent normal $(0, 1)$ random variables. Determine the covariance of $x(t)$ and $x(s)$.

3. SOME SPECIAL STOCHASTIC PROCESSES

The definition of a stochastic process, given in the previous section, is very general. In order to obtain a manageable theory, it is therefore necessary to specialize heavily. Such specializations of the theory which make it possible to characterize the distribution of $x(t_1), x(t_2), \ldots, x(t_k)$ in a simple way are particularly attractive. In this section we will discuss some special processes that are of particular interest to control theory.

Some Special Stochastic Processes

Stationary Processes

If the distribution of $x(t_1), x(t_2), \ldots, x(t_k)$ is identical to the distribution of $x(t_1 + \tau), x(t_2 + \tau), \ldots, x(t_k + \tau)$ for all τ such that $t_i \in T$ and $t_i + \tau \in T$, $i = 1, 2, \ldots, k$, the stochastic process $\{x(t), t \in T\}$ is said to be *stationary*. If only the first and second moments of the distributions are the same, the process is *weakly stationary*.

A stationary process is said to be *ergodic* if the ensemble average equals the time-average of the sample functions, that is,

$$Ex(t) = \int_\Omega x(t, \omega) \, P(d\omega) = \lim_{T \to \infty} \frac{1}{2T} \int_{-T}^{T} x(t, \omega) \, dt \qquad (3.1)$$

for almost all ω. The symbol E denotes mathematical expectation, that is, integration with respect to the measure P.

Normal Processes

A stochastic process is said to be *normal* or *Gaussian* if the joint distribution of $x(t_1), x(t_2), \ldots, x(t_k)$ is normal for every k and all $t_i \in T$, $i = 1, 2, \ldots, k$. It is well known that a normal process is completely characterized by the mean value

$$m_i = Ex(t_i) \qquad i = 1, 2, \ldots, k$$

and the covariances

$$r_{ij} = \text{cov}\,[x(t_i), x(t_j)] = E[x(t_i) - m_i][x(t_j) - m_j]^T$$
$$i, j = 1, 2, \ldots, k$$

If we introduce the vector m and the matrix R defined by

$$m = \begin{bmatrix} m_1 \\ m_2 \\ \vdots \\ m_k \end{bmatrix} \qquad R = \begin{bmatrix} r_{11} & r_{12} & \cdots & r_{1k} \\ r_{12} & r_{22} & \cdots & r_{2k} \\ \vdots & & & \\ r_{1k} & r_{2k} & \cdots & r_{kk} \end{bmatrix}$$

where R is assumed to be nonsingular, we find that the joint distribution of $x(t_1), x(t_2), \ldots, x(t_k)$ can be characterized by the frequency function

$$f(\xi) = (2\pi)^{-k/2} (\det R)^{-1/2} \exp\left[-\frac{1}{2}(\xi - m)^T R^{-1}(\xi - m)\right] \qquad (3.2)$$

It thus follows from Kolmogorov's theorem that a normal process can be defined if we know the mean value and the covariance for all possible $t_1, t_2, \ldots, t_k \in T$. A normal process is thus completely determined by the two functions

$$m(t) = Ex(t)$$
$$r(s, t) = \text{cov}\,[x(s), x(t)] = E[x(s) - m(s)][x(t) - m(t)]^T$$

which are referred to as the *mean value function* and the *covariance function*. For a stationary normal process the mean value is constant and the covariance function depends on s-t only.

Markov Processes

Let t_i and t be elements of the index set T such that $t_1 < t_2 < \cdots < t_k < t$. A stochastic process $\{x(t), t \in T\}$ is called a Markov process if

$$P\{x(t) \leq \xi \mid x(t_1), x(t_2), \ldots, x(t_k)\} = P\{x(t) \leq \xi \mid x(t_k)\} \tag{3.3}$$

where $P\{\cdot \mid x(t_k)\}$ denotes the conditional probability given $x(t_k)$. If the probability distribution of $x(t_1)$, the *initial probability distribution*

$$F(\xi_1; t_1) = P\{x(t_1) \leq \xi_1\} \tag{3.4}$$

and the *transition probability distribution*

$$F(\xi_t, t \mid \xi_s, s) = P\{x(t) \leq \xi_t \mid x(s) = \xi_s\} \tag{3.5}$$

are given, it follows from Baye's rule that the distribution function of the random variable $x(t_1), x(t_2), \ldots, x(t_k)$ is given by

$$F(\xi_1, \xi_2, \ldots, \xi_k; t_1, t_2, \ldots, t_k)$$
$$= F(\xi_k, t_k \mid \xi_{k-1}, t_{k-1}) \cdots F(\xi_2, t_2 \mid \xi_1, t_1) F(\xi_1; t_1) \tag{3.6}$$

A Markov process is thus defined by two functions, the absolute probability distribution $F(\eta; s)$ and transition probabilities $F(\xi, t \mid \eta, s)$.

Processes of Second Order

A stochastic process $\{x(t), t \in T\}$ is said to be of *second order* if $Ex^2(t) < \infty$ for all $t \in T$. For such processes the mean value function

$$m(t) = Ex(t)$$

and the covariance function

$$r(s, t) = E[x(s) - m(s)][x(t) - m(t)]^T$$

exist, and the second order properties of the distributions can thus be expressed by these two functions.

Processes with Independent Increments

Consider the stochastic process $\{x(t), t \in T\}$. Let $t_i \in T$ for $i = 1, 2, \ldots, k$, and $t_1 < t_2 < \cdots < t_k$. The process is called a process with *independent increments* if the random variables

$$x(t_k) - x(t_{k-1}), x(t_{k-1}) - x(t_{k-2}), \ldots, x(t_2) - x(t_1), x(t_1)$$

are mutually independent. If the variables are only uncorrelated, the process $\{x(t), t \in T\}$ is called a process with *uncorrelated* or *orthogonal increments*.

Some Special Stochastic Processes

A process with independent increments is specified by the distribution of the increment $x(t) - x(s)$ for arbitrary s and t and the distribution of $x(t_1)$. If the distribution of $x(t) - x(s)$ depends only on $t - s$, the process is said to have stationary increments. If $x(t) - x(s)$ are normally distributed, $\{x(t), t \in T\}$ is called a process with independent normal increments.

Let $\{x(t), t \in T\}$ be a vector-valued stochastic process with orthogonal increments. We can then define the functions F with the property

$$F(t) - F(s) = \text{cov}\,[x(t), x(t)] - \text{cov}\,[x(s), x(s)] \qquad (3.7)$$

This equation can formally be written as

$$\text{cov}\,[dx, dx] = dF(t)$$

The differential dF is called the *incremental covariance* of the process.

If the process $\{x(t), t \in T\}$ has stationary increments, the difference $F(t) - F(s)$ depends only on $s - t$. In that case denote the difference by $F_1(t - s)$, $s < t$. We then find

$$F_1(t + s) = F_1(t) + F_1(s), \qquad s \geqslant 0, t \geqslant 0 \qquad (3.8)$$

In the continuous parameter case we find that the continuous function which satisfies (3.8) is

$$F_1(t) = At \qquad (3.9)$$

Let us also discuss the properties of the covariance function for a process with orthogonal increments. For $s \geqslant t$ we have

$$\begin{aligned} r(s, t) &= \text{cov}\,[x(s), x(t)] = \text{cov}\,[x(t) + x(s) - x(t), x(t)] \\ &= \text{cov}\,[x(t), x(t)] + \text{cov}\,[x(s) - x(t), x(t)] \\ &= \text{cov}\,[x(t), x(t)] \end{aligned}$$

where the third equality follows from the definition of covariance and the last equality from the fact that the process $\{x(t), t \in T\}$ has uncorrelated increments.

We thus find that the covariance function of a process with orthogonal increments has the property

$$r(s, t) = \begin{cases} r(s, s) & \text{if } s \leqslant t \\ r(t, t) & \text{if } t \leqslant s \end{cases} = \text{cov}\,[x(\min{(s, t)}), x(\min{(s, t)})] \qquad (3.10)$$

The Wiener Process

The *Wiener process* or the *Brownian motion process* is a normal process which has been very important for the development of the theory of the stochastic process. As will be shown in later chapters, many of the disturbances affecting a control system can be modeled by processes generated from Wiener processes.

The botanist Robert Brown noticed in 1827 that small particles (with diameters of the order of 0.001 mm) submerged in a fluid have a very irregular motion. In 1905 Einstein showed that the motion could be explained by assuming that it was caused by collisions with molecules of the fluid. Einstein also made a mathematical model of the phenomenon and presented an analysis which enabled him to determine the Avogadro number by analyzing the Brownian motion. A rigorous mathematical analysis of the process was given by Wiener in 1923. Heuristically, the Brownian motion can be explained as follows: Consider a specific particle submerged in a fluid. Let $x(t)$ denote one of the coordinates of the particle with the origin chosen so that $x(0) = 0$. The motion of the particle over a time interval which is long enough is the result of the change of impulse due to many collisions. It is thus reasonable to assume that the central limit theorem applies and that the displacement is normal. It is further reasonable to assume that statistical properties of the displacement over the interval $(t, t + \tau)$ are the same as those over the interval $(s, s + \tau)$, that the displacements over nonoverlapping time intervals are independent, and that there is no trend in the motion. Axiomatically, the Wiener process can be defined by the following conditions:

1. $x(0) = 0$
2. $x(t)$ normal
3. $Ex(t) = 0$ for all $t > 0$
4. The process has independent stationary increments

Since the Wiener process is normal, it can be completely characterized by the mean value function and the covariance function. The third property implies that

$$m(t) = Ex(t) = 0$$

Since the Wiener process has independent stationary increments and $x(0) = 0$, it follows from (3.9) that

$$\text{var } x(t) = ct \tag{3.11}$$

and from (3.10) that

$$r(s, t) = \text{cov}\,[x(s), x(t)] = c \min\,(s, t) \tag{3.12}$$

The parameter c is called the variance parameter.

With some abuse of language, a vector-valued process with zero mean value and independent normal increments is also called a Wiener process. The sample functions of a Wiener process have interesting properties. They can be shown to be continuous with probability one. Their derivative exists nowhere and the paths are of infinite length.

Some Special Stochastic Processes

Singular or Purely Deterministic Processes

The concept of a singular process is of interest in connection with prediction theory. It also illustrates that it is not completely trivial to introduce meaningful mathematical models of random phenomena.

Let the column vector x be an n-dimensional stochastic variable. The distribution of x is called *singular* or more precisely *linearly singular* if there exists an n-vector, a, such that

$$P\{\omega;\ a^T x(\omega) \neq 0\} = 0 \tag{3.13}$$

Similarly a stochastic process $\{x(t, \omega), t \in T\}$ is called *(linearly) singular* if there exists a linear operator $\mathscr{L}$ such that

$$P\{\omega;\ \mathscr{L}x \neq 0\} = 0 \tag{3.14}$$

The process

$$x(t, \omega) = a(\omega) \quad \text{for all } t \tag{3.15}$$

where a is a random variable is a simple example of a singular process.

We have apparently $x(t_1, \omega) - x(t_0, \omega) = 0$ for all t and all ω. This means, for example, that the process can be predicted exactly over arbitrary time intervals. The process (3.15) is therefore also called a *purely deterministic process*. Processes of the form

$$x(t) = \sum_{i=1}^{n} a_i b_i(t) \tag{3.16}$$

where a_i are random variables and $b_i(t)$ known functions which were often used in early attempts to describe random processes, are more general examples of purely deterministic processes.

Exercises

1. Let $\{x(t), t \in T\}$ be a normal process with zero mean value. Show that
$$E[x(t_1)x(t_2)x(t_3)x(t_4)]$$
$$= r(t_1, t_2)r(t_3, t_4) + r(t_1, t_3)r(t_2, t_4) + r(t_1, t_4)r(t_2, t_3)$$

2. Let $\{x(t), t \in T\}$ be a normal process with zero mean value. Show that
$$E[x(t_1)x(t_2) \cdots x(t_{2n})] = \sum [Ex(t_{i_1})x(t_{i_2})] \cdots [Ex(t_{i_{2n-1}})x(t_{i_{2n}})]$$
where the sum is taken over the $\binom{2n}{n}$ ways to arrange the terms.

3. Let x and y be column-vectors of arbitrary dimensions and assume that the vector $\begin{bmatrix} x \\ y \end{bmatrix}$ is normal with mean $\begin{bmatrix} m_x \\ m_y \end{bmatrix}$ and covariance

$$\begin{bmatrix} R_x & R_{xy} \\ R_{yx} & R_y \end{bmatrix}$$

Show that the conditional distribution of x given y is normal with the mean

$$m = m_x + R_{xy}R_y^{-1}(y - m_y)$$

and the covariance

$$R = R_x - R_{xy}R_y^{-1}R_{yx}$$

4. Let x be an n-dimensional column-vector which is normal with mean m and covariance R_0. Let S be a symmetric matrix and v the quadratic form

$$v = x^T S x$$

Show that

$$Ev = m^T S m + \operatorname{tr} R_0 S$$

5. Consider the first order moving average

$$x(t) = e(t) + ce(t - 1)$$

where $\{e(t), t = \ldots, -1, 0, 1, \ldots\}$ is a sequence of independent normal $(0, 1)$ random variables. Is the process stationary, normal, Markovian, ergodic, or singular? Has the process independent increments?

6. Consider the first order autoregression

$$x(t + 1) + ax(t) = e(t) \qquad t = t_0, t_0 + 1, \ldots$$

where $|a| < 1$, $\{e(t)\}$ is a sequence of independent normal $(0, 1)$ random variables, and the initial state $x(t_0)$ is normal $(0, \sigma)$. The sequence $\{e(t)\}$ is also assumed to be independent of $x(t_0)$. Is the process stationary, normal, Markovian, ergodic, or singular? Has the process independent increments?

7. Consider the autoregression of Exercise 6 but assume that $x(t_0)$ and $e(t_0)$ are jointly normal with the correlation ρ. Is the process Markovian?

8. Consider the singular stochastic process $\{x(t), 0 \leqslant t < \infty\}$ defined by

$$\frac{dx}{dt} = 0$$

where the initial state is normal $(0, 1)$. Is the process ergodic? Give a predictor for the process which predicts $x(t + h)$ based on measurements of $x(t)$.

9. Consider the stochastic process

$$\frac{dx}{dt} = \begin{bmatrix} 0 & 1 \\ -1 & 0 \end{bmatrix} x$$

where the initial state is normal with zero mean value and the covariance

$$\text{cov}\,[x(0), x(0)] = \begin{pmatrix} 1 & 0 \\ 0 & 1 \end{pmatrix}$$

Is the process ergodic? Give a predictor for the process which predicts $x(t + h)$ based on observation of $\{x_1(t), t_0 \leqslant s \leqslant t\}$.

10. Let $\{x(t), t \in T\}$ be a Wiener process with unit variance parameter. Show that

$$P\{\omega;\ \max_{0 \leqslant t \leqslant T_1} x(t, \omega) \geqslant a\} = 2P\{\omega;\ x(T_1) \geqslant a\}$$

$$= \frac{2}{\sqrt{2\pi T_1}} \int_a^\infty \exp\,(-\xi^2/2T_1)\,d\xi$$

4. THE COVARIANCE FUNCTION

This section will summarize the properties of the covariance functions. Let $\{x(t), t \in T\}$ and $\{y(t), t \in T\}$ be stochastic processes of second order. The *covariance function* of the processes $\{x(t), t \in T\}$ and $\{y(t), t \in T\}$ was previously defined by

$$r_{xy}(s, t) = \text{cov}\,[x(s), y(t)] = E[x(s) - Ex(s)][y(t) - Ey(t)]^T \quad (4.1)$$

In particular if the processes are weakly stationary, $r_{xy}(s, t)$ is a function of the difference of the arguments $s - t$. Hence

$$r_{xy}(s, t) = r_{xy}(s - t) \quad (4.2)$$

When the processes $\{x(t), t \in T\}$ and $\{y(t), t \in T\}$ are the same, we refer to $r_{xx}(s, t)$ as the *autocovariance function*. To simplify notation, $r_{xx}(s, t)$ is also denoted $r_x(s, t)$.

The *cross-correlation function* of the processes $\{x(t), t \in T\}$ and $\{y(t), t \in T\}$ is defined as

$$\rho_{xy}(s, t) = \frac{r_{xy}(s, t)}{\sqrt{r_{xx}(s, s) r_{yy}(t, t)}}$$

The *auto-correlation* function of the process $\{x(t), t \in T\}$ is defined by

$$\rho_{xx}(s, t) = \frac{r_{xx}(s, t)}{\sqrt{r_{xx}(s, s) r_{xx}(t, t)}}$$

For stationary processes we have in particular

$$\rho_{xy}(\tau) = \frac{r_{xy}(\tau)}{\sqrt{r_{xx}(0)\, r_{yy}(0)}}$$

$$\rho_{xx}(\tau) = \frac{r_{xx}(\tau)}{r_{xx}(0)}$$

Since it is cumbersome to carry out the normalization, the correlation functions are seldom used. Notice that the name "autocorrelation function" is sometimes, in the literature, used to denote $Ex(t)x(t+\tau)$.

If x and y are vector valued, say n-dimensional vectors, the covariance function is defined by

$$R_{xy}(s, t) = \text{cov}[x(s), y(t)] = E[x(s) - Ex(s)][y(t) - Ey(t)]^T$$

$$= \begin{bmatrix} \text{cov}[x_1(s), y_1(t)] & \text{cov}[x_1(s), y_2(t)] & \cdots & \text{cov}[x_1(s), y_n(t)] \\ \text{cov}[x_2(s), y_1(t)] & \text{cov}[x_2(s), y_2(t)] & \cdots & \text{cov}[x_2(s), y_n(t)] \\ \vdots & & & \\ \text{cov}[x_n(s), y_1(t)] & \text{cov}[x_n(s), y_2(t)] & \cdots & \text{cov}[x_n(s), y_n(t)] \end{bmatrix}$$

Notice that all definitions hold for both continuous time and discrete time processes. The covariance function has the following important property.

THEOREM 3.1

Let $\{x(t), t \in T\}$ be a real second order random process with the covariance function $r_x(s, t)$. Then

1. $r_x(s, t) = r_x(t, s)$ (4.3)
2. The quadratic form in z_i, $\sum_{i,j=1}^n z_i z_j r_x(t_i, t_j)$, is nonnegative definite for all integers n and every choice of division points $t_i \in T$, $i = 1, 2, \cdots, n$.
3. $|r_x(s, t)|^2 \leqslant r_x(s, s)\, r_x(t, t)$ (4.4)
4. If $r_x(s, t)$ is continuous along the diagonal $s = t$ then it is continuous for all t.

Proof

The first statement follows directly from the definition of a covariance function

$$r_x(s, t) = \text{cov}[x(s), x(t)] = \text{cov}[x(t), x(s)] = r_x(t, s)$$

To prove the the second statement we assume that $Ex = 0$ and we form

$$E\{[\sum_{i,j=1}^n z_i x(t_i)]^2\} = \sum_{i,j=1}^n z_i z_j E[x(t_i)x(t_j)] = \sum_{i,j=1}^n z_i z_j r_x(t_i, t_j)$$

Since the left-hand side is the expectation of nonnegative quantity, it is nonnegative and statement 2 is proven.

The Covariance Function

Statement 3 follows from the Schwartz inequality

$$E|xy| \leq \sqrt{Ex^2 Ey^2} \tag{4.5}$$

To prove (4.5) we let α be a real constant and consider the inequality

$$(|x| + \alpha|y|)^2 \geq 0$$

Taking mean values of both sides we get

$$Ex^2 + 2\alpha E|xy| + \alpha^2 Ey^2 = (Ey^2)\left[\alpha + \frac{E|xy|}{Ey^2}\right]^2 + Ex^2 - \frac{(E|xy|)^2}{Ey^2} \geq 0$$

If the left-hand side should remain nonnegative for all α we get

$$[E|xy|]^2 \leq Ex^2 Ey^2$$

which is identical to (4.5).

To prove property 4 we form

$$|r_x(s+h, t+k) - r_x(s,t)|$$
$$= |\operatorname{cov}[x(s+h), x(t+k) - x(t)] + \operatorname{cov}[x(s+h) - x(s), x(t)]|$$
$$\leq \{r_x(s+h, s+h)[r_x(t+k, t+k) - 2r_x(t+k, t) + r_x(t, t)]\}^{1/2}$$
$$+ \{[r_x(s+h, s+h) - 2r_x(s+h, s) + r_x(s, s)]r_x(t, t)\}^{1/2}$$

where the inequality follows from (4.4). Now let $h, k \to 0$ then $r_x(t+k, t) \to r_x(t, t)$ and $r_x(t+k, t+k) \to r_x(t, t)$ because $r_x(s, t)$ is continuous for $s = t$. The right member of (4.6) will thus converge to zero and the statement is proven.

Specializing, we find that the covariance of a weakly stationary process has the properties

- $r_x(\tau) = r_x(-\tau)$
- $\sum z_i z_j r_x(t_i - t_j) \geq 0$
- $|r_x(\tau)| \leq r_x(0)$
- If $r_x(\tau)$ is continuous for $\tau = 0$, then $r_x(\tau)$ is continuous for all τ.

Exercises

1. Discuss whether the following functions can be covariance functions of a stationary stochastic process

$$r(\tau) = \text{constant}$$
$$r(\tau) = \cos \tau$$
$$r(\tau) = \begin{cases} 1 & |\tau| < 1 \\ 0 & |\tau| \geq 1 \end{cases}$$

$$r(\tau) = \begin{cases} 1 - |\tau| & |\tau| < 1 \\ 0 & |\tau| \geq 1 \end{cases}$$

$$r(\tau) = \frac{1}{1 + 2\xi|\tau| + \tau^2}$$

$$r(\tau) = \begin{cases} 2 & \tau = 0 \\ e^{-|\tau|} & \tau \neq 0 \end{cases}$$

2. Let $\{e(t), t = \ldots, -1, 0, 1, \ldots\}$ be a sequence of independent normal (0, 1) random variables. Consider the stochastic process defined by

$$x(t) + ax(t-1) = e(t) + ce(t-1)$$

where $|a| < 1$. Determine the covariance function of the process.

3. A random telegraph wave is a continuous time stochastic process $\{x(t), -\infty \leq t \leq \infty\}$ with the following properties: x does only assume the values $+1$ or -1. The probability that x changes its value in the interval $(t, t+h)$ is $\lambda h + o(h)$. Determine the covariance function of the process. Hint: The probability that the process changes n times in an interval of length t is $\lambda t^n/n! \exp(-\lambda t)$.

4. Let $\mathscr{L}$ be a linear operator which commutes with the operation of taking mathematical expectations. Show that a necessary condition that the stochastic process $\{x(t), t \in T\}$ is singular (or purely deterministic) due to the linear relation $\mathscr{L}x = 0$ is that the covariance function $r(s, t)$ of the process has the properties

$$\mathscr{L}r(\cdot, t) = 0, \quad t = \text{constant}$$
$$\mathscr{L}r(s, \cdot) = 0, \quad s = \text{constant}$$

6. Prove that

$$\text{cov}[Ax + a, By + b] = A\{\text{cov}[x, y]\}B^T$$

5. THE CONCEPT OF SPECTRAL DENSITY

We will now consider a weakly stationary stochastic process $\{x(t), t \in T\}$. The process can be characterized by its mean value m and its covariance function $r_x(\tau)$. We will now introduce a different characterization of the process. In *principle*, this will give us nothing new but it will enable us to give different physical interpretations and it will also simplify the writing of some formulas. In essence, we will introduce the Fourier transform of the covariance function and we will then obtain the equivalence of time domain and frequency domain viewpoints on deterministic systems.

The Concept of Spectral Density

According to Theorem 3.1 the covariance function has the following properties

$$r(\tau) = r(-\tau) \tag{5.1}$$

$$\sum z_i z_j r(t_i - t_j) \geq 0 \tag{5.2}$$

The covariance function is thus nonnegative definite. According to a theorem of Bochner, a nonnegative definite function can always be represented as

$$r(\tau) = \int_{-\infty}^{\infty} e^{i\omega\tau} \, dF(\omega) \quad \text{(continuous parameter processes)} \tag{5.3}$$

or

$$r(\tau) = \int_{-\pi}^{\pi} e^{i\omega\tau} \, dF(\omega) \quad \text{(discrete parameter processes)} \tag{5.4}$$

where F is a nondecreasing function. The function F is called the *spectral distribution function* of the stochastic process. It can be decomposed into three components

$$F(\omega) = F_a(\omega) + F_d(\omega) + F_s(\omega) \tag{5.5}$$

where F_a is absolutely continuous, F_d is a step function and F_s is a continuous function that is constant almost everywhere. The function F_s is called the singular part. The functions F_a and F_d can be represented as

$$F_a(\omega) = \int^{\omega} \phi(\omega') \, d\omega' \tag{5.6}$$

$$F_d(\omega) = \sum_{\omega_\nu \leq \omega} F(\omega_\nu) \tag{5.7}$$

where $\phi(\omega)$ is called the *spectral density function* or simply the *spectral density*. If we assume that the singular part F_s and the step function F_d are zero, we thus find the following equation relating the spectral density and the covariance function

$$\begin{cases} \phi(\omega) = \dfrac{1}{2\pi} \int_{-\infty}^{\infty} e^{-i\omega t} \, r(t) \, dt & (5.8) \\ r(t) = \int_{-\infty}^{\infty} e^{i\omega t} \phi(\omega) \, d\omega & (5.9) \end{cases} \quad \text{(continuous parameter case)}$$

$$\begin{cases} \phi(\omega) = \dfrac{1}{2\pi} \sum_{n=-\infty}^{\infty} r(n) e^{-in\omega} & (5.10) \\ r(n) = \int_{-\pi}^{\pi} e^{in\omega} \phi(\omega) \, d\omega & (5.11) \end{cases} \quad \text{(discrete parameter case)}$$

If we consider the spectral density as a distribution or a generalized function these equations will hold also when $F_d \neq 0$.

Notice that
$$\operatorname{var} x = r(0) = \int dF(\omega) \tag{5.12}$$
where the integral is taken over the interval $(-\pi, \pi)$ in the discrete time case and over $(-\infty, \infty)$ in the continuous time case. The total variation of the spectral distribution function thus equals the variance of the process Similarly the quantity
$$\int_{-\omega_2}^{-\omega_1} dF(\omega) + \int_{\omega_1}^{\omega_2} dF(\omega) \qquad \omega_2 > \omega_1 > 0 \tag{5.13}$$
can be interpreted as the contribution to the total variance of the process from frequencies in the interval (ω_1, ω_2). The function $F(\omega)$ thus expresses how the variance of the stochastic process is distributed on various frequencies. This physical interpretation explains the name spectral distribution function.

When reading the literature, the placement of the factor 2π in the Fourier transform varies. If we remember the physical interpretation given above, we have a simple rule (5.12) to memorize the convention used throughout this book. The following pairs of transforms are also often used in literature.
$$S(f) = \int_{-\infty}^{\infty} e^{-2\pi i f \tau} r(\tau) \, d\tau \tag{5.14}$$
$$r(\tau) = \int_{-\infty}^{\infty} e^{+2\pi i f \tau} S(f) \, df \tag{5.15}$$
If we introduce
$$\phi(\omega) \, d\omega = S(f) \, df \qquad \omega = 2\pi f$$
we find the following correspondence between $\phi(\omega)$ and $S(f)$
$$\phi(\omega) = \frac{1}{2\pi} S\left(\frac{\omega}{2\pi}\right) \tag{5.16}$$
Using the scales rad/sec and Hz we find that both ϕ and S have the properties that, for processes with $F_d = 0$ and $F_s = 0$, the area under the curves defined by the spectral density equals the total variance.

Decomposition of Stationary Processes

There is a decomposition of a stationary process $\{x(t), t \in T\}$ which corresponds to the decomposion (5.5) of the spectral distribution function. It can be shown that a stationary stochastic process $\{x(t), t \in T\}$ can be decomposed into three independent processes, $\{x_a(t), t \in T\}$, $\{x_d(t), t \in T\}$ and $\{x_s(t), t \in T\}$, having the spectral distribution functions F_a, F_d, and F_s respectively such that

The Concept of Spectral Density

$$x(t) = x_a(t) + x_d(t) + x_s(t) \tag{5.17}$$

If the function F_d has a finite number of jumps, the process x_d consists of a finite sum of harmonics. The process is thus of the form (3.16) where $b_i(t)$ are sine-functions, and it follows from the analysis of Section 3 that the process is purely deterministic. The process x_d will in general have a enumerable number of discontinuities. It can, however, be shown that the process is deterministic also in the general case. It is also possible to show that the component x_s is purely deterministic. The component x_a can be either deterministic or nondeterministic.

In the discrete parameter case it has been shown by Kolmogorov that a process is purely deterministic if the integral

$$I = \int_{-\pi}^{\pi} |\log F_a'(\omega)| \, d\omega \tag{5.18}$$

is infinite and that the process is nondeterministic if the integral (5.18) is finite.

The corresponding criterion for continuous parameter processes is given by the integral

$$I = \int_{-\pi}^{\pi} \frac{|\log F_a'(\omega)|}{1 + \omega^2} \, d\omega \tag{5.19}$$

This result is due to Wiener. The criterion that the integral (5.19) is finite is called the *Paley-Wiener condition*.

The Concept of White Noise

Having introduced the spectral distribution function $F(\omega)$, we will now use it to introduce special stochastic processes which are called white noise. We will thus consider a weakly stationary stochastic process $\{x(t), t \in T\}$. Without losing generality, we can assume that the mean value of the process is zero. Recalling that the spectral distribution function expressed how the variance of the process was distributed on different frequencies, [see (5.13)], and making an analogy with optics, we introduce the following definition:

DEFINITION 5.1

A weakly stationary stochastic process with $F(\omega) = \text{const} \cdot \omega$ is called *white noise*.

We immediately observe that for a white noise process, the singular part F_s and the discrete part F_d of the decomposition of the spectral distribution function will vanish. White noise thus has a constant spectral density: $\phi(\omega) = \text{constant}$. We will now investigate the implications of the definition. To do this we will consider the discrete and continuous time processes separately.

Discrete Time White Noise

To further analyze the properties of discrete time white noise we will first calculate its covariance function. Introducing $\phi(\omega) = \text{constant} = c$ into (5.11) we find

$$r(n) = \int_{-\pi}^{\pi} e^{in\omega} c \, d\omega = \frac{2c}{n} \sin n\pi \qquad (5.20)$$

Discrete time white noise thus has the property

$$r(n) = \begin{cases} 2\pi c & n = 0 \\ 0 & n = \pm 1, \pm 2, \ldots \end{cases} \qquad (5.21)$$

This implies that the values of the process at different times are uncorrelated and, if we consider normal white noise, also independent. White noise in the discrete time case is thus a process which consists of a sequence of uncorrelated (in the normal case also independent) stochastic variables. Discrete time white noise is therefore sometimes referred to as a *completely uncorrelated process* or a *pure random process*.

Continuous Time White Noise

It was mentioned in the introduction that continuous time processes are much more difficult to analyze than discrete time processes. The introduction of continuous time white noise is a good example of the sort of difficulties that will be encountered. In analogy with the discrete time case, the definition implies that

$$\phi(\omega) = \text{constant} = c$$

As the variance of the process is the integral of $\phi(\omega)$ over $(-\infty, \infty)$ we find immediately that continuous time white noise does not have finite variance. Continuous time white noise thus is not a second order random process.

As the Fourier transform of a constant equals the distribution with all mass at the origin or finally the Dirac delta function, we find that *formally* white noise has the covariance function

$$r(\tau) = 2\pi c \delta(\tau)$$

Hence, also in the continuous time case, we find that white noise has the property that $x(t)$ and $x(s)$ are uncorrelated if $t \neq s$ in complete analogy with the discrete time case. Notice, however, that in the continuous time case we have the complication that white noise does not have finite variance. If we try to bypass this difficulty by constructing a stochastic process with finite variance such that $x(t)$ and $x(s)$ are uncorrelated for $s \neq t$, we will find that such a process is zero in a certain sense. Since this will require

The Concept of Spectral Density

more theory than we have presently available, we must postpone a detailed discussion of this to Chapter 3 (see Theorem 3.4.1).

Since white noise has infinite variance we could try to introduce other processes which have essentially constant spectral densities but finite variance. There are many ways in which this could be done. One alternative is the so-called band limited white noise characterized by the spectral density

$$\phi(\omega) = \begin{cases} c & |\omega| < \Omega \\ 0 & |\omega| \geqslant \Omega \end{cases}$$

This process has the covariance function

$$r(\tau) = \int_{-\Omega}^{\Omega} ce^{i\omega\tau} d\omega = \frac{2c}{\tau} \sin \Omega\tau \qquad (5.22)$$

The correlation of two values of the process $x(t)$ and $x(s)$ separated by a given interval $|t - s| > \delta$ can thus be made arbitrarily small by choosing Ω sufficiently large. Notice, however, that for given Ω the values of the process at time t and s will always be correlated if t and s are chosen sufficiently close together.

We will now analyze what happens to the function (5.22) when $\Omega \to \infty$. For $\tau \neq 0$, the value of the function will tend to zero. As $r(0) = 2c\Omega$, we find that $r(0)$ will tend to infinity. It is easier analytically to consider the integral of r

$$R(\tau) = \int_{-\infty}^{\tau} r(s)\, ds = 2c \int_{-\infty}^{\tau} \frac{\sin \Omega s}{s}\, ds = 2c \int_{-\infty}^{\Omega\tau} \frac{\sin x}{x}\, dx$$

We find

$$\lim_{\Omega \to \infty} R(\tau) = \begin{cases} 0 & \tau < 0 \\ \pi c & \tau = 0 \\ 2\pi c & \tau > 0 \end{cases}$$

because

$$\int_0^\infty \frac{\sin x}{x}\, dx = \pi$$

The integral of the covariance function thus equals a stepfunction. Formally, the limiting covariance function then becomes a Dirac delta function

$$r(\tau) \to 2\pi c \delta(\tau)$$

Apart from bandlimited noise the random process with the covariance function

$$r(\tau) = \frac{2\pi \cdot a}{2} \cdot e^{-a|\tau|}$$

and the spectral density

$$\phi(\omega) = \frac{a^2}{\omega^2 + a^2}$$

is frequently used. In this case we have formally

$$\lim_{a \to \infty} \phi(\omega) = 1$$
$$\lim_{a \to \infty} r(\tau) = 2\pi \delta(\tau)$$

In spite of the difficulties which will naturally arise due to the infinite variance, the concept of white noise is very important in the theory and application of stochastic processes. White noise is frequently used to model random processes, having essentially constant spectral density over a certain frequency range, in cases where it is immaterial how the spectral density goes to zero outside the frequency range of interest.

The reason why white noise is convenient to use is that the values of the white noise process at distinct values of time are uncorrelated (or independent if the process is also Gaussian). The price we pay when using band limited noise is that we get correlation between the values of the process at neighboring points which often complicates the analysis. The use of white noise in the theory of stochastic process resembles in many ways the use of the Dirac delta function in the analysis of linear systems.

Exercises

1. A stationary stochastic process has the covariance function

$$r(\tau) = e^{-\alpha|\tau|}$$
$$r(\tau) = e^{-a^2\tau^2}$$
$$r(\tau) = A + B \cos \omega\tau$$
$$r(\tau) = e^{-\alpha|\tau|} \cos \beta\tau$$

   Determine the corresponding spectral distribution functions and their decompositions.

2. Determine the spectral densities and the covariance functions for the following stochastic processes

$$x(t) = e(t) + ce(t - 1)$$
$$x(t) + ax(t - 1) = e(t - 1)$$
$$x(t) + ax(t - 1) = e(t) + ce(t - 1)$$

   where $\{e(t), t = \ldots -1, 0, 1, \ldots\}$ is a sequence of independent normal $(0, 1)$ random variables and $|a| < 1$.

3. A more accurate description of the Brownian motion is given by the

following model (the Langevin equation)

$$\frac{dv}{dt} + \alpha v = e(t)$$

where v is the velocity of the particle and $\{e(t)\}$ is bandlimited white noise with the covariance function given by (5.22). Determine the covariance function of the velocity and show that the covariance function converges to $r(\tau) = \text{const} \cdot \exp(-\alpha\tau)$, if the noise bandwidth goes to infinity.

6. ANALYSIS OF STOCHASTIC PROCESSES

In order to analyze dynamical systems whose inputs are stochastic processes, it is necessary to develop the analysis of stochastic processes. We will thus need concepts such as continuity, derivative, and integral of a stochastic process. This section will present the foundation of the analysis of stochastic processes. We will first introduce the notion of convergence of sequences of random variables. It turns out that the theory will be richer than the corresponding results for real variables because we have a greater choice of topologies. Starting from the concept of convergence, we will then define continuity, derivative, and integral for stochastic processes.

Convergence

We will consider a sequence of stochastic variables $\{x_n(\omega), n = 1, 2, \ldots\}$ and we will discuss what we should mean by the limit of such a sequence. This can be formulated in many ways. In the following we will discuss the most common limit concepts.

DEFINITION 6.1

The sequence $\{x_n(\omega)\}$ converges with *probability one* to the stochastic variable $x(\omega)$ if $x_n(\omega) \to x(\omega)$ for all ω, except possibly for a set of ω-values having probability measure zero

$$P\{\omega;\ x_n(\omega) \to x(\omega)\} = 1 \tag{6.1}$$

DEFINITION 6.2

The sequence $\{x_n(\omega)\}$ converges to $x(\omega)$ *in probability* if for every $\varepsilon > 0$

$$\lim_{n\to\infty} P\{\omega;\ |x_n(\omega) - x(\omega)| \geq \varepsilon\} = 0 \tag{6.2}$$

DEFINITION 6.3

The sequence $\{x_n(\omega)\}$ converges to $x(\omega)$ in the *mean square* if

$$\lim_{n\to\infty} E \mid x_n - x \mid^2 = \lim_{n\to\infty} \int \mid x_n(\omega) - x(\omega) \mid^2 P(d\omega) = 0 \qquad (6.3)$$

The different convergence concepts are related. We have, for example, Theorem 6.1.

THEOREM 6.1

Convergence with probability one implies convergence in probability. Convergence in the mean square implies convergence in probability.

Choice of Convergence Concepts

Having defined convergence we can now proceed to define concepts such as continuity, differentiability, integrability and analyticity. We can, for example, say that a process is continuous with probability one at t if $x(t + h)$ converges to $x(t)$ with probability one as h tends to zero. Since we have a choice of convergence concepts, it is natural to ask which concept is the most natural to choose, from the point of view of applications to stochastic control theory. In the applications it would, for example, be highly desirable to state that almost all sample functions are continuous. This means convergence with probability one uniformly in t. There is unfortunately a fundamental difficulty in making such a statement.

It was stated in Section 2 that for continuous time processes, the sets

$$\{\omega; x(t, \omega) \leqslant c \text{ for all } t \in (a, b)\}$$
$$\{\omega, x(t, \omega) \text{ continuous for all } t \in (a, b)\}$$

are not Borel sets. Hence it is not possible to assign probability measures for such sets using finite dimensional distributions.

Also notice that continuity with probability one does not imply that all sample functions are continuous. Consider, for example, the random telegraph wave which is a process that assumes only two values, $+1$ or -1. The probability of a change in the interval $[t, t + h]$ is $\lambda h + o(h)$. The sample functions of this process are not continuous because the probability of continuity throughout an interval of length T is $\exp(-\lambda T)$ which goes to zero when T goes to infinity. The process is, however, continuous with probability one for fixed $t = t_0$ because

$$P\{\omega; x(t_0 + h, \omega) - x(t_0, \omega) \neq 0\} = 1 - e^{-\lambda h}$$

and the right hand side converges to zero as h goes to zero.

The criteria for convergence with probability one are often difficult to establish. We will therefore in the sequel use convergence in the mean square for the simple reason that it leads to very simple analysis. It will be shown in the following that there are simple criteria for continuity, differentiability and integrability in the mean square of second order processes. Having introduced the concept of limit we can now proceed in a

Analysis of Stochastic Processes

straight forward manner to introduce continuity, derivative, and integral, thus forming the foundation for the analysis of stochastic processes. It should, however, be emphasized that in many applications it would be preferable to use uniform convergence in t with probability one.

Properties of Mean Square Convergence

To investigate mean square convergence we can use the Cauchy criterion. Let $\{x_n\}$ be a sequence of random variables such that $x_n \to x$ in the mean square. We have

$$|x_n - x_m|^2 = |x_n - x - (x_m - x)|^2 \leqslant |x_n - x|^2 + |x_m - x|^2 \\ + 2|x_n - x| \cdot |x_m - x| \leqslant 2|x_n - x|^2 + 2|x_m - x|^2$$

because $|ab| < a^2 + b^2$. Taking mathematical expectations we find

$$E|x_n - x_m|^2 \leqslant 2E|x_n - x|^2 + 2E|x_m - x|^2 \qquad (6.4)$$

The right member of this inequality converges to zero as $n, m \to \infty$ because $x_n \to x$ as $n \to \infty$.

Conversely it can be shown that if $E|x_n - x_m|^2 \to 0$ as $n, m \to \infty$, then there exists x such that $x_n \to x$ in the mean square as $n \to \infty$. This is the famous Riesz-Fisher theorem. We also have the following important result.

THEOREM 6.2

Let $\{x_n\}$ be a sequence of random variables. Assume that $Ex_n^2 < \infty$ and that $x_n \to x$ in the mean square as $n \to \infty$. Then

$$\lim_{n \to \infty} Ex_n = E \lim_{n \to \infty} x_n = Ex \qquad (6.5)$$

Proof

We have

$$|E(x_n - x)| \leqslant E|x_n - x| \leqslant \sqrt{E\,1^2 \cdot E|x_n - x|^2} \qquad (6.6)$$

where the first inequality follows from $|\int f(x)\,dx| \leqslant \int |f(x)|\,dx$ and the second from Schwartz inequality (4.5). Since $x_n \to x$ in the mean square, the right hand side of (6.6) converges to zero and we thus have (6.5).

Continuity

The notion of continuity of a stochastic process is defined as follows.

DEFINITION 6.4

A second order stochastic process $\{x(t), t \in T\}$ is said to be continuous in the mean square at time t if

$$\lim_{h \to 0} E[x(t + h) - x(t)]^2 = 0$$

It is easy to find out if a stochastic process is continuous in the mean square by analyzing its covariance function.

THEOREM 6.3

A second order stochastic process $\{x(t),\ t \in T\}$ is continuous in the mean square at $t \in T$ if and only if the mean value function $m(t)$ is continuous at t and if the covariance function is continuous at $(t,\ t)$.

Proof

We have

$$[x(t+h) - x(t)]^2 = [x(t+h) - m(t+h) - x(t) + m(t)]^2$$
$$+ 2[x(t+h) - x(t)][m(t+h) - m(t)] - [m(t+h) - m(t)]^2$$

Take mathematical expectation of both members and we find

$$E[x(t+h) - x(t)]^2$$
$$= \operatorname{cov}[x(t+h) - x(t),\ x(t+h) - x(t)] + [m(t+h) - m(t)]^2$$
$$= r(t+h,\ t+h) - 2r(t+h,\ t) + r(t,\ t) + [m(t+h) - m(t)]^2 \quad (*)$$

If $r(s,\ t)$ and $m(t)$ are continuous we now get

$$\lim_{h \to 0} E\{x(t+h) - x(t)\}^2 = 0.$$

It follows from Theorem 3.1 (property 3) that

$$r(t+h,\ t+h) - 2r(t+h,\ t) + r(t,\ t) \geq 0$$

The right-hand side of (*) is thus a sum of two nonnegative terms. If the left member converges to zero, each term of the right member also converges to zero. We thus find that continuity of the process in the mean square implies that the mean value function and the covariance function are continuous.

EXAMPLE 6.1

As an application we will now investigate the continuity of the Wiener process. For this process we have

$$E[x(t+h) - x(t)]^2 = Ah$$

and it thus follows directly from the definition that the process is continuous. The covariance function of the Wiener process is

$$r(s,\ t) = A \min(s,\ t)$$

Since this function is continuous the continuity of the Wiener process also follows from Theorem 6.3. It can also be shown that the paths of a Wiener process are continuous with probability one.

Analysis of Stochastic Processes

Differentiability

We now proceed to define differentiability and derivative.

DEFINITION 6.5

A second order stochastic process $\{x(t), t \in T\}$ is said to be differentiable in the mean square at $t_0 \in T$ if the limit

$$\lim_{h \to 0} \frac{x(t_0 + h) - x(t_0)}{h} = x'(t_0)$$

exists in the sense of mean square convergence, that is, if

$$\lim_{h \to 0} E\left\{\frac{x(t_0 + h) - x(t_0)}{h} - x'(t_0)\right\}^2 = 0$$

If a process is differentiable for all $t \in T$, it is said to be a differentiable stochastic process.

We have the following criterion for differentiability.

THEOREM 6.4

A second order stochastic process $\{x(t), t \in T\}$ is differentiable in the mean square at $t_0 \in T$ if and only if the mean value function $m(t)$ is differentiable at t_0 and the generalized second order derivative of the covariance function

$$\frac{\partial^2 r(s, t)}{\partial s \, \partial t}$$

exists at $s = t = t_0$.

Proof

We will first prove the if statement. The generalized second order derivative is defined as the limit of

$$\frac{r(s + h, t + k) - r(s, t + k) - r(s + h, t) + r(s, t)}{hk}$$

when $h, k \to 0$.

To prove the existence of a limit we form the Cauchy sequence

$$\left[\frac{x(t_0 + h) - x(t_0)}{h} - \frac{x(t_0 + k) - x(t_0)}{k}\right]^2$$

$$= \frac{x(t_0 + h) - x(t_0)}{h} \cdot \frac{x(t_0 + h) - x(t_0)}{h}$$

$$- 2\frac{x(t_0 + h) - x(t_0)}{h} \cdot \frac{x(t_0 + k) - x(t_0)}{k}$$

$$+ \frac{x(t_0 + k) - x(t_0)}{k} \cdot \frac{x(t_0 + k) - x(t_0)}{k}$$

Taking mathematical expectation for each term we find

$$E \frac{x(t_0 + h) - x(t_0)}{h} \cdot \frac{x(t_0 + k) - x(t_0)}{k}$$
$$= \text{cov}\left[\frac{x(t_0 + h) - x(t_0)}{h} \cdot \frac{x(t_0 + k) - x(t_0)}{k}\right]$$
$$+ \frac{m(t_0 + h) - m(t_0)}{h} \cdot \frac{m(t_0 + k) - m(t_0)}{k}$$
$$= \frac{r(t_0 + h, t_0 + k) - r(t_0 + h, t_0) - r(t_0, t_0 + k) + r(t_0, t_0)}{h \cdot k}$$
$$+ \frac{m(t_0 + h) - m(t_0)}{h} \cdot \frac{m(t_0 + k) - m(t_0)}{k} \quad (6.7)$$

According to the assumptions, the mean value function is differentiable and the mixed second derivative of $r(s, t)$ exists. Hence

$$\lim_{h, k \to 0} E \frac{x(t_0 + h) - x(t_0)}{h} \cdot \frac{x(t_0 + k) - x(t_0)}{k}$$
$$= \frac{\partial^2 r(s, t)}{\partial s\, \partial t}\bigg|_{s=t=t_0} + m'(t_0) \cdot m'(t_0)$$

The right member of (6.7) is thus finite. Evaluating all three terms we find that

$$E\left[\frac{x(t_0 + h) - x(t_0)}{h} - \frac{x(t_0 + k) - x(t_0)}{k}\right]^2 \to 0$$

as $h, k \to 0$. We have thus proven the if statement. To prove the only if statement we put $h = k$ and observe that the right member of (6.7) is a sum of two nonnegative terms. If the left member converges to zero, each of the terms of the right member will also converge to zero.

We thus have very simple criteria for differentiability. We will now give a result which is very useful when making formal calculations on stochastic processes.

Let the stochastic process $\{x(t), t \in T\}$ be differentiable at $s \in T$. It then follows from Theorem 6.2 that

$$E\left[\frac{d}{dt}x(t)\right] = \frac{d}{dt}Ex(t) = \frac{dm(t)}{dt} \quad (6.8)$$

$$\text{cov}\left[\frac{d}{ds}x(s), \frac{d}{dt}x(t)\right] = \frac{d}{ds} \cdot \frac{d}{dt}\text{cov}\,[x(t), x(t)] = \frac{\partial^2 r(s, t)}{\partial s\, \partial t} \quad (6.9)$$

$$\text{cov}\left[\frac{d}{ds}x(s), x(t)\right] = \frac{d}{ds}\text{cov}\,[x(s), x(t)] = \frac{\partial r(s, t)}{\partial s} \quad (6.10)$$

Analysis of Stochastic Processes

Let us now specialize to weakly stationary processes. For such processes we have

$$r(s, t) = r(s - t)$$

Hence

$$\frac{\partial^2 r(s, t)}{\partial s\, \partial t} = \frac{\partial}{\partial s} \cdot \frac{\partial}{\partial t} r(s - t) = - r''(s - t)$$

We thus find that a stationary process is differentiable in the mean square if the covariance function is twice differentiable at the origin. It further follows from (6.9) that the differentiated process is weakly stationary with the covariance function $- r''(\tau)$.

EXAMPLE 6.2

As an application we will now investigate the differentiability of the Wiener process. We will first use the definition directly

$$\lim_{h \to 0} \frac{x(t + h) - x(t)}{h}$$

If $\{x(t), t \in T\}$ is a Wiener process we have

$$E[x(t + h) - x(t)]^2 = h$$

Hence

$$E\left[\frac{x(t + h) - x(t)}{h}\right]^2 = h^{-1} \quad (6.11)$$

As $h \to 0$ we thus find that (6.11) diverges and we thus find that the Wiener process is not differentiable.

We will now obtain the same result by applying Theorem 6.4. The Wiener process has the covariance function

$$r(s, t) = \min(s, t)$$

We find

$$\frac{\partial r(s, t)}{\partial s} = \begin{cases} 1 & s < t \\ 0 & s > t \end{cases}$$

which implies that the mixed second derivative does not exist. According to Theorem 6.4, the Wiener process thus is not differentiable. However, if we calculate formally we find

$$\frac{\partial^2 r(s, t)}{\partial s\, \partial t} = - \delta(s - t)$$

where $\delta(r)$ is Dirac's delta function. Formally we thus find that the derivative of the Wiener process is white noise.

Since the Wiener process was introduced as a mathematical model for the motion of a particle in Brownian motion it might seem strange that the Wiener process does not have a derivative. Going back to the physical interpretation, this implies that it is not possible to define the velocity of the particle. To resolve this difficulty we will discuss how the model was derived. The equation of motion for a particle submerged in a fluid is

$$m\frac{d^2x}{dt^2} + D\frac{dx}{dt} = F$$

where m is the mass, D the coefficient of viscous damping, and F the resultant of the forces acting on the particle. Einstein assumed $m = 0$, that is, he neglected the mass forces in comparison with the viscous forces. The equation of motion then becomes

$$D\frac{dx}{dt} = F$$

With constant force, neglecting the mass implies that the velocity instantaneously assumes the equilibrium value. Using idealized shock theory we only compute the change of momentum due to the impact and we do not analyze the details of the change in velocity during the shock. If we now assume that the time interval between the shocks is infinitely small we find it natural that it is not possible to define the velocity.

Integrability

We will now introduce integrals of stochastic processes. Let $\{x(t), t \in T\}$ be a stochastic process of second order. Consider an interval $[a, b] \in T$. Let $a = t_0 < t_1 < \cdots < t_n = b$ be a subdivision of $[a, b]$. Consider the sum

$$I_n = \sum_{k=1}^{n} x(\tau_k)[t_k - t_{k-1}] \qquad (6.12)$$

where $t_{k-1} \leqslant \tau_k \leqslant t_k$. The process $\{x(t), t \in T\}$ is said to be Riemann integrable if I_n converges to a limit in the mean square as $n \to \infty$ in such a way that

$$\max_{1 \leqslant k \leqslant n} |t_k - t_{k-1}| \to 0$$

The limit is called the mean square Riemann integral of x over $[a, b]$ and is denoted by

$$I = \int_a^b x(t)\, dt \qquad (6.13)$$

We have the following result.

Analysis of Stochastic Processes

THEOREM 6.5

Let $\{x(t), t \in T\}$ be a stochastic process of second order with mean value function $m(t)$ and covariance function $r(s, t)$. The process is Riemann integrable if the integrals

$$\int_a^b m(t)\, dt$$

and

$$\int\int_a^b r(s, t)\, ds\, dt$$

exist. In that case we have

$$E\int_a^b x(t)\,dt = \int_a^b Ex(t)\,dt = \int_a^b m(t)\, dt \qquad (6.14)$$

$$E\int\int_a^b x(t)x(s)\, ds\, dt = \int\int_a^b \{Ex(t)x(s)\}\, ds\, dt$$

$$= \left[\int_a^b m(t)\, dt\right]^2 + \int\int_a^b r(s, t)\, ds\, dt \qquad (6.15)$$

A different type of integral will be introduced in Section 5 of Chapter 3.

Exercises

1. Consider stationary stochastic processes with the covariance functions

    $$r(\tau) = e^{-\alpha|\tau|}$$
    $$r(\tau) = \begin{cases} 2 & \tau = 0 \\ e^{-|\tau|} & \tau \neq 0 \end{cases}$$

    Are the processes continuous in the mean square?

2. Consider stationary stochastic processes with the covariance functions

    $$r(\tau) = e^{-\alpha|\tau|}$$
    $$r(\tau) = \frac{\alpha}{\alpha^2 + \tau^2}$$
    $$r(\tau) = \frac{\sin \alpha\tau}{\tau}$$

    Are the processes differentiable in the mean square?

3. Consider the random telegraph wave. Are the sample functions differentiable with probability one? Is the process differentiable with probability one or in the mean square?

4. Consider a normal stationary process with the covariance function

$$r(\tau) = (1 + |\tau|)e^{-|\tau|}$$

Show that the process is differentiable in the mean square and prove that x and dx/dt are independent.

5. Let $\{x(t), t \geqslant 0\}$ be a Wiener process with unit variance parameter. Show that the process is integrable in the mean square. Determine the covariance function of the integrated process

$$z(t) = \int_0^t x(s)\,ds$$

$\left(\text{Answer: } E\,z(s)\,z(t) = \dfrac{s^2(3t-s)}{6}\right)$

6. Consider a stochastic process $\{x(t), -\infty < t < \infty\}$ with the following properties: The sample functions assume the values $+1$ and -1 only. The probability that x changes in the interval $(t, t+h)$ is $\lambda h + o(h)$. Answer the following questions.

Are the sample functions continuous?
Is the stochastic process continuous in the mean square?
Is the stochastic process continuous with probability one?
Is the stochastic process differentiable in the mean square?
Are the sample functions differentiable?
Is the stochastic process integrable in the mean square?
Are the sample functions integrable?

Hint: The probability that the sample functions changes n times in the interval $(0, t)$ is $p_n = ((\lambda t)^n/n!)\exp(-\lambda t)$.

7. BIBLIOGRAPHY AND COMMENTS

There are several good textbooks available on stochastic processes. The book

Karlin, S., *A First Course in Stochastic Processes*, Academic Press, New York, 1966.

is an excellent introduction. Among other books on an elementary level we can mention

Cox, D. R. and Miller, H. D., *The Theory of Stochastic Processes*, Methuen, London, 1965.
Parzen, E., *Stochastic Processes*, Holden-Day, San Fransisco, 1962.
Prabhu, N. U., *Stochastic Processes—Basic Theory and Its Applications*, MacMillan, New York, 1965.

A more advanced presentation is given in the excellent textbook

Gikhman, I.I. and Skorokhod, A.V., *Introduction to the Theory of Random Processes*, W. B. Saunders, Philadelphia, 1969.

Bibliography and Comments

The books

Doob, J. L., *Stochastic Processes*, Wiley, New York, 1953.
Loeve, M., *Probability Theory*, Van Nostrand, Princeton, New Jersey, 1963.

are no easy reading but Doob's book in particular is a good reference work with particular emphasis on the measure theoretical aspects.

The original work on Kolmogorov's theorem

Kolmogorov, A. N., *Foundations of the Theory of Probability*, Chelsea, New York, 1931.

is still very readable.

The recent book

Cramér, H. and Leadbetter, M. R., *Stationary and Related Stochastic Processes*, Wiley, New York, 1967.

is of particular interest because it gives an elegant way to bypass the difficulty discussed in Section 2 and to introduce continuity of sample functions with probability one.

There are also many books available which deal with stochastic processes of special types, e.g.,

Yaglom, A. M., *An Introduction to the Theory of Stationary Random Functions*, Prentice Hall, Englewood Cliffs, New Jersey, 1962.
Bharucha-Reid, A. T., *Elements of the Theory of Markov Processes and Their Applications*, McGraw-Hill, New York, 1960.
Levy, P., *Processus Stochastiques et Mouvement Brownian*, Gauthier-Villars, Paris, 1948.
Ito, K. and McKean, H. P., *Diffusion Processes and Their Sample Paths*, Springer-Verlag, Berlin, 1965.

CHAPTER 3

STOCHASTIC STATE MODELS

1. INTRODUCTION

State models, i.e., systems of first order difference or differential equations, are very convenient for the analysis of deterministic systems. This chapter will show how the notion of state can be carried over to stochastic systems. This leads to the introduction of stochastic difference equations and stochastic differential equations. The processes defined by such equations are Markov processes.

The discrete time case is straigtforward and we can get a stochastic state model simply by adding a disturbance term to the right member of an ordinary difference equation. Roughly speaking, the disturbance should be a sequence of independent random variables in order to obtain a state model. The discrete time case is covered in Sections 2 and 3. The continuous time case is more involved. We first argue heuristically in Section 4. A direct attempt to generalize the discrete time case by adding a disturbance term to the right member of an ordinary differential equation fails. The reason for this is that a continuous random process, which has the property that $x(t)$ and $x(s)$ are independent for $t \neq s$, is zero in the mean square. This means that when a disturbance term of this nature is added to an ordinary differential equation we will not get any effect. This analysis is carried out in Section 4. Having understood the nature of the difficulty we will then arrive at the concept of a stochastic differential equation in a natural way. In the Sections 5 to 8 we will then develop the machinery required to handle stochastic differential equations. We will thus give a precise meaning to such equations and show how they are interpreted and manipulated. Modeling of physical processes by stochastic state models is discussed in Section 9, and in Section 10 we consider the approximation of a stochastic differential equation by a stochastic difference equation.

2. DISCRETE TIME SYSTEMS

The concept of state has its roots in the cause-and-effect relations of classical mechanics. For example, the motion of a system of particles is uniquely determined for all future time by the future forces and by the present positions and momenta of all particles. It is immaterial for the future motion how the actual positions and momenta were attained. Intuitively speaking, the state is the minimal amount of information about the past history of a system which is required to predict the future motion. For a deterministic system in free motion, the future motion is uniquely determined by the actual value of the state.

For a stochastic system we naturally can not require that the future motion be uniquely determined by the actual state x. A natural extension of the concept of state to stochastic systems would be to require that the *probability distributions* of the state variable x at future times should be uniquely determined by the actual value of the state. Recalling the definition of a Markov process (Section 3 of Chapter 2) we would thus require that the system can be described as a Markov process.

This section will consider discrete time systems only. The index set T will thus be the set of integers $\{\ldots, -1, 0, 1, \ldots\}$. An n-dimensional deterministic discrete time system can be described by the difference equation

$$x(t+1) = g(x(t), t), \quad t \in T \tag{2.1}$$

where x is an n-dimensional state vector and t denotes time. The future motion of the system is thus uniquely determined by the value of x at time t. It does not depend on the manner in which $x(t)$ was attained.

We will now discuss how the model (2.1) can be converted to a stochastic state model. One way to do this is to assume that $x(t+1)$ is not uniquely given by $x(t)$ as expressed by (2.1), but that $x(t+1)$ is a random variable which depends on $x(t)$ and t. We can then write

$$x(t+1) = g(x(t), t) + v(x(t), t), \quad t \in T \tag{2.2}$$

where g is the conditional mean of $x(t+1)$ given $x(t)$ and v is a random variable with zero mean.

If (2.2) should be a stochastic state model, we must require that the conditional probability distribution of $x(t+1)$ given $x(t)$ does not depend on the past values of x. This implies that the conditional distribution of $v(t)$ given $x(t)$ must not depend on $x(s)$ for $s < t$. A model (2.2) with this property is called a *stochastic difference equation*. The process $\{x(t), t \in T\}$ defined by (2.2) is then also a Markov process.

If we specialize further and assume that the conditional distribution of $v(t)$ given $x(t)$ is normal, the stochastic variable v can always be normalized by its variance $\sigma^2(x, t)$ so that $v(t)/\sigma(x, t)$ is normal $(0, 1)$. We thus get

$$v = v(x, t) = \sigma(x, t) e(t) \tag{2.3}$$

where $e(t)$ is independent of x. It is thus possible to take $\{e(t), t \in T\}$ as a sequence of independent equally distributed normal $(0, 1)$ random variables. Equation (2.2) then becomes

$$x(t + 1) = g(x(t), t) + \sigma(x(t), t) e(t), \quad t \in T \tag{2.4}$$

If we specialize even further and assume that g is linear in x and that σ does not depend on x we obtain the *linear stochastic difference equation*

$$x(t + 1) = \Phi(t + 1; t)x(t) + \Gamma(t)e(t), \quad t \in T \tag{2.5}$$

Notice that we can easily repeat the argument for the case when x is an n-vector. We will then arrive at (2.5) where x is a n-vector and $\{e(t), t \in T\}$ a sequence of independent equally-distributed Gaussian random vectors with zero mean values and covariance matrix R_0.

Equation (2.5) can also be written as

$$x(t + 1) = \Phi(t + 1; t)x(t) + v(t) \tag{2.6}$$

where $\{v(t), t \in T\}$ is a sequence of independent Gaussian vectors with zero mean values and covariance matrices $R_1 = \Gamma(t) R_0 \Gamma^T(t)$. If (2.6) should be a state model we must also require that $\{v(t), t \in T\}$ is independent of the initial state.

If only one component of the state vector is of interest we can always write (2.6) as an nth order difference equation

$$\begin{aligned}x_1(t) + a_1(t)x_1(t-1) + \cdots + a_n(t)x_1(t-n) \\= c_1(t)e(t-1) + c_2(t)e(t-2) + \cdots + c_n(t)e(t-n)\end{aligned} \tag{2.7}$$

For discrete time systems we can thus easily introduce a stochastic state model in the form of a stochastic difference equation. In the linear case such an equation reduces to a linear difference equation with a forcing function which is discrete time white noise.

Special cases of (2.7) have been of much interest in the statistical literature. Section 2 of Chapter 2 discussed the *autoregressive process*

$$x_1(t) + a_1 x_1(t-1) + \cdots + a_n x_1(t-n) = e(t)$$

and the *moving average process*

$$x_1(t) = c_1 e(t-1) + c_2 e(t-2) + \cdots + c_n e(t-n)$$

3. SOLUTION OF STOCHASTIC DIFFERENCE EQUATIONS

An ordinary difference equation is solved if we know the values of x for all t. As an ordinary difference equation defines each $x(t)$ recursively from the previous value, the difference equation as such can be regarded as an algorithm for producing the solution. Similarly we will consider a stochastic difference equation solved if we know the joint probability distribution of values of the state variables at arbitrary values of time.

Consider, for example, the stochastic difference equation (2.2). Assume that we would like to know the joint probability distribution of $x(t_0)$ and $x(t_0 + 1)$. As $x(t_0 + 1)$ is given as a function of $x(t_0)$ and $v(x(t_0), t)$ we can calculate the conditional distribution of $x(t_0 + 1)$ given $x(t_0)$, using the ordinary rules for calculating the distribution of a function of a random variable. If the distribution of $x(t_0)$ is known, the joint distribution of $x(t_0)$ and $x(t_0 + 1)$ can then be obtained using Bayes rule. We thus find that the joint probability distribution of successive values of the state variable can be written as

$$F[x(t), x(t-1), \ldots, x(t_0)]$$
$$= F[x(t) \mid x(t-1)]F[x(t-1) \mid x(t-2)] \ldots F[x(t_0+1) \mid x(t_0)]F[x(t_0)]$$

where $F[x(t+1) \mid x(t)]$ denotes the conditional distribution of $x(t+1)$ given $x(t)$ (the transition probabilities), and $F[x(t_0)]$ is the distribution of $x(t_0)$. It is clear from this expression that the solution of a stochastic difference equation is a Markov process. Compare the definition in Section 3 of Chapter 2. To carry out the computation outlined above in detail will naturally be far from trivial in the general situation. We will therefore specialize to linear normal systems where all details are easily handled analytically.

Linear Equations

Consider the linear normal stochastic differential equation

$$x(t+1) = \Phi(t+1; t)x(t) + e(t) \qquad (3.1)$$

where x is an n-dimensional state vector, $\{e(t), t \in T\}$ is a sequence of n-dimensional normal independent random vectors, Φ is an $n \times n$ matrix with time varying elements. The vectors $e(t)$ and $e(s)$ are thus independent if $t \neq s$. The vector $e(t)$ is also independent of $x(t)$. The normal distribution of $e(t)$ is specified by the first and second moments:

$$Ee(t) = \begin{bmatrix} 0 \\ 0 \\ \vdots \\ 0 \end{bmatrix}, \quad Ee(t)e^T(t) = \begin{bmatrix} r_{11}(t) & r_{12}(t) & \cdots & r_{1n}(t) \\ r_{12}(t) & r_{22}(t) & \cdots & r_{2n}(t) \\ \vdots & & & \\ r_{1n}(t) & r_{2n}(t) & \cdots & r_{nn}(t) \end{bmatrix} = R_1(t) \quad (3.2)$$

The initial state $x(t_0)$ is also assumed normal with mean m_0 and covariance matrix R_0.

The stochastic process $\{x(t), t \in T\}$ is a normal process since the values of x at particular times are linear combinations of normal variables. The stochastic process can thus be completely characterized by its mean value function and its covariance function. These two functions will now be determined.

The Mean Value Function

To determine the mean value function we take mathematical expectation of both members of (3.1). We then find

$$m(t + 1) = Ex(t + 1)$$
$$= E[\Phi(t + 1; t)x(t) + e(t)] = \Phi(t + 1; t)m(t) \quad (3.3)$$
$$m(t_0) = Ex(t_0) = m_0 \quad (3.4)$$

The mean value function is thus given by the linear difference equation (3.3) with the initial condition (3.4).

The Covariance Function

To evaluate the covariance function we assume that m_0 is zero. This is no loss in generality since we can always introduce the variables $z(t) = x(t) - m(t)$, but it will make the calculations simpler. Assume $s \geqslant t$. Then

$$x(s) = \Phi(s; t)x(t) + \Phi(s; t + 1)e(t) + \cdots$$
$$+ \Phi(s; s - 1)e(s - 2) + e(s - 1)$$

where

$$\Phi(s; t) = \Phi(s; s - 1)\Phi(s - 1; s - 2) \cdots \Phi(t + 1; t) \quad (3.5)$$

Hence

$$Ex(s)x^T(t) = E\{[\Phi(s; t)x(t) + \Phi(s; t + 1)e(t) + \cdots + e(s - 1)]x^T(t)\}$$

As $e(s)$ has zero mean and also is independent of $x(t)$ if $s \geqslant t$, we find that all terms of the right member except the first one will vanish. Hence

$$R(s, t) = \Phi(s; t)R(t, t) = \Phi(s; t)P(t), \quad s \geqslant t \quad (3.6)$$

It now remains to determine the covariance of x

$$P(t) = \operatorname{cov}[x(t), x(t)] = Ex(t)x^T(t)$$

We form

$$x(t + 1)x^T(t + 1) = [\Phi(t + 1; t)x(t) + e(t)][\Phi(t + 1; t)x(t) + e(t)]^T$$
$$= \Phi(t + 1; t)x(t)x^T(t)\Phi^T(t + 1; t) + \Phi(t + 1; t)x(t)e^T(t)$$
$$+ e(t)x^T(t)\Phi^T(t + 1; t) + e(t)e^T(t)$$

Solution of Stochastic Difference Equations

Taking mathemical expectation and observing that $x(t)$ and $e(t)$ are independent, we find the following difference equation for $P(t)$.

$$P(t + 1) = \Phi(t + 1; t)P(t)\Phi^T(t + 1; t) + R_1(t) \tag{3.7}$$

The initial condition is given by

$$P(t_0) = Ex(t_0)x^T(t_0) = R_0 \tag{3.8}$$

Notice the physical interpretation of the different terms of (3.7). The first term of the right member shows how the variance $P(t)$ of the state at time t is transformed through the system dynamics. The second term R_1 gives the increase in variance due to the disturbance $e(t)$. We summarize the result as Theorem 3.1.

THEOREM 3.1

The solution of the normal and linear stochastic difference equation (3.1) is a normal random process characterized by the mean value function $m(t)$ given by the difference equation

$$m(t + 1) = \Phi(t + 1; t)m(t) \tag{3.3}$$

with the initial condition

$$m(t_0) = m_0 \tag{3.4}$$

and the covariance function

$$R(s, t) = \Phi(s; t)P(t) \qquad s \geqslant t \tag{3.6}$$

where $P(t)$ satisfies

$$P(t + 1) = \Phi(t + 1; t)P(t)\Phi^T(t + 1; t) + R_1 \tag{3.7}$$

with initial condition

$$P(t_0) = R_0 \tag{3.8}$$

When the matrices Φ and R_1 are constant it follows from (3.7) and (3.8) that

$$P(t) = \Phi^t R_0 (\Phi^T)^t + \sum_{s=0}^{t-1} \Phi^s R_1 (\Phi^T)^s \tag{3.9}$$

If all the eigenvalues of the matrix Φ have magnitudes strictly less than one, the above series will converge and we find that the limit

$$P_\infty = \lim_{t \to \infty} P(t)$$

exists. Taking limits in (3.7) we also find that P_∞ satisfies the following equation

$$P_\infty = \Phi P_\infty \Phi^T + R_1 \tag{3.10}$$

For linear discrete time stochastic difference equations, we thus find that conditional distributions of future states given $x(t)$ are normal. The mean values and covariances of the distribution are easily computed from recursive equations.

Exercises

1. A dynamical system is described by the difference equation
$$x(t+1) = \begin{pmatrix} \cos h & \sin h \\ -\sin h & \cos h \end{pmatrix} x(t)$$
where $h = \pi/(4n)$. The initial state $x(0)$ is normal with the mean value
$$Ex(0) = \begin{pmatrix} 1 \\ 0 \end{pmatrix}$$
and the covariance
$$\text{cov}[x(0), x(0)] = \begin{pmatrix} 1 & -1 \\ -1 & 1 \end{pmatrix}$$
Determine the smallest t^* such that the components x_1 and x_2 are independent for $t = t^*$. Determine the distribution of $x(t^*)$.

2. A dynamical system is governed by the stochastic difference equation
$$x(t+1) = \begin{pmatrix} 1.5 & 1 \\ -0.7 & 0 \end{pmatrix} x(t) + \begin{pmatrix} 1.0 \\ 0.5 \end{pmatrix} e(t)$$
where $\{e(t), t \in T\}$ is a sequence of independent normal $(0, 1)$ stochastic variables. Determine the covariance of the steady state distribution.

3. A stochastic process is defined by the stochastic difference equation
$$x(t+1) = ax(t) + e(t) \qquad |a| < 1$$
where $\{e(t), t \in T\}$ is a sequence of independent normal $(0, \sigma)$ random variables. The initial state $x(t_0)$ is normal $(0, \sigma_0)$, and the variables $\{e(t), t \in T\}$ are independent of $x(t_0)$. Determine the variance of $x(t)$ and the limit of the variance as $t \to \infty$ or $t_0 \to -\infty$. Show that if σ_0 is chosen as $\sigma_0^2 = \lim_{t \to \infty} P(t)$ and if the index set is taken as $T = \{t_0, t_0 + 1, \ldots\}$ the process $\{x(t), t \in T\}$ is stationary. Determine the covariance function and the spectral density for this stationary process.

4. Consider a stationary stochastic process which satisfies the stochastic difference equation

$$x(t+1) = \Phi x(t) + e(t)$$

where $\{e(t), t \in T\}$ is a sequence of independent equally distributed vectors with zero mean values and the covariance matrix R_1. Let the characteristic polynomial of Φ be

$$\det[\lambda I - \Phi] = \lambda^n + a_1\lambda^{n-1} + \cdots + a_n$$

Show that the covariance function, $r(t)$, of an arbitrary linear combination of the components of the state variable satisfies

$$r(t) + a_1 r(t-1) + \cdots + a_n r(t-n) = 0 \qquad t \geqslant n$$

4. CONTINUOUS TIME SYSTEMS

This section will now discuss continuous time systems. In analogy with the discrete time case, we will try to construct a state model by adding a stochastic disturbance to an ordinary differential equation. When we pursue this in detail, we will encounter some difficulties. To overcome these, we introduce the notion of stochastic differential equations. This is done heuristically in this section. The following sections of this chapter will then give precise definitions of such equations and their solutions.

A Nonsuccessful Construction

In analogy with the treatment of discrete time systems in Section 2, we start with a deterministic state model described by an ordinary differential equation.

$$\frac{dx}{dt} = f(x, t) \tag{4.1}$$

This equation means that the rate of change of the state is uniquely determined by the current value of the state. In order to obtain a stochastic state model, we then assume that dx/dt is a random variable whose probability distribution is uniquely determined by time and the current value of the state vector. Hence

$$\frac{dx}{dt} = f(x, t) + v(x, t) \tag{4.2}$$

where $\{v(x, t), t \in T\}$ is a zero mean random process. There is no generality lost by this assumption because a nonzero mean can always be accounted for by changing the function f appropriately. It now remains to determine the suitable properties of the random process $\{v(x, t)\ t \in T\}$. If (4.2) should be a state model we must require that $v(x, t)$ and $v(y, s)$ are independent for every x and y if $t \neq s$, otherwise the probability distribution of dx/dt would depend not only on the actual state but also on the manner in which the actual state is attained. If (4.2) should have any meaning we must impose some regularity conditions. If we use the mean square topology, it is natural to require that dx/dt has finite variance, that is, that v has finite variance. It might also be necessary to impose some continuity assumptions, for example, that v is continuous in the sense of mean square convergence. Before proceeding we will now analyze the character of a stochastic process which has these properties.

THEOREM 4.1

Let $\{v(t), t \in T\}$ be a continuous time second order stochastic process with the properties

1. $v(t)$ is independent of $v(s)$ for $t \neq s$
2. $v(t)$ continuous in mean square for all $t \in T$ and has bounded variance
3. $v(t)$ has zero mean

Then $Ev^2(t) = 0$.

Proof

As the process $\{v(t), t \in T\}$ is continuous in the mean square it can be integrated. We can thus form

$$u(t) = \int_0^t v(s)\, ds \qquad (4.3)$$

where the integral is understood to be defined in the sense of mean square. We find

$$Eu(t) = E\int_0^t v(s)\, ds = \int_0^t Ev(s)\, ds = 0$$

where the second equality follows from Theorem 6.2 of Chapter 2. We have further

$$Eu^2(t) = E\left(\int_0^t v(s)ds\right)^2 = E[(\lim \sum v(t_i)(t_{i+1} - t_i))]^2$$
$$= E \lim \sum_i \sum_j v(t_i)v(t_j)(t_{i+1} - t_i)(t_{j+1} - t_j)$$
$$= \lim \sum_i \sum_j Ev(t_i)v(t_j)(t_{i+1} - t_i)(t_{j+1} - t_j)$$
$$= \lim \sum_i Ev^2(t_i)(t_{i+1} - t_i)^2$$

where $(0 = t_0, t_1, \cdots, t_N = t)$ is a subdivision of the interval $(0, t)$ and the last equality follows from the property 1. Since $\{v(t)\}$ has bounded variance we have $Ev^2 \leqslant c$.

Continuous Time System

Now let max $(t_{i+1} - t_i) \to 0$ in such a way that max $(t_{i+1} - t_i) \leqslant a/N$. Then

$$Eu^2(t) = \lim \sum Ev^2(t_i)(t_{i+1} - t_i)^2 \leqslant \lim_{N\to\infty} c\left(\frac{a}{N}\right)^2 N = 0$$

Hence

$$Eu^2(t) = 0$$

Equation (4.3) implies that the mean square derivative of u exists. Hence

$$Ev^2 = E\left(\frac{du}{dt}\right)^2 = E \lim_{h\downarrow 0} \lim_{k\downarrow 0} \frac{u(t+h) - u(t)}{h} \cdot \frac{u(t+k) - u(t)}{k}$$

$$= \lim_{h\downarrow 0}\lim_{k\downarrow 0} \frac{1}{hk} E[u(t+h) - u(t)][u(t+k) - u(t)] = 0$$

where the last equality follows from Schwartz inequality

$$| Eu(t)u(s) |^2 \leqslant Eu^2(s) Eu^2(t) = 0$$

Compare (4.5) of Chapter 2. We thus find that $v(t)$ is zero in the mean square sense and the theorem is proven.

We will now discuss the implications of this theorem with respect to the proposed model (4.2). As the process $\{v(x, t), t \in T\}$ with the desired properties is zero in the sense of mean square, it will have no influence on the solution of (4.2). Integrating the proposed stochastic state model (4.2) and comparing it with the deterministic model (4.1), we find that the difference between the solutions is zero in the mean square, and we have thus failed in our attempts to construct a stochastic state model with desired properties.

In order to obtain a sensible stochastic state model in the continuous case, we must thus change the requirements on the process $\{v(t), t \in T\}$. Apart from the assumption of zero mean, which is not essential, we are left with two conditions

1. $v(s)$ and $v(t)$ independent
2. v has finite variance

Hence if we insist to have a state model condition 2 must be relaxed. The relaxation of this condition must naturally raise questions as to the validity of (4.2) because if v does not have finite variance neither has dx/dt. For a stochastic state model we thus cannot expect dx/dt to exist !

Back to Basics

We have thus failed in our initial attempt to construct a stochastic state model for continuous time systems, but we have gained some insight and we will now make another attempt. We first observe that an ordinary differential equation can be obtained by a limit process. Starting with the difference

$$x(t + h) - x(t) = f(x, t)h + o(h) \tag{4.4}$$

we get (4.1) by dividing by h and letting $h \to 0$. As (4.4) is a difference equation we can easily obtain a stochastic difference equation by adding a disturbance to the right member as was done in Section 2. Let $\{v(t), t \in T\}$ be a random process with independent increments. Consider the following model

$$x(t + h) - x(t) = f(x, t)h + v(t + h) - v(t) + o(h) \tag{4.5}$$

As $\{v(t), t \in T\}$ is a process with independent increments, (4.5) is obviously a state model for all values of h. In analogy with the previous analysis we will now specialize and assume that the conditional distribution of $v(t + h) - v(t)$ given $x(t)$ is normal. Hence

$$v(t + h) - v(t) = \sigma(x,t) [w(t + h) - w(t)] \tag{4.6}$$

where $\{w(t), t \in T\}$ is a Wiener process with unit variance parameter. We thus obtain the following stochastic state model

$$x(t + h) - x(t) = f(x, t)h + \sigma(x, t)[w(t + h) - w(t)] + o(h) \tag{4.7}$$

We have

$$E[x(t + h) - x(t)] = f(x, t)h + o(h) \tag{4.8}$$

$$\operatorname{var}[x(t + h) - x(t)] = \sigma^2(x, t)E[w(t + h) - w(t)]^2 + o(h)$$
$$= h\sigma^2(x,t) + o(h) \tag{4.9}$$

Notice in particular that the variance of the increment is proportional to h (and not h^2). It is thus clear that we cannot divide (4.7) by h and let h tend to zero, because the derivative of the Wiener process does not exist. Compare Section 6 of Chapter 2. We can, however, let h go to zero in (4.7). Doing so we obtain the following formal expression

$$dx = f(x, t) dt + \sigma(x, t) dw \tag{4.10}$$

which is called a *stochastic differential equation*. The model (4.7) implies that the increment of the state variable is a sum of two terms. One is a deterministic term which is the product of a function of the state and a time increment. The other is a stochastic term which is a function of the state multiplied by the increment of a Wiener process.

A stochastic differential equation is called *linear* if f is linear in x and if σ does not depend on x. The argument can easily be generalized to the vector case. A linear stochastic differential equation is thus

$$dx = Ax\, dt + dv \tag{4.11}$$

where x is a n-vector and $\{v(t), t \in T\}$ an n-dimensional Wiener process with incremental covariance $R_1 dt$. The elements of the matrices A and R_1 may be continuous functions of time. Notice in (4.10) that $E(dw)^2 = dt$ which means that dw is of the magnitude $\sqrt{dt}$ in the mean square metric. This implies that some care must be exercised when making formal manipulations with expressions involving dw.

Also notice that if we accept the notion of continuous time white noise, (4.10) can be interpreted as

$$\frac{dx}{dt} = f(x, t) + \sigma(x, t)e(t) \qquad (4.12)$$

where $\{e(t), t \in T\}$ is continuous time white noise.

If we compare (4.12) with the discussion in the beginning of this section, we thus find that if we want to obtain a nontrivial stochastic state model by adding a disturbance to the right member of an ordinary stochastic differential equation, the disturbance cannot have finite variance but must be of white noise character.

Using the notion of a stochastic differential equation, it will be possible to handle differential equations driven by white noise rigorously. It should be mentioned that formal manipulations of equations like (4.12) can easily lead to wrong results.

Before proceeding let us also briefly discuss the use of a stochastic differential equation as a model for physical processes. Having an equation like (4.10) we know that the derivative dx/dt does not exist in any reasonable sense. Using (4.10) as a model of a physical process thus looks strange. There are, however, many practical problems both in control and communications where we would not like to take derivatives of certain signals. Modeling these signals as stochastic differential equations is then appropriate.

Backward Differences

If we interpret the stochastic differential (4.10) as the limit in the mean square of a difference equation we must exercise some care. In the preceding analysis we have considered (4.10) as the limit of

$$x(t + h) - x(t) = f(x(t), t)h + \sigma(x(t), t)[w(t + h) - w(t)] + o(h) \qquad (4.13)$$

Notice, however, that we will get different results if we consider (4.10) as the limit of

$$x(t) - x(t - h) = f(x(t), t)h + \sigma(x(t), t)[w(t) - w(t - h)] + o(h) \qquad (4.14)$$

even if f and σ are continuous. To show this we will compute the first two moments of the increment of the process (4.14) in the case when f is continuously differentiable and σ twice continuously differentiable. We observe that the difference $[w(t) - w(t - h)]$ is independent of $x(t - h)$ but that it depends on $x(t)$. A Taylor series expansion of the right member of (4.14) about the point $(x(t - h), t - h)$ now gives

$$x(t) - x(t - h) = f(x(t - h), t - h)h + \sigma(x(t - h), t - h)[w(t) - w(t - h)]$$
$$+ \sigma_x(x(t - h), t - h)[w(t) - w(t - h)][x(t) - x(t - h)] + o(h) \quad (4.15)$$

The third term of the right member arises from the fact that $[x(t) - x(t - h)]$ and $[w(t) - w(t - h)]$ are both of the magnitude $\sqrt{h}$ in the mean square metric. Since only the term $\sigma[w(t) - w(t - h)]$ of the right member of (4.14) is of magnitude $\sqrt{h}$ we get

$$x(t) - x(t - h) = f(x(t - h), t - h)h$$
$$+ \sigma(x(t - h), t - h)[w(t) - w(t - h)]$$
$$+ \sigma_x(x(t - h), t - h)\sigma(x(t - h), t - h)[w(t) - w(t - h)]^2 + o(h)$$

Taking mathematical expectation we find

$$E[x(t) - x(t - h)]$$
$$= f(x(t - h), t - h)h + \sigma_x(x(t - h), t - h)\sigma(x(t - h), t - h)h + o(h) \quad (4.16)$$
$$\text{var}\,[x(t) - x(t - h)] = \sigma^2(x, t)h + o(h) \quad (4.17)$$

A comparison of (4.8) and (4.16) shows that the mean of the increment of the process depends on the type of difference that is used. We get

$$Edx = \begin{cases} f(x, t)dt & \text{(forward difference)} \\ f(x, t)dt + \sigma_x(x, t)\sigma(x, t)dt & \text{(backward difference)} \end{cases} \quad (4.18)$$

If we use a mixed difference

$$\Delta x(t) = (1 - \lambda)[x(t + h) - x(t)] + \lambda[x(t) - x(t - h)] \quad (4.19)$$

we get

$$Edx = f(x, t)dt + \lambda\sigma_x(x, t)\sigma(x, t)dt \quad (4.20)$$

We also find that the variance of the increment is the same in all cases.

$$\text{var}\,(dx) = \sigma^2(x, t)dt \quad (4.21)$$

Although the results obtained are not greatly different, it is naturally very important to keep in mind the type of difference that is used when defining the stochastic differential. In the heuristic arguments which led to the stochastic differential equation, we actually introduced $f(x, t)h$ as the mean value of the increment. To retain this intuitive idea we are thus forced to interpret the stochastic differential equation as a limit of a forward difference equation of type (4.13).

We have thus given a heuristic motivation for stochastic differential equations and we will now proceed to give a precise meaning of this concept. Borrowing the ideas from the theory of ordinary differential equations, we can obtain the desired result in at least two different ways; by taking limits of a difference equation or by converting the differential equation to an integral equation, which can be handled by successive approximations. In this particular case we can thus try to define (4.10) as a limit of difference equations, or we can first try to show existence and uniqueness of the integral equation

$$x(t) = x(t_0) + \int_0^t f(x(s), s)\, ds + \int_0^t \sigma(x(s), s)\, dw(s) \qquad (4.22)$$

The first approach has been followed by Bernstein, Levy, and Gikhman. The integral equation has been studied by Ito. To handle the integral (4.22) we must first give a meaning to integrals of the type which are represented by the last term of the right member. This will be done in the following sections.

Exercise

1. Let $\{w(t),\ t \in T\}$ be a Wiener process with unit variance parameter. Put

$$\Delta w(t) = w(t + h) - w(t)$$

Show that

$$\int_0^h (dw)^2 = \lim \sum [w(t_{i+1}) - w(t_i)]^2 = h$$

with probability one.

5. STOCHASTIC INTEGRALS

In the previous section we have found that integrals of the type

$$\int f(t)\, dy(t) \qquad (5.1)$$

where $\{y(t),\ t \in T\}$ is a process with independent normal increments, are required in order to give a precise meaning to a stochastic differential equation.

In special cases y can be a Wiener process. Such a process is continuous with probability one. However, almost all sample functions have unbounded variation. The integral (5.1) thus cannot immediately be interpreted as an ordinary Stieltjes integral. In this section we will outline how to define the integral (5.1). We will also discuss some properties which are of interest for the control and modeling problem. Many results will be stated without proof. For details, refer to Doob's book on Stochastic processes.*

We will first consider the case when f is a deterministic function and later consider the case when f is a random process. In that case we find that the integral (5.1) has some properties that are drastically different from an ordinary Stieltjes integral.

The Integral of a Deterministic Function

When f is a deterministic function there are at least two ways to define the integral (5.1). If f is sufficiently smooth, the integral can be defined as follows

$$\int_a^b f(t)\, dy(t) = f(b)y(b) - f(a)y(a) - \int_a^b y(t)\, df(t) \tag{5.2}$$

Since the sample functions of the process $\{y(t), t \in T\}$ are continuous with probability one, the integral of the right member exists for almost all sample functions if f has bounded variation,

This way of defining the integral has the nice feature that the integral can be interpreted as an integral of sample functions. The definition can not, however, be extended to the case when f is a stochastic process, for example, a Wiener process. Neither does the definition preserve the intuitive idea that the integral is a limit of sums of independent random variables.

For this reason we will therefore give another definition of the integral (5.1) using standard methods of integration theory. We will first define the integral when f is a class of piecewise constant functions, and we will then extend the integral to more general classes of functions. Assume that f is constant over the intervals $[t_i, t_{i+1})$. The integral (5.1) is then defined as

$$I = \int f(t)\, dy(t) = \sum f(\tau_i)[y(t_{i+1}) - y(t_i)] \tag{5.3}$$

where $t_i \leqslant \tau_i < t_{i+1}$. The integral has the following properties

$$EI = \sum f(\tau_i) E[y(t_{i+1}) - y(t_i)] = \int f(t)\, dm_y(t) \tag{5.4}$$

where

$$m_y(t) = Ey(t) \tag{5.5}$$

This follows from Theorem 6.2 of Chapter 2.
Furthermore

$$\operatorname{var} I = \operatorname{var} \sum_i f(\tau_i)[y(t_{i+1}) - y(t_i)]$$

* See Bibliography and Comments.

$$= \sum_i \sum_j f(\tau_i) f(\tau_j) \operatorname{cov}\{[y(t_{i+1}) - y(t_i)], [y(t_{j+1}) - y(t_j)]\}$$

$$= \sum_i f^2(\tau_i)[r(t_{i+1}) - r(t_i)] = \int f^2(\tau) \, dr(\tau) \qquad (5.6)$$

Now let the function f be a limit of a sequence of piecewise constant functions $\{f_n\}$ in the sense that

$$\max\left[\int (f_n - f)^2 dr, \int |f_n - f| \, dm\right] \to 0$$

The integral of the function f can now be defined using standard extension technique. We thus define the integral of the function f as

$$I = \lim \int f_n(t) \, dy(t) \qquad (5.7)$$

where the limit is taken in the sense of mean squares

$$E[I - \int f_n(t) \, dy(t)]^2 \to 0 \qquad (5.8)$$

It is also possible to define the integral as any random variable which equals I with probability one. In this way it is possible to extend the definition of the integral to functions f for which the integrals $\int f^2 \, dr$ and $\int f \, dm$ exist. The extension will preserve the properties (5.4) and (5.6). Hence

$$E \int f(t) \, dy(t) = \int f(t) \, dm(t) \qquad (5.9)$$

$$\operatorname{var} \int f(t) \, dy(t) = \int f^2(t) \, dr(t) \qquad (5.10)$$

A detailed proof is given in Doob's book on Stochastic Processes.

Integrals of Stochastic Processes

We will now consider the case when f is a stochastic process. The integral (5.1) can then be defined as a limit in the quadratic mean as was done previously. If f does not depend on y the generalization is straightforward and no particularly interesting results are obtained. If f depends on y the integral will, however, have some strange properties. In this case the integral will for example depend on the choice of τ_i. We illustrate this by an example.

EXAMPLE 5.1
Consider the integral

$$\int_0^t w(s) \, dw(s) \qquad (5.11)$$

where $\{w(t), t \in T\}$ is a Wiener process with unit variance parameter. Let the interval $(0, t)$ be divided into N subintervals by the points $0 = t_1, t_2, \ldots, t_N, t_{N+1} = t$. Using the outlined procedure the integral (5.11) can for example be defined by either of the following expressions

$$I_0 = \lim \sum_{i=1}^{N} w(t_i)[w(t_{i+1}) - w(t_i)] \quad (5.12)$$

$$I_1 = \lim \sum_{i=1}^{N} w(t_{i+1})[w(t_{i+1}) - w(t_i)] \quad (5.13)$$

where the limit is taken in the sense of mean square.

If the integrals could be defined as ordinary Stieltjes integrals, they would be identical because the integrand is continuous. The stochastic integrals, however, do not have this property because

$$I_1 - I_0 = \lim \sum_{i=1}^{N} [w(t_{i+1}) - w(t_i)]^2 = t \quad (5.14)$$

Compare the exercise in Section 4.

We thus find from the Example 5.1 that the choice of τ_i in the formula (5.3) is crucial for the definitions of stochastic integrals. Having made this observation, we also find that it is possible to define a continuum of stochastic integrals by the formula

$$I_\lambda = (1 - \lambda)I_0 + \lambda I_1 \quad 0 \leqslant \lambda \leqslant 1$$

$$= \lim \sum_{i=1}^{N} [(1 - \lambda)w(t_i) + \lambda w(t_{i+1})][w(t_{i+1}) - w(t_i)] \quad (5.15)$$

Some of these integrals have special names. I_0 is called the *Ito integral* and $I_{0.5}$ is called the *Stratonovich integral*. To find out whether there is any preference for any of these integrals we will investigate their properties closer.

Let $\{y(t), t \in T\}$ be a normal process with independent increments. Denote the mean value function by $m(t)$ and the variance function by $r(t)$. The Ito integral defined by

$$\int f(t) \, dy(t) = \lim \sum f(t_i)[y(t_{i+1}) - y(t_i)] \quad (5.16)$$

then has the properties

$$E \int f(t) \, dy(t) = \int \{E f(t)\} \, dm(t) \quad (5.17)$$

$$\text{cov}\left[\int f(t) \, dy(t), \int g(t) \, dy(t)\right] = \int [E f(t)g(t)] \, dr(t) \quad (5.18)$$

A rigorous proof of this is found in Doob's book on Stochastic processes.

Stochastic Integrals 61

The equations (5.17) and (5.18) imply that the operations of taking mathematical expectation and integration in the sense of Ito can be interchanged.

As was illustrated in Example 5.1 the formula (5.17) will not hold for the integrals I_λ if $\lambda \neq 0$.

Integration by Parts

In ordinary calculus we have the following well-known formula for integration by parts

$$\int_0^t f(s)\, dy(s) = f(s)\, y(s) \Big|_0^t - \int_0^t y(s)\, df(s) \tag{5.19}$$

The formula can also be written as

$$f(t)\, y(t) - f(0)\, y(0) = \int_0^t f(s)\, dy(s) + \int_0^t y(s)\, df(s) \tag{5.19'}$$

We will now investigate if there is a corresponding formula for stochastic integrals.

Let $0 = t_1, t_2, \ldots, t_{N+1} = t$ be a subdivision of the interval $(0, t)$. Consider the identity

$$f(t_{k+1}) y(t_{k+1}) - f(t_k) y(t_k)$$
$$= f(t_{k+1})[y(t_{k+1}) - y(t_k)] + y(t_k)[f(t_{k+1}) - f(t_k)]$$
$$= y(t_{k+1})[f(t_{k+1}) - f(t_k)] + f(t_k)[y(t_{k+1}) - y(t_k)]$$

Summing over k and taking limits in the sense of mean square as $\max_i |t_{i+1} - t_i|$ tends to zero we find

$$f(t)\, y(t) - f(0)\, y(0) = I_1(f, dy) + I_0(y, df)$$
$$= I_1(y, df) + I_0(f, dy) \tag{5.20}$$

where

$$I_0(f, dy) = \lim \sum_{i=1}^N f(t_i)[y(t_{i+1}) - y(t_i)] \tag{5.21}$$

$$I_1(f, dy) = \lim \sum_{i=1}^N f(t_{i+1})[y(t_{i+1}) - y(t_i)] \tag{5.22}$$

We thus find that there is a correspondence to the formula for integration by parts. Notice, however, that if f is a stochastic process with independent increments we need two integral concepts I_0 and I_1 to obtain the formula.

Also notice that if we introduce the symmetric integral

$$I_{0,s} = \frac{1}{2}(I_0 + I_1) \tag{5.23}$$

we get from (5.20)

$$f(t) y(t) - f(0) y(0) = I_{0.5}(f, dy) + I_{0.5}(y, df) \qquad (5.24)$$

which is the well-known formula (5.19') for integration by parts. The symmetric integral $I_{0.5}$ defined by

$$I_{0.5} = \lim \frac{1}{2} \sum_{i=1}^{N} [f(t_i) + f(t_{i+1})][y(t_{i+1}) - y(t_i)] \qquad (5.25)$$

was previously called the Stratonovich integral. We can thus conclude that the ordinary formula for integration by parts is valid also for stochastic integrals, if they are interpreted as Stratonovich integrals.

Comparison with Formal Integration

To get further insight into the properties of the stochastic integral we will make comparisons with formal integration in cases where this is possible. We will first consider the integral $\int w \, dw$. If w is an ordinary function we get

$$\int_0^t w(s) \, dw(s) = \frac{1}{2}(w^2(t) - w^2(0))$$

Now if $\{w(t)\}$ is a Wiener process with unit variance parameter we can interprete the integral in many different ways. Consider

$$I_\lambda(w, dw) = \lim \sum_{i=1}^{N} [\lambda w(t_{i+1}) + (1 - \lambda) w(t_i)][w(t_{i+1}) - w(t_i)]$$

The summand can be rewritten as

$$[\lambda w(t_{i+1}) + (1 - \lambda) w(t_i)][w(t_{i+1}) - w(t_i)]$$
$$= w(t_i) w(t_{i+1}) - w^2(t_i) + \lambda [w(t_{i+1}) - w(t_i)]^2$$
$$= \frac{1}{2}[w^2(t_{i+1}) - w^2(t_i)] + \left(\lambda - \frac{1}{2}\right)[w(t_{i+1}) - w(t_i)]^2$$

As the sum

$$\sum_{i=1}^{N} [w(t_{i+1}) - w(t_i)]^2$$

converges to t with probability one as $N \to \infty$ we have

$$I_\lambda(w, dw) = \frac{1}{2}[w^2(t) - w^2(0)] + \left(\lambda - \frac{1}{2}\right) \cdot t$$

In this particular example, the stochastic integral is thus equivalent to formal integration if we interpret the stochastic integral as the Stratonovich integral.

Linear Stochastic Differential Equations

Exercise

1. Consider the integral

$$I = \int_0^t f(s)\, dy(s)$$

where $\{y(t),\, t \in T\}$ is a process with independent normal increments with mean m and incremental covariance dr. Assume $y(0) = 0$. Let f, m, and r be continuously differentiable. Denote the derivatives by f', m', and r'. Define the integral I by (5.2)

$$I = f(t)\, y(t) - \int_0^t f'(s)\, y(s)\, ds$$

Show that

$$EI = \int_0^t f(s) m'(s)\, ds$$

$$\operatorname{var} I = \int_0^t f^2(s) r'(s)\, ds$$

Compare with the (5.9) and (5.10).

6. LINEAR STOCHASTIC DIFFERENTIAL EQUATIONS

This section will give a precise meaning to the linear stochastic differential equation introduced in the Section 4. Thus consider

$$dx = A(t) x\, dt + dv \tag{6.1}$$

where x is an n-vector, $\{v(t),\, t \in T\}$ an n-dimensional Wiener process with incremental covariance $R_1\, dt$. The matrix A is $n \times n$. The elements of A and R_1 are continuous functions of time. It is assumed that the initial value $x(t_0)$ is a normal random variable with mean m_0 and covariance R_0.

Comparing with the heuristic arguments of Section 4, we find that $\sigma = I$ does not depend on x. If (6.1) is interpreted as a limit in mean square of a difference equation, it is immaterial whether we use forward or backward differences. Compare (4.15).

If v in (6.1) would have bounded variation, the solution of (6.1) can be written as

$$x(t) = \Phi(t;\, t_0)\, x(t_0) + \int_{t_0}^t \Phi(t;\, s)\, dv(s) \tag{6.2}$$

where Φ satisfies the differential equation

$$\frac{d\Phi(t;\, t_0)}{dt} = A(t)\, \Phi(t;\, t_0) \tag{6.3}$$

with the initial condition

$$\Phi(t_0; t_0) = I \qquad (6.4)$$

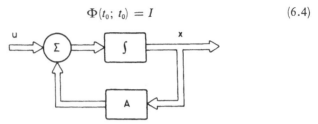

Fig. 3.1. Block diagram of the system $dx/dt = Ax + u$.

Equation (6.1) is often used to model a system whose input u is bandlimited white noise. Compare the diagram in Fig. 3.1. To preserve such an interpretation, it would be desirable to represent (6.1) by a block diagram similar to that of Fig. 3.1 where the input is the *derivative* of v. Since v does not have a derivative this cannot be done. However, if the integral of (6.2) is defined by (5.2), the solution of (6.1) can be interpreted as an integral of sample functions and it can also be represented by a block diagram, where the input is v. See Fig. 3.2. Notice that such a representation can, strictly speaking, not be given if (6.1) is interpreted using stochastic integrals, since the integration symbol of the usual block diagram notation represents integration of time functions.

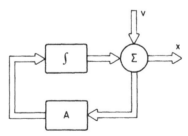

Fig. 3.2. Block diagram of the stochastic differential equation (6.1) when the solution is interpreted using the integral definition given by (5.2).

The properties of the solution of (6.2) will now be investigated. Since x is a linear function of a normal process, it is also normal and can be characterized completely by the mean value function and the covariance function. To compute these, we proceed as follows

$$Ex(t) = \Phi(t; t_0) Ex(t_0) + E\int_{t_0}^{t} \Phi(t; s)\, dv(s)$$

Using the property (5.9) of the stochastic integral we find that the second term of the right member will vanish. Hence

Linear Stochastic Differential Equations

$$m_x(t) = Ex(t) = \Phi(t; t_0) Ex(t_0) = \Phi(t; t_0) m_0 \qquad (6.5)$$

Taking derivatives we get

$$\frac{dm_x}{dt} = \frac{d}{dt} \Phi(t; t_0) m_0 = A(t)\Phi(t; t_0) m_0 = A(t) m_x \qquad (6.6)$$

where the second equality follows from (6.3). Equation (6.6) can also be derived directly by taking mathematical expectation of (6.1) and using Theorem 6.2 of Chapter 2.

The initial value of (6.6) is obtained from (6.5). We get

$$m_x(t_0) = m_0 \qquad (6.7)$$

To compute the covariance of x we assume $m_0 = 0$. This can always be achieved by subtracting m_x from x. Hence $Ex(t) = 0$. We get for $s \geqslant t$

$$R(s, t) = \text{cov}[x(s), x(t)] = Ex(s) x^T(t)$$
$$= E[\Phi(s; t)x(t) + \int_t^s \Phi(s; s') \, dv(s')] x^T(t)$$
$$= \Phi(s; t) Ex(t) x^T(t) = \Phi(s; t) R(t, t) = \Phi(s; t) P(t) \qquad (6.8)$$

The first equality follows from $Ex(t) = 0$ and the third equality follows from the fact that $v(s)$ is independent of $x(t)$ for $s \geqslant t$. Analogously we find

$$R(s, t) = R(s, s)\Phi^T(s; t) = P(s)\Phi^T(s; t) \qquad (6.8')$$

for $s \leqslant t$. To compute the covariance $P(t) = R(t, t)$ of $x(t)$ we form

$$\text{cov}[x(t), x(t)] = Ex(t) x^T(t)$$
$$= E[\Phi(t; t_0)x(t_0) + \int_{t_0}^t \Phi(t; s) \, dv(s)] \cdot [\Phi(t; t_0)x(t_0) + \int_{t_0}^t \Phi(t; s) \, dv(s)]^T$$
$$= \Phi(t; t_0) Ex(t_0) x^T(t_0) \Phi^T(t; t_0)$$
$$+ E\left[\int_{t_0}^t \Phi(t; s) \, dv(s)\right]\left[\int_{t_0}^t \Phi(t; s) \, dv(s)\right]^T$$
$$= \Phi(t; t_0) R_0 \Phi^T(t; t_0) + \int_{t_0}^t \Phi(t; s) R_1(s) \Phi^T(t; s) \, ds \qquad (6.9)$$

The third equality follows from $v(s)$ and $x(t)$ being independent for $s \geqslant t$ and the last equality follows from the property (5.10) of the stochastic integral.

Taking derivatives of (6.9) we find

$$\frac{dP}{dt} = \left[\frac{d}{dt}\Phi(t; t_0)\right] R_0 \Phi^T(t; t_0) + \Phi(t; t_0) R_0 \frac{d}{dt} \Phi^T(t; t_0)$$
$$+ \Phi(t; t) R_1(t) \Phi^T(t; t) + \int_{t_0}^t \left[\frac{d}{dt}\Phi(t; s)\right] R_1(s) \Phi^T(t; s) \, ds$$

$$+ \int_{t_0}^{t} \Phi(t; s) R_1(s) \left[\frac{d}{dt} \Phi^T(t; s) \right] ds \tag{6.10}$$

It follows from (6.3) that

$$\frac{d}{dt} \Phi^T(t; t_0) = \Phi^T(t; t_0) A^T(t) \tag{6.11}$$

Equations (6.10) and (6.11) now give

$$\frac{dP}{dt} = A(t)P + PA^T(t) + R_1(t) \tag{6.12}$$

$$P(t_0) = R_0 \tag{6.13}$$

The differential equation (6.12) can also be derived directly from the stochastic differential equation.

Consider the difference

$$P(t + h) - P(t) = Ex(t + h)x^T(t + h) - Ex(t)x^T(t)$$
$$= E\{[x(t + h) - x(t)][x(t + h) - x(t)]^T + x(t)[x(t + h) - x(t)]^T$$
$$\quad + [x(t + h) - x(t)]x^T(t)\}$$
$$= E[(Axh + \Delta v)(Axh + \Delta v)^T + x(Axh + \Delta v)^T$$
$$\quad + (Axh + \Delta v)x^T] + o(h)$$
$$= R_1 h + (Exx^T)A^T h + AhE(xx^T) + o(h)$$
$$= (PA^T + AP + R_1)h + o(h)$$

The fourth equality follows from x and Δv being independent and $E \Delta v (\Delta v)^T = R_1 h + o(h)$. Now divide by h and let $h \to 0$ and we get (6.12). Summarizing we find Theorem 6.1.

THEOREM 6.1

The solution of the stochastic differential equation is a normal process with mean value $m_x(t)$ and covariance $R(s, t)$ where

$$\frac{dm_x}{dt} = A(t) m_x \tag{6.6}$$

$$m_x(t_0) = m_0 \tag{6.7}$$

$$R(s, t) = \begin{cases} \Phi(s; t) P(t) & s \geqslant t \\ P(s) \Phi^T(s; t) & s \leqslant t \end{cases} \tag{6.8}$$

$$\frac{dP}{dt} = AP + PA^T + R_1 \tag{6.12}$$

$$P(t_0) = R_0 \tag{6.13}$$

Linear Stochastic Differential Equations

Remark

Notice that we will obtain the same formulas for the mean and the covariance of the process $\{x(t), t \in T\}$ if it is assumed that $\{v(t), t \in T\}$ is a stochastic process whose increments are uncorrelated with zero mean values and the incremental covariance $R_1 \, dt$.

Having obtained this result we might ask whether it was really worthwhile to go through the machinery of developing the stochastic differential equation in the linear case. Would it not be possible to obtain the same result by formal manipulation of the equality

$$\frac{dx}{dt} = Ax + e \qquad (6.14)$$

where $\{e\}$ is continuous time white noise, that is, a stationary stochastic process with the covariance function

$$\text{cov}\,[e(t), e(s)] = R_1 \delta(t - s)$$

To demonstrate that erroneous results are easily obtained, we will show some results of formal manipulations of (6.14). Consider, for example, the evaluation of the covariance matrix

$$P(t) = Ex(t)\,x^T(t)$$

we get

$$\frac{dP}{dt} = E\frac{dx}{dt}x^T + Ex\frac{dx^T}{dt} = E(Ax + e)x^T + Ex(Ax + e)^T$$
$$= A\,Exx^T + (Exx^T)A^T = AP + PA^T$$

This is apparently wrong. We only get the first two terms of (6.12). Notice that the result would be the same as if $e \equiv 0$. The erroneous result originates from the fact that the derivative dx/dt does not exist. Taking into account that dx is of magnitude $\sqrt{dt}$, we find that the ordinary rules for differentiation do not apply. Consider

$$\Delta xx^T = (x + \Delta x)(x + \Delta x)^T - xx^T = x(\Delta x)^T + (\Delta x)x^T + (\Delta x)(\Delta x)^T$$

If Δx is of magnitude Δt as is the case in ordinary calculus, we find that the last term is of magnitude $(\Delta t)^2$ and small compared to $x\,\Delta x^T$. When x is a Wiener process the last term will, however, be of magnitude Δt. Taking mathematical expectation we get

$$\Delta Exx^T = Ex(Ax\,\Delta t + \Delta v)^T + E(Ax\,\Delta t + \Delta v)x^T$$
$$+ E(Ax\,\Delta t + \Delta v)(Ax\,\Delta t + \Delta v)^T$$
$$= [(Exx^T)A^T + AExx^T]\Delta t + E(\Delta v)(\Delta v)^T + o(\Delta t)$$

Hence

$$dP = (PA^T + AP + R_1)\,dt$$

which gives the correct result.

Exercises

1. Variations in thrust in a rocket engine can be described approximatively as a white noise variation in thrust vector angle. Assume that the spectral density is

   $$N = 0.0004 \text{ rad}^2/\text{Hz}$$

   and that the rocket acceleration has the magnitude 3 m/sec.^{2}. Determine the variance of lateral error and lateral velocity after 100 seconds acceleration.
   Hint: The lateral motion can be described by $\ddot{x} = a\theta$ where θ is the thrust vector angle.

2. Evaluate the steady state covariance matrix for the state of the system

   $$dx = \begin{bmatrix} -a_1 & -a_2 \\ 1 & 0 \end{bmatrix} x\,dt + \begin{bmatrix} 1 \\ 0 \end{bmatrix} dv$$

   where $a_1 > 0$, $a_2 > 0$ and $\{v(t),\ t \in T\}$ is a Wiener process with unit variance parameter.

3. A scalar stochastic process $\{x(t),\ t \in T\}$ satisfies the stochastic differential equation

   $$dx = \alpha x\,dt + dv$$

   where $\{v(t),\ t \in T\}$ is a Wiener process with variance parameter r_1. The initial state x_0 is normal with mean m_0 and covariance r_0. The process $\{v(t),\ t \in T\}$ is independent of $x(t_0)$. Determine the mean value function $m(t)$ and the covariance function for $x(t)$. Give conditions which ensure that the limit of $Ex^2(t)$ exists as $t \to \infty$ or $t_0 \to -\infty$. Show that if we choose $m_0 = 0$, $r_0 = \lim_{t \to \infty} Ex^2(t)$, then the process $\{x(t),\ t_0 \leqslant t < \infty\}$ is stationary. Compute the spectral density and the covariance function for this stationary process.

4. Consider the stationary stochastic process which satisfies the stochastic differential equation

   $$dx = Ax\,dt + dv \qquad (*)$$

   where the characteristic polynomial of the A matrix

   $$\det[\lambda I - A] = \lambda^n + a_1\lambda^{n-1} + \cdots + a_n$$

   has all its zeros in the left half plane and $\{v(t),\ t \in T\}$ is a Wiener process with incremental covariance $R_1 dt$. Show that the covariance

Linear Stochastic Differential Equations

function of an arbitrary linear combination of the state variables of (*) satisfies the differential equation

$$\frac{d^n r}{dt^n} + a_1 \frac{d^{n-1} r}{dt^{n-1}} + \cdots + a_n r = 0, \qquad t > 0$$

5. Consider the stochastic differential equation

$$dx = Ax\, dt + b\, dv$$

where A is a constant $n \times n$ matrix, b a constant n-vector and $\{v(t), t \in T\}$ a Wiener process with unit variance parameter. Assume that all eigenvalues of A have negative real parts. Show that the steady state covariance of x is given by

$$R = \int_0^\infty z(t)\, z^T(t)\, dt$$

where z is the solution of the differential equation

$$\frac{dz}{dt} = Az$$

with the initial condition

$$z(0) = b$$

6. Consider the stationary stochastic process

$$\begin{cases} dx = Ax\, dt + b\, dv \\ y = x_1 \end{cases}$$

where A is a constant $n \times n$ matrix, b a constant n-vector, and $\{v(t), t \in T\}$ a Wiener process with unit variance parameter. Show that the following procedure can be used when evaluating the covariance function of y using an analog computer.

Integrate the differential equation

$$\frac{dz}{dt} = Az$$

with initial condition

$$z(0) = b$$

and compute the components of the vector r defined by

$$r_i = \int_0^\infty z_1(t)\, z_i(t)\, dt \qquad i = 1, 2, \ldots, n$$

The covariance function $r_y(t)$ is then given by

$$\frac{dz}{dt} = Az$$
$$z(0) = r$$
$$r_y(t) = z_1(t)$$

7. Show that the problem of evaluating the covariance matrix of the state of (6.1) is the dual of the problem of evaluating the quadratic loss function

$$V = \int_0^\infty x^T(t) R_1 x(t) \, dt$$

for the dynamical system

$$\frac{dx}{dt} = A^T x$$

with initial condition

$$x(0) = b$$

8. A free particle in a liquid can be described by the so-called Langevin equation

$$m \frac{dv}{dt} + fv = K(t)$$

where m is the mass, f the coefficient of viscous friction, and K is the fluctuating force due to the collisions with molecules of the liquid. The force K has zero average and covariance function which tends to zero very fast in comparison with m/f. The force K can thus be regarded as a white noise process. It follows from the equipartition law of statistical mechanics that the intensity of the fluctuating force is such that

$$\frac{1}{2} mE(v^2) = \frac{1}{2} kT$$

where k is the Boltzman constant and T the temperature of the surrounding medium. Assuming that the particle is initially at rest, determine the probability distribution of the velocity as a function of time.

9. Derive the result corresponding to Theorem 6.1 when the solution of the stochastic differential equation (6.1) is defined by (6.2) and the integral of (6.2) is defined as

$$\int_{t_0}^t \Phi(t;s) \, dv(s) = v(t) - \Phi(t;t_0) v(t_0) - \int_{t_0}^t \left[\frac{d}{ds} \Phi(t;s) \right] v(s) \, ds$$

7. NONLINEAR STOCHASTIC DIFFERENTIAL EQUATIONS

We will now consider the nonlinear stochastic differential equation

$$dx = f(x, t)dt + \sigma(x, t)dw \qquad (7.1)$$

where $\{w(t), t \in T\}$ is a Wiener process with incremental covariance $I\,dt$. By the formal expression (7.1) we mean that x is the solution of the stochastic integral equation

$$x(t) = x(t_0) + \int_{t_0}^{t} f(x(s), s)\,ds + \int_{t_0}^{t} \sigma(x(s), s)\,dw(s) \qquad (7.2)$$

To interpret this equation, we must first define the integral of the right member. This was discussed in Section 5. As was shown in that section, the definition could be done in many different ways. We can, for example, use Ito integrals, Stratonovich integrals, or the integral I_λ defined by (5.15). Notice that no matter what integral concept is used, the integral in (7.2) cannot be interpreted as an ordinary Stieltjes integral of the sample functions because almost all sample functions of a Wiener process have unbounded variation.

No matter what integral concept we use, we must investigate when (7.2) has a unique solution. This can be decided by the standard technique of successive approximations. We thus consider the sequence of stochastic processes

$$x^{n+1}(t) = x^n(t_0) + \int_{t_0}^{t} f(x^n(s), s)\,ds + \int_{t_0}^{t} \sigma(x^n(s), s)\,dw(s) \qquad (7.3)$$

where

$$x^0(t) = x(t_0) \qquad \text{for all } t$$

In the scalar case it has been shown by Ito that if the functions f and σ are such that

$$|f(x, t)| \leq K[1 + x^2]$$
$$0 \leq \sigma(x, t) \leq K[1 + x^2]$$
$$|f(x, t) - f(y, t)| \leq K|x - y|$$
$$|\sigma(x, t) - \sigma(y, t)| \leq K|x - y|$$

then there exists a stochastic process $\{x(t), t \in T\}$ whose sample functions are continuous with probability one. The process $\{x(t)\}$ also satisfies (7.2) with probability one. The multivariate case is treated by Skorokhod.

The solution of (7.2) will, of course, depend on the integral concept that is used to define the stochastic integral. We observe that if the Ito integral is used we get

$$Ex(t) = Ex(t_0) + E\int_{t_0}^{t} f(x(s), s)\, ds$$

because of the property (5.17) of the Ito integral. If (7.2) is defined, using the Ito integral we thus find

$$E[x(t + h) - x(t) \mid x(t)] = f(x, t)h + o(h) \tag{7.4}$$

If we instead interpret the integral in (7.2) as the integral I_λ given by the equation (5.15) we get

$$E[x(t + h) - x(t) \mid x(t)] = [f(x, t) + \lambda \sigma_x(x, t)\sigma(x, t)]h + o(h) \tag{7.5}$$

In all cases we find that the variance of the increment equals

$$\text{var}\,[x(t + h) - x(t) \mid x(t)] = \sigma(x, t)\sigma^T(x, t)h + o(h) \tag{7.6}$$

Recalling the heuristic development of the stochastic state model, we find that $f(x, t)h$ was introduced as the mean value of the increment $x(t + h) - x(t)$. In order to retain this intuitive property of the state model we will thus have to define the stochastic integral as an Ito integral.

When analyzing linear stochastic differential equations in Section 6, we found that the conditional probability distributions of future states $x(t)$, given the actual state $x(t_0)$, were Gaussian. Differential equations for the mean value and the covariance of the distributions were given in Theorem 6.1. For nonlinear systems the corresponding conditional distributions will not be normal. It is, however, possible to derive a partial differential equation for the conditional distribution.

Consider the stochastic process given by (7.1). Let $p(x, t; x_0, t_0)$ denote the probability density of being in state x at time t given that the process is in state x_0 at t_0. Under suitable regularity conditions, it can be shown that p satisfies the following parabolic partial differential equation

$$\frac{\partial p}{\partial t} = \mathscr{L}p = -\sum_{i=1}^{n} \frac{\partial}{\partial x_i}(pf_i) + \frac{1}{2}\sum_{i,j,k=1}^{n} \frac{\partial^2}{\partial x_i\, \partial x_j}(p\sigma_{ik}\sigma_{jk}) \tag{7.7}$$

with initial condition

$$p(x, t_0; x_0, t_0) = \delta(x - x_0) \tag{7.8}$$

This equation is known as the *Fokker-Planck equation* or the *Kolmogorov forward equation*. The operator $\mathscr{L}$ is called the *Kolmogorov forward operator* and its adjoint

$$\mathscr{L}^* = \sum_{i=1}^{n} f_i \frac{\partial}{\partial x_i} + \frac{1}{2}\sum_{i,j,k=1}^{n} \sigma_{ik}\sigma_{jk}\frac{\partial^2}{\partial x_i\, \partial x_j} \tag{7.9}$$

is called the *Kolmogorov backward operator*.

Nonlinear Stochastic Differential Equations

Exercises

1. Consider the differential equation
$$\frac{dx_1}{dt} = e$$
$$\frac{dx_2}{dt} = x_1 e$$
with zero initial condition. Let $\{e(t), t \in T\}$ be a stationary stochastic process with zero mean value and the covariance function $r(\tau)$. Evaluate the mean value of x_2. Compute the limit of the mean value as $r(\tau)$ tends to the Dirac distribution.

2. Consider the stochastic differential equation
$$dx_1 = dw$$
$$dx_2 = x_1 \, dw$$
where $\{w(t), t \in T\}$ is the Wiener process and all initial conditions are zero. The solution to the equation is
$$x_1(t) = w(t)$$
$$x_2(t) = \int_0^t w(s) \, dw(s)$$
Calculate the mean value of x_2 when the integral above is interpreted as the integral I_λ. Consider in particular $\lambda = 0$, 0.5 and 1.

3. Consider the stochastic difference equation
$$\Delta x_1(t) = \Delta w(t)$$
$$\Delta x_2(t) = x_1(t) \Delta w(t)$$
Calculate the mean values of x_1 and x_2 when
$$\Delta f(t) = f(t+h) - f(t)$$
$$\Delta f(t) = f(t) - f(t-h)$$
$$\Delta f(t) = \frac{1}{2}\{f(t+h) - f(t-h)\}$$

4. Let $\{e(t), t \in T\}$ be a stationary stochastic process with covariance function $r(\tau)$. Consider the difference equations
$$x_1(t+h) = x_1(t) + he(t)$$
$$x_2(t+h) = x_2(t) + hx_1(t)\,e(t), \quad t = 0, h, 2h, \cdots$$
with $x_1(0) = x_2(0) = 0$. Evaluate $x_1(Nh)$ and $x_2(Nh)$ and their mean values. Compute the limit of $Ex_2(Nh)$ when $h \to 0$ such that $Nh =$ const and also the limit of $Ex_2(Nh)$ when

$$r(\nu h - \mu h) \to \begin{cases} r(0) & \nu = \mu \\ 0 & \nu \neq \mu \end{cases}$$

as $h \to 0$.

8. STOCHASTIC CALCULUS – THE ITO DIFFERENTIATION RULE

It has been shown in previous sections that the stochastic integrals and the stochastic differential equations do not obey the rules of ordinary calculus. We have also found that the main reason for this is that the increment of a Wiener process dw is of magnitude $\sqrt{dt}$ in the mean square metric. This implies that when differentials of processes which are functions of Wiener processes are computed, we must take into account all terms which are quadratic in dw as these will be of magnitude dt. Keeping this in mind, it is possible to develop the stochastic calculus. The key to this is the following theorem of Ito.

THEOREM 8.1 (The Ito Differentiation Rule)
Let the n-vector x satisfy the stochastic differential equation

$$dx = f(x, t)\, dt + \sigma(x, t)\, dw \qquad (8.1)$$

where $\{w(t), t \in T\}$ is a Wiener process with incremental variance $I\, dt$. Let the function $y(x, t)$ be continuously differentiable in t and twice continuously differentiable in x. Then y satisfies the following stochastic differential equation

$$dy = \frac{\partial y}{\partial t}\, dt + \sum_{i=1}^{n} \frac{\partial y}{\partial x_i}\, dx_i + \frac{1}{2} \sum_{i,j,k=1}^{n} \frac{\partial^2 y}{\partial x_i \partial x_j} \sigma_{ik}\sigma_{jk}\, dt$$

$$= \left[\frac{\partial y}{\partial t} + \sum_{i=1}^{n} \frac{\partial y}{\partial x_i} f_i + \frac{1}{2} \sum_{i,j,k=1}^{n} \frac{\partial^2 y}{\partial x_i \partial x_j} \sigma_{ik}\sigma_{jk} \right] dt + \sum_{i=1}^{n} \frac{\partial y}{\partial x_i} (\sigma\, dw)_i \qquad (8.2)$$

Equation (8.2) can be written in a slightly more compact form if suitable notations are introduced. Let y_t denote the partial derivative $\partial y/\partial t$, y_x the gradient, and y_{xx} the matrix of second order partial derivatives. We then find that (8.2) reduces to

$$dy = [\, y_t + y_x^T f + \frac{1}{2} \operatorname{tr}(y_{xx} \sigma\sigma^T)\,]\, dt + y_x^T \sigma\, dw \qquad (8.3)$$

Theorem 8.1 will not be proven. We will, however, indicate how it can be obtained formally. A Taylor series expansion of y gives

$$\Delta y = y_t \Delta t + y_x^T \Delta x + 1/2\, (\Delta x)^T y_{xx} (\Delta x) + o(\Delta t) \qquad (8.4)$$

But
$$\Delta x = f(x, t)\Delta t + \sigma(x, t)\Delta w$$
Hence
$$(\Delta x)^T y_{xx} \Delta x = (\Delta w)^T \sigma^T y_{xx} \sigma \Delta w + o(\Delta t) = \mathrm{tr}(\Delta w)^T \sigma^T y_{xx} \sigma \Delta w + o(\Delta t)$$
$$= \mathrm{tr}\, y_{xx} \sigma \Delta w \Delta w^T \sigma^T + o(\Delta t) \quad (8.5)$$

We thus find
$$E\Delta y = [y_t + y_x^T f(x, t) + \frac{1}{2} \mathrm{tr}\, y_{xx}\sigma\sigma^T]\Delta t + o(\Delta t)$$
$$\mathrm{var}\,(\Delta y) = y_x^T \sigma\sigma^T y_x \cdot \Delta t + o(\Delta t)$$
because $E\,||\,\Delta w\,||^4 \simeq (\Delta t)^2$.

Notice that in ordinary calculus the term $1/2\,(\Delta x)^T y_{xx}\Delta x$ would be of the order $(\Delta t)^2$. We will now give a few examples which illustrate the application of the Ito differentiation rule.

EXAMPLE 8.1

Let $\{w(t), t \in T\}$ be a Wiener process with unit variance parameter. Consider the function
$$y(t) = e^{w(t)} \quad (8.6)$$
Applying Theorem 8.1 we find
$$dy = e^{w(t)}dw + \frac{1}{2} e^{w(t)} dt$$

The function y thus satisfies the following stochastic differential equation
$$\begin{cases} dy = y(1/2\, dt + dw) \\ y(0) = 1 \end{cases} \quad (8.7)$$
or, conversely, the stochastic differential Eq. (8.7) has the solution (8.6).

EXAMPLE 8.2

Consider the stochastic differential equation
$$dx = Ax\, dt + dv$$
where $\{v(t), t \in T\}$ is a Wiener process with covariance parameter $R_1 dt$. Put
$$y(x, t) = x^T S(t) x$$
Theorem 8.1 then gives
$$dy = d(x^T S x) = \left[x^T \frac{dS}{dt} x + x^T A^T S x + x^T S A x + \mathrm{tr}\, SR_1 \right] dt$$
$$+ dv^T S x + x^T S\, dv \quad (8.8)$$

Evaluation of Loss Functions

When evaluating the performance of a system governed by a stochastic differential equation, we are led to analyze expressions of the following type

$$V(x, t) = E\left[\int_t^{t_1} G(x(s), s)\, ds \mid x(t) = x\right] \tag{8.9}$$

where the stochastic process $\{x(t), t \in T\}$ satisfies (8.1) and G is a scalar function. If G is twice continuously differentiable, it is possible to derive a partial differential equation for V. We get

$$V(x, t) = E\left[\int_t^{t+h} G(x(s), s)\, ds + V(x(t+h), t+h) \mid x(t) = x\right]$$
$$= E[G(x, t)h + V(x, t) + dV(x, t) + o(h) \mid x(t) = x] \tag{8.10}$$

Evaluating dV using Theorem 8.1 and rearranging the terms we get

$$\frac{\partial V}{\partial t} + G + V_x^T f + \frac{1}{2} \text{tr}\, V_{xx}\sigma\sigma^T = 0 \tag{8.11}$$

The boundary value is obtained directly from (8.9)

$$V(x, t_1) = 0 \tag{8.12}$$

We thus find that the evaluation of the loss function (8.9) can be reduced to the solution of the partial differential equation (8.11) with the boundary value (8.12). Notice that (8.11) can be written as

$$\frac{\partial V}{\partial t} + G + \mathscr{L}^* V = 0 \tag{8.13}$$

where $\mathscr{L}^*$ is the Kolmogorov backward operator defined by (7.9).

In quantum mechanics this result has also been used in order to obtain numerical solutions to partial differential eqautions of type (8.11). Trajectories of (8.1) with initial condition x_0 have thus been generated and the value of V at x_0 has been obtained by evaluating (8.9) along the trajectories and averaging, using Monte Carlo techniques.

Exercises

1. Let $\{w(t), t \in T\}$ be a Wiener process with unit variance parameter. Show that the stochastic differential equation

$$\begin{cases} dx = \sqrt{x}\, dw + \frac{1}{4}\, dt \\ x(0) = 1 \end{cases}$$

has the solution

$$x(t) = (1 + \frac{1}{2} w(t))^2$$

2. Let $\{w(t), t \in T\}$ be a Wiener process with unit variance parameter. Show that the stochastic differential equation

$$dx_1 = x_2 dw - \frac{1}{2} x_1 dt \qquad x_1(0) = 0$$

$$dx_2 = -x_1 dw - \frac{1}{2} x_2 dt \qquad x_2(0) = 1$$

has the solution

$$\begin{cases} x_1(t) = \sin w(t) \\ x_2(t) = \cos w(t) \end{cases}$$

3. Find the solution to the stochastic differential equation

$$\begin{cases} dx_1 = x_2 dt \\ dx_2 = dw \end{cases}$$

with the initial condition

$$x_1(0) = 1$$
$$x_2(0) = 0$$

where $\{w(t), t \in T\}$ is a Wiener process with unit variance parameter.

4. Use the Ito differentiation rule to derive Theorem 6.1.

5. The Ito differentials have the property

$$E[x(t+h) - x(t) - d_0 x(t)]^2 = o(h)$$

Compare Section 4. It was also shown in Section 4 that it is possible to introduce other differentials, for example, the backward differential d_1 with the property

$$E[x(t) - x(t-h) - d_1 x(t)]^2 = o(h)$$

or

$$d_\lambda x = (1 - \lambda) d_0 x + \lambda d_1 x$$

Derive the formulas corresponding to the Ito differentiation rule for the differential $d_\lambda x$.

6. Consider the evaluation of the loss function

$$V(a, t) = E\left\{\int_t^{t_1} x^T(s) Q x(s)\, ds \mid x(t) = a\right\}$$

for the linear stochastic differential equation

$$dx = Ax\, dt + dw$$

where $\{w(t), t \in T\}$ is a Wiener process with incremental covariance $R\,dt$. Give a partial differential equation for V and show that it has the solution

$$V(a, t) = a^T S(t) a + \int_t^{t_1} tr S(s) R(s)\, ds$$

where

$$\frac{dS}{dt} + A^T S + SA + Q = 0, \quad S(t_1) = 0$$

7. Consider the linear stochastic differential equation

$$dx = Ax\,dt + dv$$

where $\{v(t), t \in T\}$ is a Wiener process with incremental covariance $R\,dt$. Let

$$y(t, x) = x^T Sx$$

Show that (8.8) holds when the solution of the stochastic differential equation is defined as (6.2) where the integral is defined by (5.2).

9. MODELING OF PHYSICAL PROCESSES BY STOCHASTIC DIFFERENTIAL EQUATIONS

If the stochastic differential equations are going to be of any practical value, it is necessary to know how physical processes should be modeled by such equations. As the result of such a modeling procedure can be judged only by comparison with practical experiments, there is not much we can say in general. As an illustration we will discuss the modeling of a Brownian motion. The motion of a small particle submerged in a fluid can be described by the equation

$$m\frac{d^2 x}{dt^2} + c\frac{dx}{dt} = f \tag{9.1}$$

where x is one coordinate, m is the mass, c the coefficient of viscous friction, and f the forces acting on the particle. For convenience we introduce the state variables $x_1 = x$ and $x_2 = dx/dt$. Equation (9.1) then becomes

$$\frac{dx_1}{dt} = x_2$$

$$\frac{dx_2}{dt} = -\frac{c}{m} x_2 + \frac{1}{m} f \tag{9.2}$$

The forces are due to collisions with molecules of the fluid in thermal motion. If the particle is large, compared to the mean free path of the molecules of the fluid, and, if the impact time is considered infinitesimal, the force will be of white noise character. The mean time between collisions in a liquid of room temperature is of the order 10^{-21} seconds. It thus looks reasonable to model the motion by the following linear stochastic differential equation

$$dx_1 = x_2\, dt$$
$$dx_2 = -\frac{c}{m} x_2\, dt + \frac{1}{m} dw \qquad (9.3)$$

where $\{w(t), t \in T\}$ is a Wiener process with variance parameter r. Equation (9.3) is the so-called Langevin equation.

Applying Theorem 6.1 we find that the covariance function of the process is given by

$$R(s, t) = \Phi(s - t)\, P(t) \qquad (9.4)$$

where

$$\Phi(t) = \exp\begin{bmatrix} 0 & t \\ 0 & -ct/m \end{bmatrix} = \begin{bmatrix} 1 & m/c\,(1 - \exp(-ct/m)) \\ 0 & \exp(-ct/m) \end{bmatrix} \qquad (9.5)$$

and

$$P = \begin{pmatrix} p_1 & p_2 \\ p_2 & p_3 \end{pmatrix} \qquad (9.6)$$

is the solution of the equation (6.12). Writing this equation in terms of components we get

$$\frac{dp_1}{dt} = 2p_2$$
$$\frac{dp_2}{dt} = -\frac{c}{m} p_2 + p_3 \qquad (9.7)$$
$$\frac{dp_3}{dt} = -\frac{2c}{m} p_3 + \frac{r}{m^2}$$

Assuming that the velocity distribution is in steady state and the position of the particle is known exactly at $t = 0$, we get the following initial condition

$$p_1(0) = 0$$
$$p_2(0) = 0 \qquad (9.8)$$
$$p_3(0) = \frac{r}{2mc}$$

The coefficient r can be determined from the equipartition law of statistical mechanics, which implies that in equilibrium the mean energy of each degree of freedom equals $1/2\,kT$ where k is Boltzmann's constant and T the absolute temperature.

Hence

$$\frac{1}{2} m\, Ex_2{}^2 = \frac{m}{2} \cdot \frac{r}{2mc} = \frac{1}{2} kT \qquad (9.9)$$

which gives

$$r = 2kTc \qquad (9.10)$$

Solving (9.7) with the initial conditions in (9.8) we get

$$p_1(t) = \frac{r}{c^2}\left[t - \frac{m}{c}(1 - e^{-ct/m})\right] = \frac{2kT}{c}\left[t - \frac{m}{c}(1 - e^{-ct/m})\right]$$

$$p_2(t) = \frac{r}{2c^2}[1 - e^{-ct/m}] = \frac{kT}{c}[1 - e^{-ct/m}] \qquad (9.11)$$

$$p_3(t) = \frac{r}{2cm} = \frac{kT}{m}$$

The covariance functions of position and velocity are then given by

$$r_{11}(s, t) = p_1(t) + \frac{m}{c}[1 - e^{-(c(s-t)/m)}]p_2(t), \quad s \geqslant t \qquad (9.12)$$

$$r_{22}(s, t) = \frac{kT}{m} e^{-(c(s-t)/m)}, \qquad s \geqslant t \qquad (9.13)$$

The relevance of the model (9.3) can now be tested, for example, by observing the motion of the particle and observing whether statistics of the displacements of the particle can be described as a normal process with the covariance function (9.12).

Before proceeding, we will consider the orders of magnitude involved. For particles of colloidal platinum in water with a radius of 2.5×10^{-8} m we get $m = 2.5 \times 10^{-18}$ kg and $c = 7.5 \times 10^{-12}$ kg/sec. Hence $c/m = 3 \times 10^6$ sec^{-1} !

If the particles are observed at time intervals not closer than one second apart, we find that the correlation between the velocities at the observation times is smaller than $e^{-3\,000\,000}$! It also follows from (9.12) that the covariance function of the displacement is very well approximated by

$$r_{11}(s, t) \sim r(s, t) = \frac{2kT}{c} \min(s, t) \qquad (9.14)$$

The error in this approximation is less than 10^{-6} !

The function $r(s, t)$ in (9.14) is, however, the covariance function of a Wiener process with variance parameter $2kT/c$. Notice that we would get the covariance function (9.14) exactly if the " mass term " m

Modeling of Physical Processes by Stochastic Differential Equations 81

d^2x/dt^2 in (9.1) is neglected or if we assume $dx_2 = 0$ in (9.3). This gives the generally accepted model

$$dx = (1/c)\, dw \qquad (9.15)$$

for the Brownian motion. Notice, however, that this approximation only holds when t is large in comparison with $m/c = 3.3 \times 10^{-7}$ seconds. Hence for problems, where it is sufficient to consider values of the process at times which are not closer than $\theta = m/c$, the model (9.15) will give a very good description of the situation. Also notice that the model (9.15) is of lower order than the original model. The fact that we have a model which is not valid if we consider motions over too small time intervals explains why it is not possible to define the velocity of a particle described by (9.15).

Exercises

1. The motion of a galvanometer in thermal equilibrium with its surroundings can be described by the following equation

$$J\frac{d^2\varphi}{dt^2} + D\frac{d\varphi}{dt} + C\varphi = M$$

   where the torque M is due to collisions on the mirror by molecules of the air. Determine the variance and the spectral density of the galvanometer's deflection if the torque M can be modeled by white noise. Hint: The equipartition law gives

$$\frac{1}{2} JE\left(\frac{d\varphi}{dt}\right)^2 = \frac{1}{2} CE\varphi^2 = \frac{1}{2} kT$$

2. A single-axis gyro-stabilized platform can be described by the following crude model

$$D\frac{d\varphi}{dt} = m + H\dot{\theta}$$

$$\dot{\theta} = -c\varphi$$

   where φ is the output signal of the gyro, θ the angular deflection of the platform, D the viscous damping coefficient of the gyro, H the angular momentum of the gyro, and m the disturbing torques on the floated gyro due to thermal motion of the molecules. Assume that m can be modeled by white noise. Determine the variance of the platform deflection angle due to the thermal fluctuations. Show that for values of t which are large in comparison with D/Hc we get $E\theta^2 \sim 2kT\,DH^{-2}$ where T is absolute temperature and k Boltzmanns constant. Hint: The equipartition law gives $1/2\, cHE\,\varphi^2 = 1/2\, kT$. Put in the following numerical values, $D = 0.03 \text{ kg m}^2/\text{sec}$, $H = 0.01 \text{ kg m}^2/\text{sec}$, and evaluate the standard deviation of platform drift for $t = 1$ sec, and $t = 1$ hour.

10. SAMPLING A STOCHASTIC DIFFERENTIAL EQUATION

There are many situations when predictors and regulators for continuous time processes are implemented using a digital computer. As the digital computer operates sequentially in time, the output signals will be entered into core memory of the computer at discrete times only. The predictions and control signals will also be generated only at discrete times. The theory required to describe the processes can then be simplified considerably if we choose to describe the value of the state variable at the sampling times only. The essential simplification is that the stochastic differential equations are reduced to stochastic difference equations. We will now show how the reduction is carried out in a specific case.

Assume that the process is described by the equations

$$dx = Ax\,dt + dv \tag{10.1}$$

$$dy = Cx\,dt + de \tag{10.2}$$

where x is an n-dimensional state vector, y an r-dimensional vector of measured output signals, and $\{v(t), -\infty \leqslant t \leqslant \infty\}$ and $\{e(t), -\infty, \leqslant t \leqslant \infty\}$ are n- and r-dimensional Wiener-processes with the incremental covariances $R_1\,dt$ and $R_2\,dt$ respectively. It is assumed that the processes e and v are independent. Equation (10.1) represents the process and (10.2) represents how the measurements are related to the state variables. The process $\{e(t)\}$ can thus be considered as the measurement error. Now assume that the output is sampled at discrete times $t_1, t_2, t_3, \cdots$ and that we want equations which relate the values of the state variable x and the measured output y at the sampling times.

Integration of (10.1) and (10.2) gives

$$x(t_{i+1}) = \Phi(t_{i+1}; t_i)x(t_i) + \tilde{v}(t_i) \tag{10.3}$$

$$y(t_{i+1}) = y(t_i) + \int_{t_i}^{t_{i+1}} dy(s)$$

$$= y(t_i) + \left[\int_{t_i}^{t_{i+1}} C(s)\,\Phi(s; t_i)\,ds\right]x(t_i) + \tilde{e}(t_i) \tag{10.4}$$

where the $n \times n$ matrix Φ is defined by

$$\begin{cases} \dfrac{d\Phi(t; t_i)}{dt} = A(t)\,\Phi(t; t_i) & t_i \leqslant t \leqslant t_{i+1} \\ \Phi(t_i; t_i) = I \end{cases} \tag{10.5}$$

and
$$\tilde{v}(t_i) = \int_{t_i}^{t_{i+1}} \Phi(t_{i+1};\, t)\, dv(t) \tag{10.6}$$

$$\tilde{e}(t_i) = \int_{t_i}^{t_{i+1}} C(s) \int_{t_i}^{s} \Phi(s;\, t)\, dv(t)\, ds + \int_{t_i}^{t_{i+1}} de(s) \tag{10.7}$$

Rewriting the first integral of the right member of (10.7) by changing the order of the integrations, we get

$$\int_{t_i}^{t_{i+1}} C(s) \int_{t_i}^{s} \Phi(s;\, t)\, dv(t)\, ds = \int_{t_i}^{t_{i+1}} \left(\int_{t}^{t_{i+1}} C(s)\, \Phi(s;\, t)\, ds \right) dv(t)$$

$$= \int_{t_i}^{t_{i+1}} \theta(t_{i+1};\, t)\, dv(t) \tag{10.8}$$

where
$$\theta(t_{i+1};\, t) = \int_{t}^{t_{i+1}} C(s)\, \Phi(s;\, t)\, ds \tag{10.9}$$

Hence
$$\tilde{e}(t_i) = \int_{t_i}^{t_{i+1}} \theta(t_{i+1};\, t)\, dv(t) + e(t_{i+1}) - e(t_i) \tag{10.10}$$

Using the properties of the stochastic integral we now find that $\tilde{v}(t_i)$ and $\tilde{v}(t_k)$ are independent if $i \neq k$. We also get from (10.6) and (10.10)

$$E\tilde{v}(t_i) = E \int_{t_i}^{t_{i+1}} \Phi(t_{i+1};\, t)\, dv(t) = 0$$

$$E\tilde{e}(t_i) = E\left[\int_{t_i}^{t_{i+1}} \theta(t_{i+1};\, t)\, dv(t) + e(t_{i+1}) - e(t_i) \right] = 0$$

Furthermore (10.6) and (10.10) give

$$E\tilde{v}(t_i)\tilde{v}^T(t_i) = E\left[\iint_{t_i}^{t_{i+1}} \Phi(t_{i+1};\, t)\, dv(t)\, dv^T(s)\, \Phi^T(t_{i+1};\, s) \right]$$

$$= \int_{t_i}^{t_{i+1}} \Phi(t_{i+1};\, t)\, R_1(t)\, \Phi^T(t_{i+1};\, t)\, dt \tag{10.11}$$

$$E\tilde{v}(t_i)^T\tilde{e}(t_i) = E\left[\iint_{t_i}^{t_{i+1}} \Phi(t_{i+1};\, t)\, dv(t)\, dv^T(s)\, \theta^T(t_{i+1};\, s) \right]$$

$$= \int_{t_i}^{t_{i+1}} \Phi(t_{i+1};\, t)\, R_1(t)\, \theta^T(t_{i+1};\, t)\, dt \tag{10.12}$$

$$E\tilde{e}(t_i)\tilde{e}^T(t_i) = E\left[\iint_{t_i}^{t_{i+1}} \theta(t_{i+1};\, t)\, dv(t)\, dv^T(s)\, \theta^T(t_{i+1};\, s) \right.$$
$$\left. + (e(t_{i+1}) - e(t_i))^2 \right]$$
$$= \int_{t_i}^{t_{i+1}} \theta(t_{i+1};\, t)\, R_1(t)\, \theta^T(t_{i+1};\, t)\, dt + \int_{t_i}^{t_{i+1}} R_2(t)\, dt \tag{10.13}$$

The mixed terms which contain both de and dv will vanish because the processes v and e are independent. Summing up we get Theorem 10.1.

THEOREM 10.1

The values of state variables and outputs of the stochastic differential equations (10.1) and (10.2) at discrete times t_i are related through the stochastic difference equations

$$x(t_{i+1}) = \Phi x(t_i) + \tilde{v}(t_i) \tag{10.3}$$

$$z(t_{i+1}) = y(t_{i+1}) - y(t_i) = \theta x(t_i) + \tilde{e}(t_i) \tag{10.14}$$

where $\Phi = \Phi(t_{i+1}; t_i)$ is the fundamental matrix of (10.5); $\theta = \theta(t_{i+1}; t_i)$ defined by (10.9), $\{\tilde{v}(t_i), i = 1, 2, \cdots\}$ and $\{\tilde{e}(t_i), i = 1, 2, \cdots\}$ are sequences of independent normal random variables with zero mean values and the covariances

$$E\tilde{v}(t_i)\tilde{v}^T(t_i) = \tilde{R}_1(t_i) = \int_{t_i}^{t_{i+1}} \Phi(t_{i+1}; s) R_1(s) \Phi^T(t_{i+1}; s) \, ds \tag{10.11}$$

$$E\tilde{v}(t_i)\tilde{e}^T(t_i) = \tilde{R}_{12}(t_i) = \int_{t_i}^{t_{i+1}} \Phi(t_{i+1}; s) R_1(s) \theta^T(t_{i+1}; s) \, ds \tag{10.12}$$

$$E\tilde{e}(t_i)\tilde{e}^T(t_i) = \tilde{R}_2(t_i) = \int_{t_i}^{t_{i+1}} [\theta(t_{i+1}; s) R_1(s) \theta^T(t_{i+1}; s) + R_2(s)] \, ds \tag{10.13}$$

The stochastic difference equations (10.3) and (10.4) are referred to as the sampled versions of (10.1) and (10.2).

It follows from the derivation that the statistical properties of (10.1), (10.2), (10.3), and (10.4) are identical at the sampling intervals. This means that the sampled version can be conveniently used to approximate the continuous time problem. Notice that the "measurement errors" $\tilde{e}$ and the "process disturbances" $\tilde{v}$ of (10.3) and (10.4) can be dependent even if e and v are independent in (10.1) and (10.2).

An Application

We will now discuss some implications of Theorem 10.1 for the practical implementation of digital predictors and regulators. In practice we use (10.1) and (10.2) to model the process

$$\frac{dx}{dt} = Ax + \dot{v} \tag{10.15}$$

$$\dot{y} = Cx + \dot{e} \tag{10.16}$$

where $\{\dot{v}(t), -\infty < t < \infty\}$ and $\{\dot{e}(t), -\infty < t < \infty\}$ are stationary Gaussian processes with spectral densities which are constant up to high frequencies. (Compare Section 9.) If we model (10.15) and (10.16), and design a digital filter based on the corresponding sampled models, it means that the signal which is fed into the control computer is given by

$$y(t_{i+1}) - y(t_i) = \int_{t_i}^{t_{i+1}} \dot{y}(t) \, dt \tag{10.17}$$

Sampling a Stochastic Differential Equation

Interpreting this physically, we thus find that the measurement signal $\dot{y}$ which contains high frequency noise is not sampled directly. The measurement signal $\dot{y}$ is first integrated using an analog integrator which is reset to zero at the beginning of each sampling interval. From a practical point of view, this makes a lot of sense because the integration will reduce the high frequency noise effectively. Also notice that, if some other type of analog prefiltering is used, we must either take the dynamics of the prefilter into account in the discrete model or choose the prefilter in very special ways, if the sampled version is going to be a stochastic difference equation.

Exercises

1. Consider the stochastic differential equation

$$dx = \begin{pmatrix} 0 & 1 \\ 0 & 0 \end{pmatrix} x \, dt + \begin{pmatrix} 0 \\ 1 \end{pmatrix} dv$$

$$dy = x_1 \, dt + de$$

   where $\{v(t), t \in T\}$ is a Wiener process with unit variance parameter and $\{e(t), t \in T\}$ is a Wiener process with variance parameter r. Determine the sampled version when the sampling interval is h.

2. Consider the stochastic differential equations (10.1) and (10.2) where the Wiener processes $\{v(t), t \in T\}$ and $\{e(t), t \in T\}$ are correlated with the joint incremental covariance

$$E \begin{pmatrix} dv \\ de \end{pmatrix} (dv^T \, de^T) = \begin{pmatrix} R_1 & R_{12} \\ R_{12}^T & R_2 \end{pmatrix} dt$$

   Show that the sampled version of the stochastic differential equation are given by (10.3) and (10.4) where $\{\tilde{e}(t_i)\}$ and $\{\tilde{v}(t_i)\}$ are sequences of independent normal variables with zero mean values and the covariances

$$E v(t_i) v^T(t_i) = \tilde{R}_1(t_i) = \int_{t_i}^{t_{i+1}} \Phi(t_{i+1}; s) R_1(s) \Phi^T(t_{i+1}; s) \, ds$$

$$E v(t_i) e^T(t_i) = \tilde{R}_{12}(t_i) = \int_{t_i}^{t_{i+1}} \Phi(t_{i+1}; s) [R_1(s) \theta^T(t_{i+1}; s) + R_{12}(s)] \, ds$$

$$E e(t_i) e^T(t_i) = \tilde{R}_2(t_i) = \int_{t_i}^{t_{i+1}} [\theta(t_{i+1}; s) R_1(s) \theta^T(t_{i+1}; s) + \theta(t_{i+1}; s) R_{12}(s)$$
$$+ R_{12}^T(s) \theta^T(t_{i+1}; s) + R_2(s)] \, ds$$

3. Consider the system with

$$A = \begin{pmatrix} 0 & 0 & 0 \\ 1 & 0 & 0 \\ 0 & 1 & 0 \end{pmatrix}$$

$$C = \begin{pmatrix} 0 & 1 & 0 \\ 0 & 0 & 1 \end{pmatrix}$$

$$R_1 = \begin{pmatrix} q_1 & 0 & 0 \\ 0 & q_2 & 0 \\ 0 & 0 & r \end{pmatrix}, \quad R_{12} = \begin{pmatrix} 0 & 0 \\ 0 & 0 \\ r & 0 \end{pmatrix}, \quad R_2 = \begin{pmatrix} r & 0 \\ 0 & r \end{pmatrix}$$

Show that the sampled version with sampling interval h is given by

$$\Phi = \begin{pmatrix} 1 & 0 & 0 \\ h & 1 & 0 \\ \frac{1}{2}h^2 & h & 1 \end{pmatrix}, \quad \theta = \begin{pmatrix} \frac{1}{2}h^2 & h & 0 \\ \frac{1}{6}h^3 & \frac{1}{2}h^2 & h \end{pmatrix}$$

$$\tilde{R}_1 = \begin{pmatrix} q_1 h & \frac{1}{2} q_1 h^2 & \frac{1}{6} q_1 h^3 \\ \frac{1}{2} q_1 h^2 & \frac{1}{3} q_1 h^3 + q_2 h & \frac{1}{8} q_1 h^4 + \frac{1}{2} q_2 h^2 \\ \frac{1}{6} q_1 h^3 & \frac{1}{8} q_1 h^4 + \frac{1}{2} q_2 h^2 & \frac{1}{20} q_1 h^5 + \frac{1}{3} q_1 h^3 + rh \end{pmatrix}$$

$$\tilde{R}_{12} = \begin{pmatrix} \frac{1}{6} q_1 h^3 & \frac{1}{24} q_1 h^4 \\ \frac{1}{8} q_1 h^4 + \frac{1}{2} q_2 h^2 & \frac{1}{30} q_1 h^5 + \frac{1}{6} q_2 h^3 \\ \frac{1}{20} q_1 h^5 + \frac{1}{3} q_2 h^3 + rh & \frac{1}{72} q_1 h^6 + \frac{1}{8} q_1 h^4 + \frac{1}{2} rh^2 \end{pmatrix}$$

$$\tilde{R}_2 = \begin{pmatrix} \frac{1}{40} q_1 h^5 + \frac{1}{3} q_2 h^3 + rh & \frac{1}{72} q_1 h^6 + \frac{1}{8} q_2 h^4 + \frac{1}{2} rh^2 \\ \frac{1}{72} q_1 h^6 + \frac{1}{8} q_2 h^4 + \frac{1}{2} rh^2 & \frac{1}{252} q_1 h^7 + \frac{1}{20} q_2 h^5 + \frac{1}{3} rh^3 \end{pmatrix}$$

4. Derive the result which corresponds to Theorem 10.1 when the output of the sampled system is defined by

$$z(t_{i+1}) = \int_{\tau_i}^{t_{i+1}} dy(s) \qquad t_i < \tau_i < t_{i+1}$$

instead of (10.14).

11. BIBLIOGRAPHY AND COMMENTS

The idea of characterizing stochastic processes by stochastic difference equation is due to Yule who introduced the auto-regressive process in

Yule, G. U., "On a Method of Investigating Periodicities in Disturbed Series with Special Reference to Wolfer's Sunspot Numbers," *Phil. Trans. Roy. Soc. A* **226** 267-298 (1927).

Stochastic difference equations are also discussed in

Wold, H., *Stationary Time Series*, Almqvist and Wiksell, Uppsala, 1938.

Stochastic differential equations have been used heuristically in physics since the beginning of this century in connection with analysis of Brownian motion. Einstein's papers which are collected in

Einstein, A., *Investigations on the Theory of the Brownian Motion*, Dover, New York, 1956.

are well worth reading.

The book

Wax, N., *Selected Papers on Noise and Stochastic Processes*, Dover, New York, 1954.

contains an excellent selection of early papers. Thermal motions related to Brownian motion also give a definite limit to the precision of all measurements. This was early observed in connection with high precision galvanometers and circuit noise (Nyquist noise) in electronic amplifiers. These phenomena are discussed in

Barnes R.B. and Silverman, S. "Brownian Motion as a Natural Limit to all Measuring Processes," *Rev. Mod. Phys.* **6**, 162-192 (1934).

McCombie, C. W., "Fluctuation Theory in Physical Measurements" *Rep. Prog. in Phys.* **16** 266-320 (1953).

A rigorous concept of stochastic differential equations was introduced in

Bernstein, S.N., "Principes de la Théorie des Equations Differentielles Stochastiques," *Trudy. Fiz.-Mat. Inst. Steklov.* **5**, 95-124 (1934).

Bernstein, S.N., "Equations Differentielles Stochastiques," *Act. Sci. et Ind.* **738** 5-31 (1938).

Stochastic differential equations were also discussed in

Levy, P., *Processus Stochastiques et Mouvement Brownien*, Gauthier-Villars, Paris, 1948.

Theorems of existence and uniqueness based on Picard iteration are presented in

Ito, K., "On Stochastic Differential Equations," *Mem. Am. Math. Soc.* No. 4 (1951).

These proofs are also available in

Doob, J. L., *Stochastic Processes*, Wiley, New York, 1953.

Gikhman, I. I. and Skorokhod, A. V. *Introduction to the Theory of Random Processes*, W. B. Saunders, Philadelphia 1969.

The book by Gikhman and Skorokhod is recommended to those who would like to pursue stochastic differential equations in full detail.

Stochastic integrals of the type

$$\int f(t)\, dw(t)$$

where $\{w(t), t \in T\}$ is a Wiener process and f a function of bounded variation, were also introduced by Wiener who defined the integral by the formula

$$\int_0^t f(s)\, dw(s) = f(t)w(t)\Big|_0^t - \int_0^t w(t)\, df(t)$$

As f is of bounded variation and the sample functions are continuous, the integral on the right is a well defined quantity.

A pedagogic exposition is given in

Wiener, N., *Nonlinear Problems in Random Theory*, MIT press, Cambridge, Massachusetts, 1958.

An approach to stochastic differential equations, which is completely different from the presentation in this chapter, can be obtained by defining white noise as a "generalized stochastic process" in analogy with the concept of generalized functions or distributions. This idea is pursued in

Gelfand, I.M. and Wilenkin, N.J. *Verallgemeinerten Funktionen*, Vol. IV, VEB Deutscher Verlag der Wissenschaften, Berlin, 1964.

A rigorous and compact treatment is also given in

Skorokhod, A. V., *Studies in the Theory of Random Processes*, Addison-Wesley, Reading, Massachusetts, 1951.

This monograph also contains generalizations of Ito's results to the vector case.

Another approach to the theory of stochastic differential equations is given in

Gikhman, I.I., On the Theory of Differential Equations of Stochastic Processes I-II *Ukr. Mat. Zh.* **2**, 37-63 (1950) and **3**, 317-339 (1951).

Gikhman introduces stochastic differential equations as limits of stochastic difference equations. The conditions required for existence and uniqueness are analogous as those given by Ito.

The stochastic integral was introduced by Ito in the previously mentioned reference. Another approach to stochastic integrals has been given in

Stratonovich, R. L., "A New Representation of Stochastic Integrals and Equations," *SIAM J. Control* **4**, 362-371 (1966).

As has been discussed there are not very great differences between the Ito and Stratonovich calculus. The formulas in the Stratonovich calculus

Bibliography and Comments 89

are sometimes simpler. The reason we prefer the Ito integral in spite of this is that we want to retain the intuitive idea of a state model. The relationships between the different integral concepts and the corresponding difference equations are based on

Åström, K. J., "On Stochastic Differential Equations," Lecture notes, Lund, June, 1965.

Stability of stochastic differential equations is discussed in

Kushner, H. J., *Stochastic Stability and Control*, Academic Press, New York, 1967.

Several problems associated with stochastic differential equations are found in

"Stochastic Problems in Control," A symposium of the American Automatic Control Council, *Am. Soc. Mech. Eng.* (1968).

This monograph also contains many additional references. The Fokker-Planck equation is, e.g., treated in

Bharucha-Reid, A. T., *Elements of the Theory of Markov Processes and Their Application*, McGraw-Hill, New York, 1960.

An exposition of the application of the Fokker-Planck equation to nonlinear systems is given in

Fuller, A. T., "Analysis of Nonlinear Stochastic Systems by means of the Fokker-Planck Equation," *Int. J. Control* **9**, 603–655 (1969).

This paper contains many references.

The relationships between quantum mechanics and stochastic differential equations are analyzed in

Gelfand, I. M. and Yaglom, A. M., "Integration in Functional Spaces and its Application in Quantum Physics," *J. Math. Phys.* **1**, 48–69 (1960).

The Ito differentiation rule was proven in

Ito, K., "On a Formula Concerning Stochastic Differentials," *Nagoya Math. J.* (Japan) **3**, 55 (1951).

A proof is also found in

Ito, K., "Lectures on Stochastic Processes," Tata Institute of Fundamental Research, Bombay, 1961.

The modeling problem discussed in Section 9 is simple since only linear systems were considered. In the nonlinear case the modeling is more difficult. Consider, e.g., the following differential equation

$$\frac{dz}{dt} = f(z) + g(z)n \qquad (*)$$

where z is a scalar and n is bandlimited white noise. It turns out that the proper Ito equation which models equation (*) is

$$d_0 x = [f(x) + \frac{1}{2} g(x) g'(x)] dt + g(x) dv$$

where $d_0 x$ denotes the Ito differential and the proper Stratonovich equation is

$$d_{0.5} x = f(x) dt + g(x) dv$$

where $d_{0.5} x$ denotes the Stratonovich differential. See, e.g.,

Stratonovich, R. L., *Topics in the Theory of Random Noise*, Vol. I, Gordon and Breach, New York, 1963.

Clark, J. M. C., "The Representation of Nonlinear Stochastic Systems with Applications to Filtering," Ph.D. thesis, Imperial College, University of London, June, 1966.

Wong, E. and Zakai, M., "On the Relation Between Ordinary and Stochastic Differential Equations and Applications to Problems in Control Theory," *3rd IFAC Congress*, London, (1966).

The modeling of equation (*) when z is a vector is even more involved. This is discussed in great detail in Clark's Thesis. The modeling problem is also closely related to the problem of simulating stochastic differential equations. A simple case is discussed in

Åström, K. J., "On a First-order Stochastic Differential Equation," *Int. J. Control* **1**, 301–326 (1965).

CHAPTER 4

ANALYSIS OF DYNAMICAL SYSTEMS WHOSE INPUTS ARE STOCHASTIC PROCESSES

1. INTRODUCTION

Up to this chapter, stochastic processes and stochastic models as such have been our main concern. This chapter will give the fundamentals of stochastic control systems. We will thus consider systems whose environments are described by disturbances in terms of stochastic processes. The basic problem is then to analyze dynamical systems whose input signals are stochastic processes. The analysis is entirely restricted to linear systems. Both continuous time and discrete time problems will, however, be discussed.

Dynamical systems can be represented in many different ways by input-output descriptions or by state models. We have found previously that stochastic processes can also be represented in many different ways, e.g., by covariance functions, spectral densities, state models. When analyzing dynamical systems whose inputs are stochastic processes, we thus have many different cases to consider.

Discrete time systems, characterized by weighting functions, and input signals as second order processes, characterized by covariance functions, are covered in Section 2. Conditions for the output to be a second order process are given. Formulas for mean values and covariances of the output are derived. It is also shown that, for stationary processes, the results can be given in a more compact form if the system is characterized by its transfer function and the processes by their spectral densities.

The results of Section 2 make it seem possible that many stochastic processes can be thought of as generated from dynamical systems whose inputs are white noise. In Section 3 this is shown to be true for discrete time processes whose spectral densities are rational functions in cos ω. The main results are referred to as the spectral factorization theorem and the representation theorem. These results are of great importance because they imply that all processes with rational spectral densities can be thought of as generated by stochastic state models, and that it is only necessary to consider the case of white noise inputs in the analysis.

Section 4 presents analysis of continuous time systems. The spectral factorization theorem and the representation theorem for continuous time processes are given in Section 5.

2. DISCRETE TIME SYSTEMS

This section will consider discrete time systems whose input signals are discrete time, second order stochastic processes. It is assumed throughout the section that the sampling interval is chosen as the time unit and that T denotes the set $\{\cdots, -1, 0, 1, \cdots\}$. Consider a discrete time dynamical system with input u and output y. See Fig. 4.1. For simplicity it is

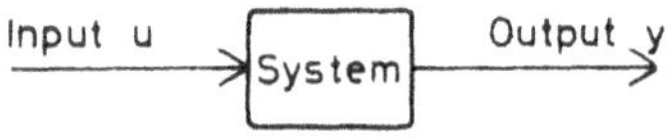

Input u → System → Output y

Fig. 4.1.

assumed that the system is time-invariant, and that it has one input and one output. These restrictions can easily be eliminated. The input u is assumed to be a stochastic process of second order with given mean value function $m_u(t)$ and given covariance function $r_u(s, t)$. The problem is then to find the stochastic properties of the output. Assuming that the system is characterized by its weighting function h, the input-output relation can then be written as

$$y(t) = \sum_{s=-\infty}^{t} h(t-s)u(s) = \sum_{s=0}^{\infty} h(s)u(t-s) \quad (2.1)$$

We must first ensure that (2.1) has a meaning. If the sum was finite, there would not be any difficulties because in that case the output y is simply a weighted sum of stochastic variables. When the sum (2.1) is infinite, we must first ensure that it converges. As was mentioned in Chapter 2, there are many concepts of convergence that can be used for stochastic variables. We will choose convergence in the mean square. To find out if the series (2.1) converges, form the Cauchy sequence

Discrete Time Systems

$$\sum_{s=n}^{m} h(s)u(t-s)$$

We have

$$E\left[\sum_{s=n}^{m} h(s)u(t-s)\right]^2 = E\sum_{s=n}^{m}\sum_{s'=n}^{m} h(s)h(s')u(t-s)u(t-s')$$
$$= \sum_{s,s'=n}^{m} h(s)h(s')r_u(t-s, t-s') \quad (2.2)$$

As $\{u(t), t \in T\}$ is a stochastic process of second order we have

$$Eu^2(t) \leqslant a < \infty. \quad \text{Hence} \quad |r_u(t,s)| \leqslant \sqrt{r_u(t,t)r_u(s,s)} = a$$

If the dynamical system is asymptotically stable we get

$$|h(s)| \leqslant \alpha^s \quad \text{with} \quad |\alpha| < 1$$

We thus find that the sum (2.2) can be made arbitrarily small by choosing m and n sufficiently large. The infinite sum (2.1) thus exists in the sense of mean square convergence, and we have thus found that $\{y(t), t \in T\}$ is a stochastic process of second order. We will now investigate the properties of this process.

To determine the mean value function we form

$$m_y(t) = Ey(t) = E\sum_{s=0}^{\infty} h(s)u(t-s)$$
$$= \sum_{s=0}^{\infty} h(s)Eu(t-s) = \sum_{s=0}^{\infty} h(s)m_u(t-s) \quad (2.3)$$

The third equality follows from Theorem 6.2 of Chapter 2. We thus find that the mean value of the output is obtained simply by sending the mean value of the input through the dynamical system.

To determine the covariance function of the output we subtract (2.3) from (2.1) and get

$$y(t) - m_y(t) = \sum_{s=0}^{\infty} h(s)[u(t-s) - m_u(t-s)]$$

The difference between the signal and its mean value thus propagates through the system in the same way as the signal itself. Because of this, we can in the future assume that the mean value of the signal is zero. This will simplify the writing.

Assuming that the input signal has zero mean value we get

$$r_y(s,t) = Ey(s)y(t) = E\sum_{k=0}^{\infty} h(k)u(s-k)\sum_{l=0}^{\infty} h(l)u(t-l)$$

$$= \sum_{k=0}^{\infty} \sum_{l=0}^{\infty} h(k)h(l) Eu(s-k)u(t-l)$$

$$= \sum_{k=0}^{\infty} \sum_{l=0}^{\infty} h(k)h(l) r_u(s-k, t-l) \qquad (2.4)$$

where the third equality follows from Theorem 6.2 of Chapter 2, which says that the operations of limit in the sense of mean squares and mathematical expectation will commute.

We also find the following equation for the covariance of the input and the output

$$r_{uy}(s, t) = Eu(s)y(t) = Eu(s) \sum_{l=0}^{\infty} h(l)u(t-l)$$

$$= \sum_{l=0}^{\infty} h(l) Eu(s)u(t-l) = \sum_{l=0}^{\infty} h(l) r_u(s, t-l) \qquad (2.5)$$

Given the system (2.1), characterized by its weighting function, the mean value $m_u(t)$ and the covariance function $r_u(s, t)$ of the input, we have thus found the mean value $m_y(t)$ and the covariance function $r_y(s, t)$ of the output. Summing up we get Theorem 2.1.

THEOREM 2.1

Consider a discrete time asymptotically stable dynamical system. Let the input signal u be a stochastic process of second order with mean value $m_u(t)$ and covariance $r_u(s, t)$. Then the output y of the system given by the sum

$$y(t) = \sum_{n=0}^{\infty} h(n)u(t-n) \qquad (2.1)$$

exists in the sense of mean square convergence. The output $\{y(t), t \in T\}$ is a stochastic process of second order with mean value

$$m_y(t) = \sum_{n=0}^{\infty} h(n) m_u(t-n) \qquad (2.3)$$

and covariance

$$r_y(s, t) = \sum_{k=0}^{\infty} \sum_{l=0}^{\infty} h(k)h(l) r_u(s-k, t-l) \qquad (2.4)$$

The covariance between input and output is given by

$$r_{uy}(s, t) = \sum_{l=0}^{\infty} h(l) r_u(s, t-l) \qquad (2.5)$$

If the input process $\{u(t), t \in T\}$ is normal, then the output $\{y(t), t \in T\}$ is also normal.

Discrete Time Systems

The statement on normality has not been proven. It follows from the fact that a sum of normal variables is a normal variable.

Stationary Processes

We will now specialize to stationary processes and rewrite the result in a slightly different form. If the input signal u is a stationary stochastic process we get

$$m_u(t) = m_u = \text{const}$$
$$r_u(s, t) = r_u(s - t)$$

Equations (2.3), (2.4), and (2.5) then reduce to

$$m_y = m_u \sum_{k=0}^{\infty} h(k) \tag{2.6}$$

$$r_y(s, t) = \sum_{k=0}^{\infty} \sum_{l=0}^{\infty} h(k)h(l)r_u(s - t + l - k) \tag{2.7}$$

$$r_{uy}(s, t) = \sum_{l=0}^{\infty} h(l)r_u(s - t + l) \tag{2.8}$$

We thus find that the mean value of the output is constant and that $r_y(s, t)$ and $r_{uy}(s, t)$ are functions of the difference $(s - t)$ only. If the input signal is (weakly) stationary and the system stable we thus find that the output is also (weakly) stationary.

Equations (2.6), (2.7), and (2.8) can be simplified further if we introduce spectral densities and pulse transfer functions. Let H denote the pulse transfer function of the system, i.e. the generating function or z-transform of the weighting function. Hence

$$H(z) = \sum_{n=0}^{\infty} z^{-n} h(n) \tag{2.9}$$

Equation (2.6) then reduces to

$$m_y = H(1) \cdot m_u \tag{2.10}$$

The spectral density ϕ_y of the output is related to r_y by

$$\phi_y(\omega) = \frac{1}{2\pi} \sum_{n=-\infty}^{\infty} e^{-in\omega} r_y(n) \tag{2.11}$$

where

$$r_y(n) = r_y(t + n, t) \tag{2.12}$$

Equation (2.7) now gives

$$\phi_y(\omega) = \frac{1}{2\pi} \sum_{n=-\infty}^{\infty} e^{-in\omega} \sum_{k=0}^{\infty} \sum_{l=0}^{\infty} h(k)h(l) r_u(n + l - k)$$

96 *Analysis of Dynamical Systems Whose Inputs Are Stochastic Processes*

$$= \frac{1}{2\pi} \sum_{n=-\infty}^{\infty} \sum_{k=0}^{\infty} \sum_{l=0}^{\infty} e^{-ik\omega} h(k) e^{il\omega} h(l) e^{-i(n+l-k)\omega} r_u(n+l-k)$$

$$= \sum_{k=0}^{\infty} e^{-ik\omega} h(k) \sum_{l=0}^{\infty} e^{il\omega} h(l) \frac{1}{2\pi} \sum_{n=-\infty}^{\infty} e^{-in\omega} r_u(n)$$

Introduce the definition of the pulse transfer function (2.9) and we find

$$\phi_y(\omega) = H(e^{-i\omega}) H(e^{i\omega}) \phi_u(\omega) \tag{2.13}$$

The spectral density function associated with $r_{uy}(\tau)$ is given by

$$\phi_{uy}(\omega) = \frac{1}{2\pi} \sum_{n=-\infty}^{\infty} e^{-in\omega} r_{uy}(n) \tag{2.14}$$

Introduce

$$r_{uy}(n) = r_{uy}(t+n, t) \tag{2.15}$$

and we find from (2.14) and (2.8)

$$\phi_{uy}(\omega) = \frac{1}{2\pi} \sum_{n=-\infty}^{\infty} e^{-in\omega} \sum_{l=0}^{\infty} h(l) r_u(n+l)$$

$$= \frac{1}{2\pi} \sum_{n=-\infty}^{\infty} \sum_{l=0}^{\infty} e^{il\omega} h(l) e^{-i(n+l)\omega} r_u(n+l)$$

$$= \sum_{l=0}^{\infty} e^{il\omega} h(l) \frac{1}{2\pi} \sum_{n=-\infty}^{\infty} e^{-in\omega} r_u(n)$$

Introduce the definition (2.9) of the pulse transfer function in this expression and we find

$$\phi_{uy}(\omega) = H(e^{-i\omega}) \phi_u(\omega) \tag{2.16}$$

Summing up we get Theorem 2.2.

THEOREM 2.2

Consider a stationary discrete time system with the pulse transfer function $H(z)$. Let the input signal be a stationary stochastic process with mean value m_u and spectral density $\phi_u(\omega)$. If the system is asymptotically stable, then the output signal given by (2.1) is a stationary stochastic process with mean value

$$m_y = H(1) m_u \tag{2.10}$$

and spectral density

$$\phi_y(\omega) = H(e^{-i\omega}) H(e^{i\omega}) \phi_u(\omega) = |H(e^{i\omega})|^2 \phi_u(\omega) \tag{2.13}$$

The cross spectral density of input and output is given by

$$\phi_{uy}(\omega) = H(e^{-i\omega}) \phi_u(\omega) \tag{2.16}$$

Discrete Time Systems

Remark 1

The result has a simple physical interpretation, $|H(e^{i\omega})|$ is the steady state amplitude of the output when the input is $\sin \omega t$. The value of the output spectral density $\phi_y(\omega)$ is then the product of the power gain $|H(e^{i\omega})|^2$ and the value of the input spectral density $\phi_u(\omega)$.

Remark 2

Equation (2.16) is often exploited to determine the transfer function of a dynamical system. For example if the input signal is white noise, that is, $\phi_u(\omega) = 1$, we get from (2.8) and (2.16)

$$r_{uy}(t) = h(-t)$$

and

$$\phi_{uy}(\omega) = H(e^{-i\omega})$$

By measuring the cross covariance or the cross spectral density of the input and output, we thus get the weighting function or the pulse transfer function of the system.

Exercises

1. Consider the linear dynamical system

$$y(t) + ay(t-1) = e(t) + ce(t-1)$$

where y is the output and the input $\{e(t)\}$ is a sequence of independent normal (0, 1) stochastic variables. Determine the covariance function of the output and the cross covariance of the input and output.

2. In this book the covariance function for stationary processes has been defined by

$$r_{xy}(\tau) = E[x(t+\tau) - Ex(t+\tau)][y(t) - Ey(t)]$$

It is also possible to define the covariance function for a stationary process by

$$r^*_{xy}(\tau) = E[x(t) - Ex(t)][y(t+\tau) - Ey(t+\tau)]$$

In that case Theorem 2.2 will change slightly. Show that (2.16) in that case changes to

$$\phi_{uy}(\omega) = H(e^{i\omega})\phi_u(\omega)$$

but that (2.10) and (2.13) remain the same.

3. Theorem 2.1 is easily generalized to time invariant systems. Consider the system

$$y(t) = \sum_{s=-\infty}^{t} g(t,s)u(s)$$

where $\{u(s), s = \cdots -1, 0, 1 \cdots\}$ is a stochastic process with covariance function $r_u(s,t)$. Show that the covariance function of the output y is given by

$$r_y(s,t) = \sum_{k=-\infty}^{s} \sum_{l=-\infty}^{t} g(s,k) r_u(k,l) g(t,l)$$

under suitable conditions. State these conditions explicitly.

4. Let $\{u(t), t \in T\}$ and $\{y(t), t \in T\}$ be stochastic processes related through the equation

$$y(t) + a(t-1)y(t-1) = u(t).$$

Show that

$$r_{uy}(s,t) + a(t-1) r_{uy}(s, t-1) = r_u(s,t)$$
$$r_{yu}(s,t) + a(s-1) r_{uy}(s-1, t) = r_u(s,t)$$
$$r_y(s,t) + a(t-1) r_y(s, t-1) = r_{yu}(s,t)$$

5. Generalize the results of Exercise 4 to the case when the processes $\{u(t), t \in T\}$ and $\{y(t), t \in T\}$ are related through

$$y(t) + a_1(t-1)y(t-1) + \ldots + a_n(t-n)y(t-n)$$
$$= u(t) + b_1(t-1)u(t-1) + \ldots + b_n(t-n)u(t-n).$$

6. Consider the stationary process $\{y(t), t \in T\}$ defined by

$$y(t) + ay(t-1) = e(t) + ce(t-1), \quad |a| < 1$$

where $\{e(t), t \in T\}$ is a stationary process with the covariance function $r(\tau) = \exp(-\alpha|\tau|)$. Determine the covariance function and the spectral density of e and y and the covariance function and the cross spectral density between e and y.

3. SPECTRAL FACTORIZATION OF DISCRETE TIME PROCESSES

Consider a time invariant discrete time system which is asymptotically stable. Let its pulse transfer function be $H(z)$ and let the input signal be white noise. It then follows from Theorem 2.2 that the output signal is a weakly stationary process with the spectral density

$$\phi_y(\omega) = H(e^{-i\omega}) H(e^{i\omega}) = |H(e^{i\omega})|^2 \quad (3.1)$$

It is then natural to ask if all spectral density functions $\phi(\omega)$ can be written

Spectral Factorization of Discrete Time Processes

as a product $H(e^{i\omega}) \cdot H(e^{-i\omega})$ and, if so, to find the function H. This is called the *spectral factorization* problem. If the spectral factorization was always possible, all stationary processes could be thought of as outputs of dynamical systems with white noise inputs. The theory of dynamical systems subject to stochastic disturbances could then be simplified considerably because it would be sufficient to analyze systems with white noise inputs. Simulations would also be simplified considerably because it would be sufcient to have generators for white noise processes.

It is not easy to solve the spectral factorization problem in general. If we limit the class of spectral density functions the problem can, however, be solved readily.

If the dynamical system is of finite order the pulse transfer function $H(z)$ is a rational function in z. The spectral density $\phi_y(\omega)$ is then a rational function of $e^{i\omega}$ or $\cos \omega$. The stochastic process is then called a process with *rational spectrum* or *rational spectral density*. We will now solve the spectral factorization problem for such processes with Theorem 3.1.

THEOREM 3.1 (Spectral Factorization Theorem)

Consider a stationary stochastic process with a rational spectral density ϕ. Then there exists a rational function H with poles inside the unit circle and zeros inside or on the unit circle such that

$$\phi(\omega) = H(e^{-i\omega})H(e^{i\omega}) = |H(e^{i\omega})|^2$$

Remark 1

Notice that it is important to require that the function H in (3.1) has no poles or zeros outside the unit circle in order to get a unique factorization. Consider for example the following spectral density

$$\phi(\omega) = \frac{1.04 + 0.4 \cos \omega}{1.25 + \cos \omega} = \frac{(e^{i\omega} + 0.2)(e^{-i\omega} + 0.2)}{(e^{i\omega} + 0.5)(e^{-i\omega} + 0.5)}$$

This spectral density can be factored in four different ways:

$$H_1(z) = \frac{z + 0.2}{z + 0.5}$$

$$H_2(z) = \frac{1 + 0.2 z}{z + 0.5} = 0.2 \frac{z + 5}{z + 0.5}$$

$$H_3(z) = \frac{z + 0.2}{1 + 0.5 z} = 2 \frac{z + 0.2}{z + 2}$$

$$H_4(z) = \frac{1 + 0.2 z}{1 + 0.5 z} = 0.4 \frac{z + 5}{z + 2}$$

Notice that only H_1 has no poles or zeros outside the unit circle. This implies that the systems with transfer functions H_1 and $1/H_1$ are both stable.

100 *Analysis of Dynamical Systems Whose Inputs Are Stochastic Processes*

As will be seen in Chapter 6, this fact is very important in filtering theory.

Proof

The proof is elementary; it is based on simple properties of rational functions. The only minor difficulty is to show that the system is stable.

As $\phi(\omega)$ is a rational function in $e^{i\omega}$ we have

$$\phi(\omega) = c' e^{i\omega\lambda} \frac{\prod_{k=1}^{n}(e^{i\omega} - \alpha'_k)}{\prod_{k=1}^{m}(e^{i\omega} - \beta'_k)} \tag{3.2}$$

No β'_k can have modulus 1 since ϕ is integrable. As ϕ is real we have

$$\phi(\omega) = \overline{\phi(\omega)}$$

where $\bar{\alpha}$ denotes the complex conjugate of the complex number α. We have

$$\overline{(e^{i\omega} - \alpha_k)} = (e^{-i\omega} - \bar{\alpha}_k) = e^{-i\omega}\bar{\alpha}_k(1/\bar{\alpha}_k - e^{i\omega})$$

Hence

$$\phi(\omega) = \bar{\phi}(\omega) = \bar{c}' e^{i\omega(m-n-\lambda)} \frac{\prod_{k=1}^{n} \bar{\alpha}'_k(1/\bar{\alpha}'_k - e^{i\omega})}{\prod_{k=1}^{m} \bar{\beta}'_k(1/\bar{\beta}'_k - e^{i\omega})}$$

To each zero α'_k of the numerator there corresponds another zero $1/\bar{\alpha}'_k$. If α'_k has modulus greater than 1, then $1/\bar{\alpha}'_k$ has modulus less than 1. The same holds for the denominator. We have further

$$|e^{i\omega} - z| = |e^{i\omega}| \cdot |z| |e^{-i\omega} - 1/z| = |z| |e^{i\omega} - 1/\bar{z}|$$

The expression (3.2) for the spectral density can thus be written as

$$\phi(\omega) = C \cdot \left| \frac{\prod_{k=1}^{n/2}(e^{i\omega} - \alpha_k)}{\prod_{l=1}^{m/2}(e^{i\omega} - \beta_l)} \right|^2 \tag{3.3}$$

where

$$0 < |\alpha_k| \leqslant 1$$
$$0 < |\beta_l| < 1$$
$$C > 0$$

Here the α_k's are the α'_k's and $1/\bar{\alpha}'_k$'s with modulus less than 1 and half of those of modulus 1. The β_k's are the β_k''s and the $1/\bar{\beta}_k$'s with modulus less than one. Summing up we find

$$\phi(\omega) = \left|\frac{A(e^{i\omega})}{B(e^{i\omega})}\right|^2 = \left|\frac{\sum_{k=0}^{n/2} a_i e^{i\omega(n-k)}}{\sum_{l=0}^{m/2} b_l e^{i(m-l)}}\right|^2 \quad (3.4)$$

where the polynomial $B(z)$ has all zeros inside the unit circle and $A(z)$ all zeros inside or on the unit circle. As the function $\phi(\omega)$ is real, we find that to each zero α_k there is another zero $\alpha_l = \bar{\alpha}_k$. The coefficients of the polynomials A and B are thus real. We have thus found a rational function

$$H(z) = \frac{A(z)}{B(z)} \quad (3.5)$$

with all poles inside the unit circle and all zeros inside or on the unit circle such that (3.1) holds. The following result is a direct consequence of Theorem 3.1.

THEOREM 3.2 (Representation Theorem)

Given a rational spectral density function $\phi(\omega)$, then there exists an asymptotically stable linear dynamical system such that the output of the system is a stationary process with spectral density $\phi(\omega)$ if the input is discrete time white noise.

Proof

It follows from Theorem 3.1 that there exists a rational function H with poles inside the unit circle such that (3.1) holds. Take a dynamical system with the pulse transfer function H. Let the input to the system be discrete time white noise. As the system is stable, it follows from Theorem 2.2 that the output is a stationary process with the spectral density ϕ.

Remark 1

Theorem 3.2 is very important. It means that if the analysis is restricted to stationary processes with rational spectral densities, we find that all such processes can be generated by sending white noise through stable dynamical systems. This means that the analysis can be restricted to systems with white noise inputs, i.e., the stochastic state models discussed in Chapter 3. Similarly, when simulating dynamical systems, it is sufficient to generate white noise. Disturbances with rational spectral densities are then easily obtained by filtering with dynamical systems.

Remark 2

Theorem 3.2 has also another consequence. It follows from the

102 *Analysis of Dynamical Systems Whose Inputs Are Stochastic Processes*

theorem that a process with rational spectral density can be represented as

$$y(t) = \sum_{k=-\infty}^{t} h(t-k)e(k) \tag{3.6}$$

where $\{e(t), t = \ldots, -1, 0, 1, \ldots\}$ is a sequence of independent equally distributed random variables. Assume that the transformation (3.6) can be inverted, i.e., there exists a function g such that

$$e(t) = \sum_{l=-\infty}^{t} g(t-l)y(l) \tag{3.7}$$

The knowledge of $y(t), y(t-1), \ldots$ is then equivalent to the knowledge of $e(t), e(t-1), \ldots$. The two sequences thus contain the same information. Now consider

$$y(t+1) = \sum_{k=-\infty}^{t+1} h(t+1-k)e(k)$$

$$= \sum_{k=-\infty}^{t} h(t+1-k)e(k) + h(0)e(t+1)$$

$$= \sum_{k=-\infty}^{t} h(t+1-k) \sum_{l=-\infty}^{k} g(k-l)y(l) + h(0)e(t+1) \tag{3.8}$$

We thus find that $y(t+1)$ can be written as the sum of two terms; one term is a linear function of $y(t), y(t-1), \ldots$, and the other term is $h(0)e(t+1)$. Under the condition that the transformation (3.6) is invertible, we thus find that $e(t+1)$ can be interpreted as the part of $y(t+1)$ which contains new information which is not available in $y(t)$, $y(t-1), \ldots$. The stochastic variables $\{e(t)\}$ in (3.6) are therefore called the *innovations* of the process $\{y(t), t \in T\}$ and the representation (3.6) is called the *innovations representation* of the process. This representation is important in connection with filtering and prediction problems.

The term

$$\sum_{k=-\infty}^{t} h(t+1-k) \sum_{l=-\infty}^{k} g(k-l) y(l)$$

is in fact the best mean square prediction of $y(t+1)$ based on $y(t)$, $y(t-1), \ldots$. This will be discussed in detail in Chapters 6 and 7.

Exercises

1. A stationary discrete time stochastic process has the spectral density

$$\phi(\omega) = \frac{2 + 2\cos\omega}{5 + 4\cos\omega}$$

Perform the spectral factorization and determine the pulse transfer function of a stable system such that the output has the spectral density ϕ when the input is white noise.

2. Consider the moving average
$$y(t) = e(t) + 4e(t-1)$$
where $\{e(t)\}$ is a sequence of independent normal (0, 1) stochastic variables. Show that a process with the same spectral density can be generated by
$$y(t) = \lambda[\varepsilon(t) + c\varepsilon(t-1)]$$
where $\{\varepsilon(t)\}$ is a sequence of independent normal (0, 1) stochastic variables and $|c| < 1$. Determine the parameters c and λ.

3. Consider a normal stationary process $\{y(t)\}$ which is generated by
$$y(t) = x_1(t) + x_2(t)$$
where
$$x_1(t+1) = -ax_1(t) + v_1(t)$$
$$x_2(t+1) = -bx_2(t) + v_2(t)$$
where $\{v_1(t)\}$ and $\{v_2(t)\}$ are sequences of independent normal (0, σ_1) respectively (0, σ_2) stochastic variables. Show that a stochastic process with the same spectral density can be represented by
$$y(t) = \lambda \frac{q+c}{(q+a)(q+b)} e(t)$$
where q is the shift operator $(qy(t) = y(t+1))$ and $\{e(t)\}$ is a sequence of independent normal (0, 1) stochastic variables. Determine the parameters λ and c.

4. Show that the stochastic processes $\{y(t), t \in T\}$ and $\{z(t), t \in T\}$ with the representations
$$x(t+1) = 0.8x(t) - 1.2e(t)$$
$$y(t) = x(t) + e(t)$$
and
$$x(t+1) = 0.8x(t) + 0.6e(t)$$
$$z(t) = x(t) + 2e(t)$$
where $\{e(t), t \in T\}$ is a sequence of independent normal (0, 1) random variables, have the same spectral density.

4. ANALYSIS OF CONTINUOUS TIME SYSTEMS WHOSE INPUT SIGNALS ARE STOCHASTIC PROCESSES

We will now turn to continuous time systems and continuous time processes. The analysis is completely analogous to the discrete time case treated in Sections 2 and 3.

Thus consider a time invariant dynamical system with one input u and one output y. Let the system be characterized by its weighting function $h(t)$. The input-output relation is then

$$y(t) = \int_{-\infty}^{t} h(t-s)u(s)\,ds = \int_{0}^{\infty} h(s)u(t-s)\,ds \qquad (4.1)$$

Now let the input signal be a stochastic process of second order. We must then first ensure that the integral in (4.1) has a meaning. To do so we will first consider a finite integration interval. It then follows from Theorem 6.5 of Chapter 2 that, if the integral is interpreted as a mean square limit of Riemann sums, it will exist if the covariance $r_u(s, t)$ of the input is continuous in both its arguments. The expression

$$\int_{0}^{a} h(s)u(t-s)\,ds$$

thus has a meaning for finite a.

To find out if the limit

$$\lim_{a \to \infty} \int_{0}^{a} h(s)u(t-s)\,ds$$

also exists, we form the Cauchy sequence

$$I(a, b) = \int_{a}^{b} h(s)u(t-s)\,ds$$

We have

$$EI^2(a, b) = E\iint_{a}^{b} h(s)u(t-s)h(s')u(t-s')\,ds\,ds'$$
$$= \iint_{a}^{b} h(s)h(s')r_u(t-s, t-s')\,ds\,ds'$$

If the dynamical system is asymptotically stable we find

$$|h(s)| \leq K \cdot e^{-\alpha s} \qquad \alpha > 0$$

As $\{u(t), t \in T\}$ is a stochastic process of second order we get

$$r_u(s, t) \leq E u^2(t) \leq \beta < \infty$$

Hence

Continuous Time Systems Whose Inputs Are Stochastic Processes

$$E\, I^2(a, b) \to 0$$

as $a, b \to \infty$. The limit thus exists and we find

$$E\, y^2(t) \leqslant \text{const} < \infty$$

The output y of the dynamical system is thus a stochastic process of second order. We will now determine its mean value function and its covariance function. We have

$$m_y(t) = E\, y(t) = E \int_0^\infty h(s) u(t - s)\, ds = \int_0^\infty h(s) m_u(t - s)\, ds \quad (4.2)$$

It follows from Theorem 6.5 of Chapter 2 that the operations E and $\int$ can be interchanged. We thus find that the mean value of the input propagates through the system as a deterministic signal.

The covariance function of the output will now be determined. We have

$$\begin{aligned} r_y(s, t) &= \text{cov}\,[y(s), y(t)] \\ &= \text{cov}\left[\int_0^\infty h(s')\, u(s - s')\, ds',\, \int_0^\infty h(s'')\, u(t - s'')\, ds''\right] \\ &= \iint_0^\infty h(s') h(s'')\, \text{cov}\,[u(s - s'), u(t - s'')]\, ds'\, ds'' \\ &= \iint_0^\infty h(s') h(s'') r_u(s - s', t - s'')\, ds'\, ds'' \end{aligned} \quad (4.3)$$

The interchange of the order of integrations follows from Theorem 6.5 of Chapter 2. We also have

$$\begin{aligned} r_{uy}(s, t) &= \text{cov}\,[u(s), y(t)] \\ &= \text{cov}\left[u(s), \int_0^\infty h(s')\, u(t - s')\, ds'\right] \\ &= \int_0^\infty h(s')\, \text{cov}\,[u(s), u(t - s')]\, ds' \\ &= \int_0^\infty h(s') r_u(s, t - s')\, ds' \end{aligned} \quad (4.4)$$

Summing up we get Theorem 4.1.

THEOREM 4.1

Consider a continuous time dynamical system which has the weighting function $h(t)$. Let the input signal u be a weakly stationary stochastic process with mean value $m_u(t)$ and the covariance function $r_u(s, t)$. If the dynamical system is asymptotically stable and if the function $r_u(s, t)$ is continuous, then the integral

$$y(t) = \int_0^\infty h(s) u(t - s)\, ds \quad (4.1)$$

exists as a mean square limit of Riemann sums. The output $\{y(t), t \in T\}$ is a stochastic process with the mean value

$$m_y(t) = \int_0^\infty h(s) m_u(t - s) \, ds \qquad (4.2)$$

and the covariance function

$$r_y(s, t) = \iint_0^\infty h(s') h(s'') r_u(s - s', t - s'') \, ds' \, ds'' \qquad (4.3)$$

The covariance function of the input and of the output is

$$r_{uy}(s, t) = \int_0^\infty h(s') r_u(s, t - s') \, ds' \qquad (4.4)$$

If the input signal is normal, then the output signal is also normal.

Stationary Processes

We will now specialize to stationary processes. If the input is a weakly stationary process we have

$$m_u(t) = m_u = \text{const}$$
$$r_u(s, t) = r_u(s - t)$$

Theorem 4.1 then gives

$$m_y = m_u \int_0^\infty h(s) \, ds \qquad (4.5)$$

$$r_y(s, t) = \iint_0^\infty h(s') h(s'') r_u(s - t - s' + s'') \, ds' \, ds'' \qquad (4.6)$$

$$r_{uy}(s, t) = \int_0^\infty h(s') r_u(s - t + s') \, ds' \qquad (4.7)$$

As these equations are convolutions, they can be simplified if we introduce Fourier or Laplace transforms. Let ϕ_u denote the spectral density of the input u, and G the transfer function of the dynamical system

$$\phi_u(\omega) = \frac{1}{2\pi} \int_{-\infty}^\infty e^{-i\omega\tau} r_u(\tau) \, d\tau \qquad (4.8)$$

$$G(s) = \int_0^\infty e^{-st} h(t) \, dt \qquad (4.9)$$

Equations (4.5), (4.6), and (4.7) then give

$$m_y = m_u \cdot G(0) \qquad (4.10)$$

$$\phi_y(\omega) = \frac{1}{2\pi} \int_{-\infty}^\infty e^{-i\omega\tau} r_y(\tau) \, d\tau$$

$$= \frac{1}{2\pi} \int_{-\infty}^\infty e^{-i\omega\tau} \iint_0^\infty h(s') h(s'') r_u(\tau - s' + s'') \, ds' \, ds'' \, d\tau$$

$$= \frac{1}{2\pi} \int_{-\infty}^{\infty} d\tau \int_0^{\infty} ds' \int_0^{\infty} ds'' \, e^{-i\omega s'} h(s') e^{i\omega s''} h(s'')$$
$$\times e^{-i\omega(\tau - s' + s'')} r_u(\tau - s' + s'')$$
$$= G(i\omega) G(-i\omega) \phi_u(\omega) \tag{4.11}$$

$$\phi_{uy}(\omega) = \frac{1}{2\pi} \int_{-\infty}^{\infty} e^{-i\omega\tau} r_{uy}(\tau) \, d\tau$$
$$= \frac{1}{2\pi} \int_{-\infty}^{\infty} e^{-i\omega\tau} \int_0^{\infty} h(s') r_u(\tau + s') \, ds' \, d\tau$$
$$= \frac{1}{2\pi} \int_{-\infty}^{\infty} d\tau \int_0^{\infty} ds' \, e^{i\omega s'} h(s') e^{-i\omega(\tau + s')} r_u(\tau + s')$$
$$= G(-i\omega) \phi_u(\omega) \tag{4.12}$$

Summing up we find Theorem 4.2.

THEOREM 4.2

A time invariant dynamical system has the transfer function G. The input signal is a weakly stationary stochastic process with the mean value m_u and the spectral density $\phi_u(\omega)$. If the dynamical system is asymptotically stable and if

$$r_u(0) = \int_{-\infty}^{\infty} \phi_u(\omega) \, d\omega \leqslant a < \infty \tag{4.13}$$

then the output signal is a weakly stationary process with mean value

$$m_y = G(0) \cdot m_u \tag{4.10}$$

and spectral density

$$\phi_y(\omega) = G(i\omega) G(-i\omega) \phi_u(\omega) \tag{4.11}$$

The input-output cross spectral density is

$$\phi_{uy}(\omega) = G(-i\omega) \phi_u(\omega) \tag{4.12}$$

Remark

Theorem 4.2 is analogous to Theorem 2.2, and the physical interpretations are identical. The condition (4.13), which has no correspondence in Theorem 2.2, ensures that the input signal has finite variance. This is a fundamental difference between continuous time and discrete time processes.

5. SPECTRAL FACTORIZATION OF CONTINUOUS TIME PROCESSES

This section is devoted to spectral factorization and representation of continuous time processes. The underlying ideas are identical to those for

108 *Analysis of Dynamical Systems Whose Inputs Are Stochastic Processes*

discrete time systems discussed in Section 3. The analysis is, however, more difficult because we will encounter the difficulties that are associated with continuous time white noise.

A continuous time stochastic process is said to have a rational spectral density if the spectral density $\phi(\omega)$ is a rational function in ω. The spectral factorization problem is to find a rational function G whose poles have negative real parts and whose zeros have nonpositive real parts such that

$$G(s)G(-s) = \phi(s) \tag{5.1}$$

where ϕ is a rational spectral density. The solution to this problem is given by Theorem 5.1.

THEOREM 5.1 (Spectral Factorization Theorem)

Let ϕ be a rational spectral density, then there exists a rational function G which has all poles in the left half plane and all zeros in the left half plane or on the imaginary axis such that

$$\phi(\omega) = G(i\omega)G(-i\omega) \tag{5.1}$$

Proof

The spectral density of a real process is an even function. Hence

$$\phi(\omega) = c \frac{\prod_{k=1}^{m}(\omega^2 - z_k'^2)}{\prod_{l=1}^{n}(\omega^2 - p_l'^2)} \tag{5.2}$$

As ϕ is integrable over $(-\infty, \infty)$, we have $m < n$ and no p_l real. As ϕ is nonnegative the real z_k' must always appear in pairs. The factors corresponding to real z_k' can then always be factored as follows

$$(\omega^2 - z_k'^2)^2 = [(i\omega)^2 + z_k'^2]^2 = (s^2 + z_k'^2)^2$$

As $\phi(\omega)$ is real we have

$$\phi(\omega) = \overline{\phi(\omega)} = \bar{c} \frac{\prod_{k=1}^{m}(\omega^2 - \overline{z_k'^2})}{\prod_{l=1}^{n}(\omega^2 - \overline{p_l'^2})}$$

where $\bar{z}$ denotes the complex conjugate of z. If z_k is a zero of ϕ then $-z_k$, $\bar{z}_k$ and $-\bar{z}_k$ are also zeros. Factors which correspond to purely imaginary z_k can be factored as

$$(\omega^2 - z_k'^2) = (-1)[(i\omega)^2 - (iz_k')^2] = (-1)[i\omega + iz_k'][i\omega - iz_k']$$
$$= (-1)(s + iz_k')(s - iz_k')$$

Spectral Factorization of Continuous Time Processes

Factors which correspond to complex z'_k can be factored as follows

$$(\omega^2 - z'^2_k)(\omega^2 - \overline{z'^2_k}) = (\omega + z'_k)(\omega - z'_k)(\omega + \overline{z'_k})(\omega - \overline{z'_k})$$
$$= (i\omega + iz'_k)(i\omega - iz'_k)(i\omega + i\overline{z'_k})(i\omega - i\overline{z'_k})$$
$$= [s + iz'_k][s - iz'_k][s - \overline{(iz'_k)}][s + \overline{(iz'_k)}]$$
$$= \{s^2 + s[iz'_k + \overline{(iz'_k)}] + |z'_k|^2\}\{s^2 - s[iz'_k + \overline{(iz'_k)}] + |z'_k|^2\}$$

Summing up we find that the spectral density can be factored as

$$\phi(\omega) = \frac{B(i\omega)\,B(-i\omega)}{A(i\omega)\,A(-i\omega)} \tag{5.3}$$

where the polynomial $A(s)$ can be chosen to have all zeros in the left half plane and $B(s)$ all zeros in the left half plane or on the imaginary axis. The rational function

$$G(s) = \frac{B(s)}{A(s)} \tag{5.4}$$

then has the desired properties.

The representation theorem is more difficult than the corresponding result for discrete time processes. Consider a time invariant dynamical system with the weighting function $h(t)$ and the transfer function $G(s)$. If Theorem 4.2 could be applied in the case when the input signal is white noise, that is $\phi_u(\omega) = 1$, we would find that the spectral density of the output signal is given by

$$\phi(\omega) = G(i\omega)\,G(-i\omega) \tag{5.1}$$

However Theorem 4.2 does not hold because when the input signal is white noise the condition given by (4.13) is not fulfilled. Also notice that the integral in the input-output relation

$$y(t) = \int_{-\infty}^{t} h(t-s)u(s)\,ds = \int_{0}^{\infty} h(s)u(t-s)\,ds \tag{5.5}$$

does not have a meaning when u is white noise. Compare Section 6 of Chapter 2. The integral (5.5) will, however, exist if the input signal u is band limited white noise. As was discussed in Section 6 of Chapter 2, we also found that such a case could be modeled by

$$y(t) = \int_{-\infty}^{t} h(t-s)\,dv(s) = \int_{0}^{\infty} h(s)\,dv(t-s) \tag{5.6}$$

where $\{v(t),\ t \in T\}$ is a stochastic process with orthogonal increments which has the incremental mean $m\,dt$ and the incremental covariance $c\,dt$.

If the weighting function h is assumed to be bounded, it follows from Section 5 of Chapter 3 that the integral exists. To show that the infinite integral exists we form

110 Analysis of Dynamical Systems Whose Inputs Are Stochastic Processes

$$E\left[\int_a^b h(t-s)\,dv(s)\right]^2 \leqslant \max_{a\leqslant s\leqslant b} h^2(t-s)\, E\left[\int_a^b dv(s)\right]^2$$

$$= c(b-a)\max_{a\leqslant s\leqslant b} h^2(t-s)$$

As the dynamical system is asymptotically stable we have

$$|h(t)| \leqslant C\cdot e^{-\alpha t} \qquad \alpha > 0$$

We thus find that

$$E\left[\int_a^b h(t-s)\,dv(s)\right]^2 \to 0 \quad \text{as} \quad t\to\infty$$

The infinite integral (5.6) then exists according to the Cauchy criterion and the stochastic process $\{y(t),\, t\in T\}$ thus is a process of second order. We will now determine the mean value and the covariance function. It follows from the properties of the stochastic integral that

$$Ey(t) = E\int_{-\infty}^t h(t-s)\,dv(s) = \int_{-\infty}^t h(t-s)\,E\,dv(s)$$

$$= \int_{-\infty}^t h(t-s)m(s)\,ds \tag{5.7}$$

and

$$r(s,t) = Ey(s)y(t) = E\int_{-\infty}^s\int_{-\infty}^t h(s-s')\,h(t-t')\,dv(s')\,dv(t')$$

$$= \int_{-\infty}^t h(s-s')h(t-s')\,c\,ds'$$

$$= c\int_0^\infty h(s-t+s')\,h(s')\,ds' \tag{5.8}$$

If m is constant, the mean value of y is also constant. The covariance function $r(s,t)$ is also a function of $(s-t)$, and we thus find that the stochastic process $\{y(t),\, t\in T\}$ is weakly stationary.

The spectral density of $\{y(t),\, t\in T\}$ is given by

$$\phi(\omega) = \frac{1}{2\pi}\int_{-\infty}^\infty e^{-i\omega\tau}\, r(\tau)\,d\tau$$

$$= \frac{c}{2\pi}\int_{-\infty}^\infty e^{-i\omega\tau}\int_0^\infty h(\tau+s')h(s')\,ds'\,d\tau$$

$$= \frac{c}{2\pi}\int_0^\infty e^{i\omega s'}\,h(s')\,ds'\int_{-\infty}^\infty e^{-i\omega(\tau+s')}h(\tau+s')\,d\tau$$

As h is a weighting function we have $h(t) = 0$ for $t \leqslant 0$. Hence

$$\phi(\omega) = \frac{c}{2\pi}\int_0^\infty e^{i\omega s'}\,h(s')\,ds'\int_0^\infty e^{-i\omega s'}\,h(s')\,ds' = \frac{c}{2\pi}G(-i\omega)\,G(i\omega) \tag{5.9}$$

Spectral Factorization of Continuous Time Processes

where G is the transfer function of the system which is equal to the Laplace transform of h. Summing up we find Theorem 5.2.

THEOREM 5.2 (Representation Theorem)

Consider a rational spectral density function $\phi(\omega)$. Then there exists an asymptotically stable, time invariant dynamical system with the weighting function h such that the stochastic process defined by

$$y(t) = \int_{-\infty}^{t} h(t - s) \, dv(s) \tag{5.6}$$

where $\{v(t), t \in T\}$ is a process with orthogonal increments, is stationary and has the spectral density ϕ.

This theorem can be given physical interpretations analogous to Theorem 3.2. When the transformation (5.6) is invertible, the representation (5.6) is called the innovations representation of the process $\{y(t), t \in T\}$.

Exercises

1. Perform the spectral factorization for the spectral density

$$\phi(\omega) = \frac{\omega^2 + 1}{\omega^4 + 8\omega^2 + 4}$$

   and give a representation of the corresponding process of the form (5.6).

2. A stationary stochastic process has the covariance function

$$r(\tau) = e^{-|\tau|} \cos 2\tau$$

   Find a representation of the process of the form

$$y(t) = \int_{-\infty}^{t} h(t - s) \, dv(s)$$

   where $\{v(t), t \in T\}$ is a process with zero mean and orthogonal increments. The incremental variance is dt. Also show how the process can be simulated on an analog computer with a band limited white noise generator.

3. The stochastic processes $\{x(t), t \in T\}$ and $\{y(t), t \in T\}$ are stationary and normal. They have the spectral densities

$$\phi_x = \frac{1}{\omega^2 + 1}$$

$$\phi_y = \frac{1}{\omega^2 + 4}$$

$$\phi_{xy} = \frac{1}{\omega^2 + i\omega + 2}$$

Give a representation of the vector process $\binom{x}{y}$ of the form (5.6).

6. BIBLIOGRAPHY AND COMMENTS

The results given by Theorems 2.1, 2.2, 4.1, and 4.2 have been known for a long time. They are available in standard texts such as

James, H. M., Nichols, N. B., and Phillips, R. S., *Theory of Servomechanisms*, McGraw-Hill, New York, 1947

Laning, J. H. and Battin, R. H., *Random Processes in Automatic Control*, McGraw-Hill, New York, 1956

Davenport, W. B. and Root, W. L., *An Introduction to the Theory of Random Signals and Noise*, McGraw-Hill, New York, 1958

Newton, G. C., Gould, L. A., and Kaiser, J. F., *Analytical Design of Linear Feedback Controls*, Wiley, New York, 1957

Solodovnikov, V.V., *Statistical Dynamics of Linear Automatic Control Systems*, Dover, New York, 1965.

The concept of spectral factorization was introduced by Wiener in

Wiener, N., *Extrapolation, Interpolation, and Smoothing of Stationary Time Series*, MIT Press Cambridge, Massachusetts and Wiley, New York, 1949.

The result is based upon a theorem by Wiener and Paley: A real non-negative function $\phi(\omega)$ which is quadratically integrable can be factorized as

$$\phi(\omega) = g(\omega)\, g(-\omega) \tag{6.1}$$

where $g(\omega)$ is the Fourier transform of a function f which vanishes for negative arguments if

$$\int_{-\infty}^{\infty} \frac{|\log \phi(\omega)|}{1 + \omega^2} \, d\omega < \infty \tag{6.2}$$

A proof of this theorem is given in

Paley, R. E. A. C. and Wiener, N., "Fourier Transforms in the Complex Domain," *Am. Math. Soc. Colleq. Publ.* Vol. 19, New York, 1934.

Wiener has also proven that the spectral factorization in the general case is given by

$$g(\omega) = \frac{1}{2\pi\, \phi(\omega)} \int_0^\infty e^{-i\omega t}\, dt \int_{-\infty}^\infty \phi(u)\, e^{iut}\, du \tag{6.3}$$

For discrete time processes the condition for spectral factorization is given by

$$\int_{-\pi}^{\pi} |\log \phi(\omega)| \, d\omega < \infty \tag{6.4}$$

The spectral factorization problem is closely connected with the filtering and prediction problem which is discussed in Chapters 6 and 7 of this book. The algorithms presented in these sections for the prediction problem are in fact a very convenient way to perform spectral factorizations in practice.

The multivariate version of the spectral factorization problem has many interesting aspects see, e.g.,

Youla, D. C., "On the Factorization of Rational Matrices," *IEEE Trans. on Information Theory* **IT-7**, 172-189 (1961).

The idea of representing a stochastic process as the output of a dynamical system whose input is white noise is old. It can be shown to be true under much more general conditions than those given by Theorems 3.2 and 5.2. It was shown in

Wold, H., *A Study in the Analysis of Stationary Time Series*, Almqvist and Wiksell, Stockholm, 1938

that a stationary discrete time process can be represented as

$$x(t) = s(t) + \sum_{n=0}^{\infty} c_n u(t-n) \tag{6.5}$$

where $\{u(t)\}$ is a sequence of orthogonal random variables, $\{s(t)\}$ is a singular process, and $\sum c_n^2$ convergent. This is the famous Wold decomposition theorem.

Cramér has shown that a stationary process which does not contain any singular component can be represented as

$$x(t) = \int_{-\infty}^{\infty} e^{i\omega t} \, dv(\omega) \tag{6.6}$$

where $\{v(t), -\infty < t < \infty\}$ is a process with orthogonal increments and the variance function $F(\lambda)$, which equals the spectral distribution of the process.

A stochastic process with the representation

$$x(t) = \int_{t_0}^{t} f(t, s) \, dv(s) \tag{6.7}$$

where $\{v(t)\}$ is a process with orthogonal increments which has the covariance function

$$r(s, t) = \int_{t_0}^{\min(s, t)} f(s, \tau) f(t, \tau) \, dR(\tau) \tag{6.8}$$

where

$$R(t) - R(s) = E[v(t) - v(s)]^2$$

It has been shown in

Karhunen, K., "Zur Spektraltheorie Stochastischer Prozesse," *Ann. Acad. Sci. Fennicae* Ser. A **34**, 7–79 (1946).

that the converse is also true, i.e., a process with the covariance function (6.8) has the representation (6.7). The representation (6.7) is called the Karhunen-Loève expansion.

The innovations concept is extensively discussed in

Kailath, T., "An Innovations Approach to Least Squares Estimation Part I: Linear Filtering in Additive White Noise," *IEEE Trans. Automatic Control* **AC-13** 646–655 (1968).

CHAPTER 5

PARAMETRIC OPTIMIZATION

1. INTRODUCTION

The previous chapters have developed the tools which are required to analyze dynamical systems whose input signals are stochastic processes. This chapter will show how these tools can be exploited to design control systems. We will thus consider systems whose environment can be described by disturbances which are stochastic processes. It is also assumed that the control system is given apart from a number of parameters which can be chosen arbitrarily. We will then analyze how to choose the parameters in order to optimize the performance of the system. It is assumed that the system can be characterized by linear equations and the performance by the expected value of a loss function, which is a quadratic function of the state variables of the system.

The problem of parametric optimization can be divided into two parts:

● evaluation of performance

● optimization of performance with respect to parameters

The optimization can occasionally be done analytically but in most cases we must use numerical methods. There are many numerical methods available; some require evaluation of the loss functions only, while other require evaluation of gradients and possibly also higher order derivatives of the loss function. It turns out that the problem of evaluating derivatives of the loss function is the same type of problem as the evaluation of the loss function itself. We will therefore concentrate on the evaluation of the loss function.

The problem can be approached in two different ways: in the time domain or the frequency domain. Analysis in the frequency domain leads to the problem of evaluating integrals such as

$$\int_{-i\infty}^{i\infty} G(s)\, G(-s)\, ds$$

or

$$\oint H(z)\, H(z^{-1})\, \frac{dz}{z}$$

where G and H are rational functions of a complex variable. The details are given in Sections 2 and 3 for discrete time systems and continuous time systems respectively. It is interesting to see that the continuous and discrete time systems will require approximately the same amount of work and the same degree of complexity. Analysis in the time domain leads to equations of the type

$$P(t+1) = \Phi P(t) \Phi^T + R_1$$

in the discrete time case or

$$\frac{dP}{dt} = AP + PA^T + R_1$$

in the continuous time case.

The frequency domain approach to the problem is discussed in Section 2 for discrete time systems and in Section 3 for continuous time systems. The time domain aspect of the problem is discussed in connection with the problem of reconstructing the state of a noisy dynamical system from noisy observations. Using heuristic arguments, we arrive at a structure for the reconstructor which seems reasonable. The reconstructor has a number of undetermined parameters. These parameters are determined in such a way as to minimize the mean square reconstruction error. It will be shown later in Chapter 7 that the structures derived heuristically are in fact optimal. The reconstructors obtained are thus Kalman filters. The discrete time case is discussed in Section 4 and the continuous time case in Section 5. It is of interest to observe that the parameters can actually be time-varying.

2. EVALUATION OF LOSS FUNCTIONS FOR DISCRETE TIME SYSTEMS

Statement of the Problem

In Chapter 4 we developed the tools which were required to analyze linear systems subject to disturbances which can be described as stochastic processes. For linear time invariant systems whose disturbances are sta-

Evaluation of Loss Functions for Discrete Time Systems

tionary stochastic processes with rational spectral densities, we found that the spectral density for any system variable can be expressed as

$$\phi(\omega) = H(z)H(z^{-1})$$

where $z = e^{i\omega}$ and H is a rational function. The variance of the system variable is then given by

$$\sigma^2 = \int_{-\pi}^{\pi} \phi(\omega)d\omega = \frac{1}{i}\int_{-\pi}^{\pi} H(e^{i\omega})H(e^{-i\omega})e^{-i\omega}d(e^{i\omega}) = \frac{1}{i}\oint H(z)H(z^{-1})\frac{dz}{z}$$

where $\oint$ denotes the integral along the unit circle in the complex plane. To compute the variance of a signal in such a case we are thus led to the problem of evaluating the integral

$$I = \frac{1}{2\pi i}\oint \frac{B(z)B(z^{-1})}{A(z)A(z^{-1})} \cdot \frac{dz}{z} \tag{2.1}$$

where A and B are polynomials with real coefficients

$$A(z) = a_0 z^n + a_1 z^{n-1} + \cdots + a_n \tag{2.2}$$
$$B(z) = b_0 z^n + b_1 z^{n-1} + \cdots + b_n \tag{2.3}$$

and $\oint$ denotes the integral along the unit circle in the positive direction. The factor $1/2\pi$ is introduced for convenience only. We also assume that a_0 is positive.

The integral (2.1) can of course be evaluated in a straightforward manner using residue calculus. It turns out, however, that the general formulas are not practical to handle for systems of high order. For this purpose we will present recursive formulas for the evaluation of the integral (2.1) which are convenient both for hand and machine calculations.

Notations and Preliminaries

We first observe that the integral (2.1) will always exist if the polynomial $A(z)$ is stable, which means that all its zeros are inside the unit circle. In such a case we can always find a stable dynamical system with the pulse transfer function $B(z)/A(z)$, and the integral (2.1) is then simply the variance of the output when the input is white noise.

If $A(z)$ has zeros on the unit circle, the integral diverges. If $A(z)$ has zeros both inside and outside the unit circle, but not on the unit circle, the integral (2.1) still exists. In such a case we can always find a polynomial $A'(z)$ with all its zeros inside the unit circle such that

$$A(z)A(z^{-1}) = A'(z)A'(z^{-1})$$

and the integral then represents the variance of the output of a stable dynamical system with the pulse transfer function $B(z)/A'(z)$.

In many practical cases, however, we obtain the integral as a result of an analysis of a dynamical system whose pulse transfer function is $B(z)/A(z)$. In such a case, it is naturally of great importance to test that the denominator $A(z)$ of the pulse transfer function has all its zeros inside the unit circle because, when this is not the case, the dynamical system will be unstable although the integral (2.1) exists.

In order to present the result in a simple form we will first introduce some notations. Let A^* denote the reciprocal polynomial defined by

$$A^*(z) = z^n A(z^{-1}) = a_0 + a_1 z + \cdots + a_n z^n \qquad (2.4)$$

Further introduce the polynomials[†]

$$A_k(z) = a_0^k z^k + a_1^k z^{k-1} + \cdots + a_k^k \qquad (2.5)$$
$$B_k(z) = b_0^k z^k + b_1^k z^{k-1} + \cdots + b_k^k \qquad (2.6)$$

which are defined recursively by

$$A_{k-1}(z) = z^{-1}\{A_k(z) - \alpha_k A_k^*(z)\} \qquad (2.7)$$
$$B_{k-1}(z) = z^{-1}\{B_k(z) - \beta_k A_k^*(z)\} \qquad (2.8)$$

where

$$\alpha_k = a_k^k / a_0^k \qquad (2.9)$$
$$\beta_k = b_k^k / a_0^k \qquad (2.10)$$

and

$$A_n(z) = A(z) \qquad (2.11)$$
$$B_n(z) = B(z) \qquad (2.12)$$

The coefficients of the polynomials A_k and B_k are thus given by the recursive equations

$$a_i^{k-1} = a_i^k - \alpha_k a_{k-i}^k \qquad i = 0, 1, \ldots, k-1 \qquad (2.13)$$
$$b_i^{k-1} = b_i^k - \beta_k a_{k-i}^k \qquad i = 0, 1, \ldots, k-1 \qquad (2.14)$$

with the initial conditions

$$a_i^n = a_i \qquad (2.15)$$
$$b_i^n = b_i \qquad (2.16)$$

If the equations given above should have any meaning, we must naturally require that all a_0^k are different from zero. The coefficient a_0^n can of course always be chosen different from zero. The following theorem gives necessary and sufficient conditions.

THEOREM 2.1

Let $a_0^k > 0$, then the following conditions are equivalent:

1. The polynomial $A_k(z)$ is stable

[†] Notice that k in a_0^k is a superscript!

2. The polynomial $A_{k-1}(z)$ is stable and a_0^{k-1} is positive.

By repeated application of this theorem, we thus find that if the polynomial $A_n(z)$ is stable, then all coefficients a_0^k are positive. In order to prove Theorem 2.1 we need the following result.

LEMMA 2.1

Let the polynomial $f(z)$, with real coefficients, have all its roots inside the unit circle, then

$$|f(z)| < |f^*(z)| \quad \text{for} \quad |z| < 1$$
$$|f(z)| = |f^*(z)| \quad \text{for} \quad |z| = 1$$
$$|f(z)| > |f^*(z)| \quad \text{for} \quad |z| > 1$$

Proof

Put

$$f(z) = \beta \prod_{i=1}^{n} (z - \alpha_i) \qquad |\alpha_i| < 1$$

Then

$$f^*(z) = \beta \prod_{i=1}^{n} (1 - \alpha_i z)$$

Introduce

$$w(z) = \frac{f(z)}{f^*(z)} = \prod_{i=1}^{n} \frac{z - \alpha_i}{1 - \alpha_i z} = \prod_{i=1}^{n} \frac{z - \alpha_i}{1 - \bar{\alpha}_i z}$$

where $\bar{\alpha}_i$ denotes the complex conjugate of α_i. The last equality follows from the fact that f has real coefficients. If α_i is a zero of f, then $\bar{\alpha}_i$ is also a zero. Now consider the transformation

$$w_i(z) = \frac{z - \alpha_i}{1 - \bar{\alpha}_i z}$$

This transformation transforms the interior of the unit circle on to itself. The unit circle is an invariant of the transformation. The transformation

$$w(z) = \prod_{i=1}^{n} w_i(z) = \prod_{i=1}^{n} \frac{z - \alpha_i}{1 - \bar{\alpha}_i z} = \frac{f(z)}{f^*(z)}$$

then also has the same properties and the lemma is proven.

Proof of Theorem 2.1

After these preliminaries we can now prove Theorem 2.1. We will first show that $1. \Longrightarrow 2$. If 1. holds, it follows from Lemma 2.1 that

$$|A_k(0)| < |A_k^*(0)|$$

But $A_k(0) = a_k{}^k$ and $A_k^*(0) = a_0{}^k$. Hence
$$|\alpha_k| = |a_k{}^k/a_0{}^k| < 1 \tag{2.17}$$
Equation (2.13) then gives
$$a_0^{k-1} = a_0{}^k - (a_k{}^k)^2/a_0{}^k = [(a_0{}^k)^2 - (a_k{}^k)^2]/a_0{}^k > 0$$
Notice that it was assumed that $a_0{}^k > 0$. As $A_k(z)$ is stable, it also follows from Lemma 2.1 that
$$|A_k(z)| \geqslant |A_k^*(z)| \quad \text{for} \quad |z| \geqslant 1$$
Combining this with (2.17) we get
$$|A_k(z)| > |\alpha_k| \cdot |A_k^*(z)| \quad \text{for} \quad |z| \geqslant 1$$
We now find from (2.7) that
$$|z| \cdot |A_{k-1}(z)| = |A_k(z) - \alpha_k A_k^*(z)|$$
$$\geqslant |A_k(z)| - |\alpha_k| \cdot |A_k^*(z)| > 0 \quad \text{for} \quad |z| \geqslant 1$$
This implies that $A_{k-1}(z)$ has no roots outside the unit circle. The first part of the theorem is thus proven.

Now assume that condition 2. holds. Then
$$a_0^{k-1} = a_0{}^k - (a_k{}^k)^2/a_0{}^k = [(a_0{}^k)^2 - (a_k{}^k)^2]/a_0{}^k > 0$$
As $a_0{}^k$ was assumed positive we get
$$|\alpha_k| = |a_k{}^k/a_0{}^k| < 1$$
It follows from (2.7) that
$$A_k(z) - \alpha_k A_k^*(z) = z A_{k-1}(z) \tag{2.18}$$
Hence
$$z^k A_k(z^{-1}) - \alpha_k z^k A_k^*(z^{-1}) = z^{k-1} A_{k-1}(z^{-1})$$
or
$$A_k^*(z) - \alpha_k A_k(z) = A_{k-1}^*(z) \tag{2.19}$$
Elimination of $A_k^*(z)$ between (2.18) and (2.19) gives
$$A_k(z) = \frac{z}{1 - \alpha_k^2} A_{k-1}(z) + \frac{\alpha_k}{1 - \alpha_k^2} A_{k-1}^*(z)$$
As $|\alpha_k| < 1$, the elimination is always possible.

For $|z| \geqslant 1$ we now have (Lemma 2.1)
$$|A_{k-1}(z)| \geqslant |A_{k-1}^*(z)|$$
Furthermore $|\alpha_k| < 1$. Hence for $|z| \geqslant 1$ we have
$$|A_k(z)| \geqslant \left|\frac{z}{1 - \alpha_k^2}\right| |A_{k-1}(z)| - \left|\frac{\alpha_k}{1 - \alpha_k^2}\right| \cdot |A_{k-1}^*(z)| > 0$$

Evaluation of Loss Functions for Discrete Time Systems

The polynomial $A_k(z)$ has no zeros outside the unit circle and the theorem is proven.

We have previously found that $a_0^k > 0$ for all k was a necessary condition for $A(z)$ to be stable. We will now show that the converse is also true. Hence assume that all a_0^k are positive. The trivial polynomial A_0 is stable as $a_0^0 > 0$. Theorem 2.1 then implies that A_1 is stable. By repeated application of Theorem 2.1, we thus find that the polynomial A_k is stable. Hence, if the polynomial $A(z)$ has all zeros inside the unit circle, then all coefficients a_0^k are positive. If any coefficient a_0^k is nonpositive, then the system with the pulse-transfer function $B(z)/A(z)$ is unstable. Summing up we get Theorem 2.2.

THEOREM 2.2

Let $a_0^n > 0$ then the following conditions are equivalent

1. $A_n(z)$ is stable
2. $a_0^k > 0$ for $k = 0, 1, \ldots, n-1$

The Main Result

We will now show that the integral (2.1) can be computed recursively. For this purpose, we introduce the integrals I_k defined by

$$I_k = \frac{1}{2\pi i} \oint \frac{B_k(z) B_k(z^{-1})}{A_k(z) A_k(z^{-1})} \cdot \frac{dz}{z} \qquad (2.20)$$

It follows from (2.1) that $I = I_n$. We now have Theorem 2.3.

THEOREM 2.3

Let the polynomial $A(z)$ have all its zeros inside the unit circle. The integrals I_k defined by (2.20) then satisfy the following recursive equation

$$[1 - \alpha_k^2] I_{k-1} = I_k - \beta_k^2 \qquad (2.21)$$
$$I_0 = \beta_0^2 \qquad (2.22)$$

Proof

As $A(z)$ has all its zeros inside the unit circle, it follows from Theorem 2.2 that all a_0^k are different from zero. It thus follows from (2.7) and (2.8) that all polynomials A_k and B_k can be defined. Furthermore it follows from Theorem 2.2 that all polynomials A_k have all zeros inside the unit circle. All integrals I_k thus exist.

To prove the theorem, we will make use of the theory of analytic functions. We will first assume that the polynomial $A(z)$ has distinct roots, which are different from zero. The integral (2.20) equals the sum of residues at the poles of the function $B_k(z) B_k(z^{-1})/\{z A_k(z) A_k(z^{-1})\}$ inside the unit circle. As the integral is invariant under the change of variables

$z \to 1/z$, we also find that the integral equals the sum of residues of the poles outside the unit circle.

Now consider
$$I_{k-1} = \frac{1}{2\pi i} \oint \frac{B_{k-1}(z) B_{k-1}(z^{-1})}{A_{k-1}(z) A_{k-1}(z^{-1})} \cdot \frac{dz}{z}$$

The poles of the integrand inside the unit circle are $z = 0$ and the zeros z_i of the polynomial $A_{k-1}(z)$. Since $A_{k-1}(z_i) = 0$, it follows from (2.7) and (2.4) that
$$A_k(z_i) = \alpha_k A_k^*(z_i) = \alpha_k z_i^k A_k(z_i^{-1})$$

Combining this equation with (2.7) and (2.4) we find
$$A_{k-1}(z_i^{-1}) = z_i[A_k(z_i^{-1}) - \alpha_k A_k^*(z_i^{-1})]$$
$$= z_i[A_k(z_i^{-1}) - \alpha_k z_i^{-k} A_k(z_i)] = (1 - \alpha_k^2) z_i A_k(z_i^{-1})$$

Further it follows from (2.4) and (2.7) that
$$A_{k-1}^*(z) = A_k^*(z) - \alpha_k A_k(z)$$

Hence
$$A_{k-1}^*(0) = A_k^*(0) - \alpha_k A_k(0) = a_0^k - \alpha_k a_k^k = a_0^k(1 - \alpha_k^2)$$

The functions
$$\frac{B_{k-1}(z) B_{k-1}(z^{-1})}{A_{k-1}(z) A_{k-1}(z^{-1})} \cdot \frac{1}{z} = \frac{B_{k-1}(z) B_{k-1}^*(z)}{A_{k-1}(z) A_{k-1}^*(z)} \cdot \frac{1}{z}$$

and
$$\frac{B_{k-1}(z) B_{k-1}(z^{-1})}{A_{k-1}(z)[z(1 - \alpha_k^2) A_k(z^{-1})]} \cdot \frac{1}{z} = \frac{B_{k-1}(z) B_{k-1}^*(z)}{A_{k-1}(z)[(1 - \alpha_k^2) A_k^*(z)]} \cdot \frac{1}{z}$$

have the same poles inside the unit circle and the same residues at these poles. Hence
$$I_{k-1} = \frac{1}{1 - \alpha_k^2} \cdot \frac{1}{2\pi i} \oint \frac{B_{k-1}(z) B_{k-1}(z^{-1})}{A_{k-1}(z) A_k(z^{-1})} \cdot \frac{dz}{z^2}$$
$$= \frac{1}{1 - \alpha_k^2} \cdot \frac{1}{2\pi i} \oint \frac{B_{k-1}(z) B_{k-1}(z^{-1})}{A_k(z) A_{k-1}(z^{-1})} dz \quad (2.23)$$

where the second equality is obtained by making the variable substitution $z \to z^{-1}$. The integrand has poles at the zeros of $A_k(z)$. It follows, however, from (2.7) that
$$A_{k-1}(z^{-1}) = z\{A_k(z^{-1}) - \alpha_k A_k^*(z^{-1})\} = z\{A_k(z^{-1}) - \alpha_k z^{-k} A_k(z)\}$$

Hence for z_i such that $A_k(z_i) = 0$, we get
$$A_{k-1}(z_i^{-1}) = z_i A_k(z_i^{-1})$$

Evaluation of Loss Functions for Discrete Time Systems

The functions

$$\frac{B_{k-1}(z)B_{k-1}(z^{-1})}{A_k(z)A_{k-1}(z^{-1})}$$

and

$$\frac{B_{k-1}(z)B_{k-1}(z^{-1})}{A_k(z)A_k(z^{-1})} \cdot \frac{1}{z} = \frac{B_{k-1}(z)B^*_{k-1}(z)}{A_k(z)A^*_k(z)}$$

thus have the same poles inside the unit circle and the same residues at these poles. The integral of these functions around the unit circle are thus the same. Equation (2.23) now gives

$$I_{k-1} = \frac{1}{1-\alpha_k^2} \cdot \frac{1}{2\pi i} \oint \frac{B_{k-1}(z)B_{k-1}(z^{-1})}{A_k(z)A_k(z^{-1})} \cdot \frac{dz}{z}$$

Now introduce (2.8) and we find

$$(1-\alpha_k^2)I_{k-1} = \frac{1}{2\pi i} \oint \frac{[B_k(z) - \beta_k A_k^*(z)][B_k(z^{-1}) - \beta_k A_k^*(z^{-1})]}{A_k(z)A_k(z^{-1})} \cdot \frac{dz}{z}$$

$$= \frac{1}{2\pi i} \oint \frac{B_k(z)B_k(z^{-1})}{A_k(z)A_k(z^{-1})} \cdot \frac{dz}{z} - \frac{\beta_k}{2\pi i} \oint \frac{B_k(z)A_k^*(z^{-1})}{A_k(z)A_k(z^{-1})} \cdot \frac{dz}{z}$$

$$- \frac{\beta_k}{2\pi i} \oint \frac{A_k^*(z)B_k(z^{-1})}{A_k(z)A_k(z^{-1})} \cdot \frac{dz}{z} + \frac{\beta_k^2}{2\pi i} \oint \frac{A_k^*(z)A_k^*(z^{-1})}{A_k(z)A_k(z^{-1})} \cdot \frac{dz}{z}$$

(2.24)

The first integral equals I_k. The second integral can be reduced as follows

$$\frac{\beta_k}{2\pi i} \oint \frac{B_k(z)A_k^*(z^{-1})}{A_k(z)A_k(z^{-1})} \cdot \frac{dz}{z} = \frac{\beta_k}{2\pi i} \oint \frac{B_k(z)A_k(z)}{A_k(z)A_k^*(z)} \cdot \frac{dz}{z}$$

$$= \frac{\beta_k}{2\pi i} \oint \frac{B_k(z)}{A_k^*(z)} \cdot \frac{dz}{z} = \beta_k \frac{B_k(0)}{A_k^*(0)} = \beta_k \frac{b_k^k}{a_0^k} = \beta_k^2$$

where the first equality follows from (2.4), the third from the residue theorem, and the fifth from (2.10). Similarly we find that the third integral of the right member of (2.24) also equals β_k^2.

Using (2.4), the fourth term of the right member of (2.24) can be reduced as follows

$$\frac{\beta_k^2}{2\pi i} \oint \frac{A_k^*(z)A_k^*(z^{-1})}{A_k(z)A_k(z^{-1})} \cdot \frac{dz}{z} = \frac{\beta_k^2}{2\pi i} \oint \frac{dz}{z} = \beta_k^2$$

Summarizing we find (2.21). When $k = 0$ we get from (2.20)

$$I_0 = \frac{1}{2\pi i} \oint \left(\frac{b_0^0}{a_0^0}\right)^2 \cdot \frac{dz}{z} = \beta_0^2$$

We have thus proven the formulas (2.21) and (2.22) when $A(z)$ has distinct roots. If A has multiple roots or roots equal to zero, we can always perturb its coefficients in order to obtain distinct and nonzero roots. Equations (2.21) and (2.22) then hold. As the numbers α_k and β_k are continuous functions of the parameters, we find that (2.20) and (2.21) hold even when A has multiple roots.

Notice that it follows from (2.13) that

$$a_0^{k-1} = a_0^k - \alpha_k a_k^k = a_0^k(1 - \alpha_k^2)$$

Equation (2.21) can then be written as

$$a_0^{k-1} I_{k-1} = a_0^k I_k - a_0^k \beta_k^2$$

or

$$a_0^k I_k = a_0^{k-1} I_{k-1} + \beta_k b_k^k = a_0^{k-1} I_{k-1} + (b_k^k)^2/a_0^k$$

COROLLARY 2.1

The integral I_k is given by

$$I_k = \frac{1}{a_0^k} \sum_{i=0}^{k} \frac{(b_i^i)^2}{a_0^i} \qquad (2.25)$$

Computational Aspects

Having obtained the recursive formula given by Theorem 2.3 we will now turn to the practical aspects of the computations. To obtain the integrals we must first compute the coefficients of the polynomials $A_k(z)$ and $B_k(z)$. This is easily done with the following tables

a_0	a_1	$\cdots$	a_{n-1}	a_n	b_0	b_1	$\cdots$	b_{n-1}	b_n
a_n	a_{n-1}	$\cdots$	a_1	a_0	a_n	a_{n-1}	$\cdots$	a_1	a_0
a_0^{n-1}	a_1^{n-1}	$\cdots$	a_{n-1}^{n-1}		b_0^{n-1}	b_1^{n-1}	$\cdots$	b_{n-1}^{n-1}	
a_{n-1}^{n-1}	a_{n-2}^{n-1}	$\cdots$	a_0^{n-1}		a_{n-1}^{n-1}	a_{n-2}^{n-1}	$\cdots$	a_0^{n-1}	
$\vdots$					$\vdots$				
a_0^1	a_1^1				b_0^1	b_1^1			
a_1^1	a_0^1				a_1^1	a_0^1			
a_0^0					b_0^0				

Each even row in the table of coefficients of A (A-table) is obtained by writing the coefficients of the proceeding row in reverse order. The even rows of the A- and B-tables are the same. The coefficients of the odd rows of both tables are obtained from the two elements above using (2.13) and (2.14)

$$a_i^{k-1} = a_i^k - \alpha_k a_{k-i}^k, \qquad \alpha_k = a_k^k/a_0^k \qquad (2.13)$$
$$b_i^{k-1} = b_i^k - \beta_k a_{k-i}^k, \qquad \beta_k = b_k^k/a_0^k \qquad (2.14)$$

Evaluation of Loss Functions for Discrete Time Systems

Using the stability criterion of Theorem 2.2, we find that the polynomial $A(z)$ has all zeros inside the unit circle if all coefficients a_0^k are positive. These elements are boldfaced in the table above. Having obtained the coefficients α_k and β_k we can now easily obtain the value of the integral from (2.25).

Notice that in order to investigate the stability of the polynomial $A(z)$ we have to form the A-table only. Hence the work required to calculate the integral I is roughly twice the work required to test the stability of the polynomial $A(z)$.

EXAMPLE

As an illustration, we will evaluate the integral for

$$A(z) = z^3 + 0.7z^2 + 0.5z - 0.3$$
$$B(z) = z^3 + 0.3z^2 + 0.2z + 0.1$$

We get the following tables

	α_k							β_k
1	0.7	0.5	−0.3	1	0.3	0.2	0.1	
−0.3	0.5	0.7	1.0	−0.3	−0.3	0.5	0.7	1.0	0.1
0.91	0.85	0.71			1.03	0.25	0.13	
0.71	0.85	0.91		0.780	0.71	0.85	0.91	0.143
0.356	0.187			0.929	0.129			
0.187	0.356			0.525	0.187	0.356		0.361
0.258				0.258	0.861			3.338

and we find $I = 2.9488$

The given formulas are well suited for machine calculations. See the FORTRAN program (p. 126) for the computation.

Exercises

1. Evaluate the integral (2.1) for

$$A(z) = z^2 + 0.4z + 0.1$$
$$B(z) = z^2 + 0.9z + 0.8$$

   (Answer $I = 1.565079$)

2. A simple inventory control system can be described by the equations

$$I(t) = I(t-1) + P(t) - S(t)$$
$$P(t) = P(t-1) + u(t-k)$$

   where I denotes inventory level, P production, and S sales. The deci-

```
      SUBROUTINE SALOSS (A, B, N, IERR, V, IN)
C
C     PROGRAM FOR EVALUATING THE INTEGRAL OF THE RATIONAL
C     FUNCTION
C         1/(2*PI*I)*B(Z)*B(1/Z)/(A(Z)*A(1/Z))*(1/Z)
C     AROUND THE UNIT CIRCLE
C
C     A—VECTOR WITH THE COEFFICIENTS OF THE POLYNOMIAL
C         A(1)*Z**N + A(2)*Z**(N − 1) + ··· + A(N + 1)
C     B—VECTOR WITH THE COEFFICIENTS OF THE POLYNOMIAL
C         B(1)*Z**N + B(2)*Z**(N − 1) + ··· + B(N + 1)
C
C         THE VECTORS A AND B ARE DESTROYED
C
C     N—ORDER OF THE POLYNOMIALS A AND B (MAX 10)
C     IERR—WHEN RETURNING IERR = 1 IF A HAS ALL ZEROS INSIDE UNIT
C         CIRCLE IERR = 0 IF THE POLYNOMIAL A HAS ANY ROOT OUTSIDE
C         OR ON THE UNIT CIRCLE OR IF A(1) IS NOT POSITIVE
C     V—THE RETURNED LOSS
C     IN—DIMENSION OF A AND B IN MAIN PROGRAM
C
C     SUBROUTINE REQUIRED
C         NONE
C
      DIMENSION A(IN), B(IN), AS(11)
C
      A0 = A(1)
      IERR = 1
      V = 0.0
      DO  10   K = 1, N
      L = N + 1 − K
      L1 = L + 1
      ALFA = A(L1)/A(1)
      BETA = B(L1)/A(1)
      V = V + BETA*B(L1)
      DO  20   I = 1, L
      M = L + 2 − I
      AS(I) = A(I) − ALFA*A(M)
   20 B(I) = B(I) − BETA*A(M)
      IF (AS(1)) 50, 50, 30
   30 DO  40   I = 1, L
   40 A(I) = AS(I)
   10 CONTINUE
      V = V + B(1)**2/A(1)
      V = V/A0
      RETURN
   50 IERR = 0
      RETURN
      END
```

sion variable is denoted by u, and the production delay is k units. Assume that the following decision rule is used to refill the inventory

$$u(t) = \alpha[I_0 - I(t)]$$

Determine the variance of the fluctuations in production and inventory level when the fluctuations in sales can be described as a sequence of independent equally distributed random variables with zero mean values and standard deviation σ.

3. Show that the integral I defined by (2.1) can be computed as the first component x_1 of the solution of the following linear equation

$$\begin{bmatrix} 2a_0 & 2a_1 & 2a_2 & 2a_3 & \cdots & 2a_n \\ a_1 & a_0+a_2 & a_1+a_3 & a_2+a_4 & \cdots & a_{n-1} \\ a_2 & a_3 & a_0+a_4 & a_1+a_5 & \cdots & a_{n-2} \\ \vdots & & & & & \vdots \\ a_{n-1} & a_n & 0 & 0 & \cdots & a_1 \\ a_n & 0 & 0 & 0 & \cdots & a_0 \end{bmatrix} \begin{bmatrix} x_1 \\ x_2 \\ x_3 \\ \vdots \\ x_n \\ x_{n+1} \end{bmatrix} = \begin{bmatrix} 2\sum_{i=0}^{n} b_i^2 \\ 2\sum_{i=0}^{n-1} b_i b_{i+1} \\ 2\sum_{i=0}^{n-2} b_i b_{i+2} \\ \vdots \\ 2\sum_{i=0}^{1} b_i b_{i+n-1} \\ 2 b_0 b_n \end{bmatrix}$$

Compare the number of computations required when evaluating the integral as a solution to this linear equation with the computations required when using Theorem 2.1.

4. When the function A_{k-1} is determined from A_k using (2.7), the constant term of A_k is eliminated. Show that a result which is similar to Theorem 2.3 can be obtained by the following reductions

$$A_{k-1}(z) = A_k(z) - \frac{a_0^k}{a_k^k} A_k^*(z)$$

$$B_{k-1}(z) = B_k(z) - \frac{b_0^k}{a_k^k} A_k^*(z)$$

which eliminate the terms of highest order in the polynomials A_k and B_k.

Hint:

$$I_k = \left[\left(\frac{a_0^k}{a_k^k}\right)^2 - 1\right] I_{k-1} + 2\frac{b_0^k b_k^k}{a_0^k a_k^k} - \left(\frac{b_0^k}{a_k^k}\right)^2$$

5. Give recursive algorithms for evaluating the integrals

$$\frac{1}{2\pi i} \oint \frac{B(z) B(z^{-1})}{A(z) A(z^{-1})} z^k \, dz$$

$$\frac{1}{2\pi i} \oint \frac{z^k B(z^{-1})}{A(z) A(z^{-1})} \, dz$$

6. Verify that the given FORTRAN program gives the desired result.

3. EVALUATION OF LOSS FUNCTIONS FOR CONTINUOUS TIME SYSTEMS

Statement of the Problem

We will now analyze the continuous time version of the problem discussed in Section 2. Consider a continuous time, linear time invariant dynamical system subject to a disturbance which is a stationary stochastic process with a rational spectral density. The variance of a system variable can be expressed by an integral of the type

$$I = \frac{1}{2\pi i} \int_{-i\infty}^{i\infty} \frac{B(s) B(-s)}{A(s) A(-s)} \, ds \tag{3.1}$$

where A and B are polynomials with real coefficients

$$A(s) = a_0 s^n + a_1 s^{n-1} + \cdots + a_{n-1} s + a_n \tag{3.2}$$

$$B(s) = \qquad b_1 s^{n-1} + \cdots + b_{n-1} s + b_n \tag{3.3}$$

The evaluation of the integral (3.1) will be discussed in this section. The integral (3.1) can also be interpreted as the variance of the signal obtained when white noise is fed through a stable filter with the transfer function $B(s)/A(s)$. The integral (3.1) will always exist if the polynomial $A(s)$ does not have any zeros on the imaginary axis. Notice that the polynomial B must be at least one degree less than the polynomial A. The physical interpretation is analogous to that of Section 2.

Notations and Preliminaries

To formulate the result we need some notations which will now be developed. A decomposition of the polynomial $A(s)$ into odd and even terms is first introduced. Hence

$$A(s) = \bar{A}(s) + \tilde{A}(s) \tag{3.4}$$

where

$$\bar{A}(s) = a_0 s^n + a_2 s^{n-2} + \cdots = \frac{1}{2}[A(s) + (-1)^n A(-s)] \tag{3.5}$$

$$\tilde{A}(s) = a_1 s^{n-1} + a_3 s^{n-3} + \cdots = \frac{1}{2}[A(s) - (-1)^n A(-s)] \tag{3.6}$$

We also introduce the polynomials $A_k(s)$ and $B_k(s)$ of lower order than n

$$A_k(s) = a_0^k s^k + a_1^k s^{k-1} + \cdots + a_k^k \tag{3.7}$$
$$B_k(s) = b_1^k s^{k-1} + b_2^k s^{k-2} + \cdots + b_k^k \tag{3.8}$$

which are defined recursively from the equations

$$A_{k-1}(s) = A_k(s) - \alpha_k s \tilde{A}_k(s) \tag{3.9}$$
$$B_{k-1}(s) = B_k(s) - \beta_k \tilde{A}_k(s) \tag{3.10}$$

where

$$\alpha_k = a_0^k / a_1^k \tag{3.11}$$
$$\beta_k = b_1^k / a_1^k \tag{3.12}$$

and

$$A_n(s) = A(s) \tag{3.13}$$
$$B_n(s) = B(s) \tag{3.14}$$

The polynomials A_{k-1} and B_{k-1} can apparently only be defined if $a_1^k \neq 0$. We will first establish necessary and sufficient conditions for this. It turns out that this problem is closely associated with the stability of the polynomials $A_k(s)$. We have the following result.

THEOREM 3.1

Let $a_0^k > 0$ then the following conditions are equivalent

1. The polynomial $A_k(s)$ has all zeros in the left half plane
2. The polynomial $A_{k-1}(s)$ has all zeros in the left half plane and a_1^k is positive.

To prove this theorem we will use the following lemma.

LEMMA 3.1

Let the real polynomial $f(s)$ with real coefficients have all zeros in the left half plane then

$$\begin{aligned} |f(s)| &< |f(-s)| \quad & \text{Re } s < 0 \\ |f(s)| &= |f(-s)| \quad & \text{Re } s = 0 \\ |f(s)| &> |f(-s)| \quad & \text{Re } s > 0 \end{aligned}$$

Proof

As f is a polynomial with zeros in the left half plane we have

$$f(s) = \beta \prod_{i=1}^{n} (s - \alpha_i), \quad \text{Re } \alpha_i < 0$$

Then

$$f(-s) = \beta \prod_{i=1}^{n}(-s - \alpha_i) = \beta \prod_{i=1}^{n}(-s - \bar{\alpha}_i)$$

Introduce

$$w(s) = \frac{f(s)}{f(-s)} = \prod_{i=1}^{n}\frac{s - \alpha_i}{-s - \bar{\alpha}_i}$$

Now consider the transformation

$$w_i(s) = \frac{\alpha_i - s}{\bar{\alpha}_i + s}$$

The transformation w_i maps the complex plane onto itself in such a way that the left half plane is mapped on the interior of the unit circle, the imaginary axis on the unit circle, and the right half plane on the exterior of the unit circle.

The transformation

$$w(s) = \prod_{i=1}^{n} w_i(s) = \prod_{i=1}^{n}\frac{\alpha_i - s}{\bar{\alpha}_i + s} = \frac{f(s)}{f(-s)}$$

has the same properties and the statements of the lemma are thus proven.

Proof of Theorem 3.1

We will first show that $1. \Longrightarrow 2$. Let $a_0^k > 0$ and let $A_k(s)$ have all its zeros in the left half plane. The proof that a_1^k is positive is by contradiction. Hence assume a_1^k nonpositive. Then take s real, positive, and sufficiently large. We then find $|A_k(s)| < |A_k(-s)|$ which contradicts Lemma 3.1. We can prove by a similar argument that a_1^k cannot be zero.

To prove that $A_{k-1}(s)$ has all its zeros in the left half plane, we observe that (3.6) and (3.9) give

$$A_{k-1}(s) = \left(1 - \frac{\alpha_k s}{2}\right)A_k(s) + (-1)^k \frac{\alpha_k s}{2} A_k(-s) \quad (3.15)$$

The polynomial $A_{k-1}(s)$ is of order $k - 1$. If we can show that the reciprocal polynomial

$$A_{k-1}^*(s) = s^{k-1}A_{k-1}(s^{-1}) = s^{-2}\left[\left(s - \frac{\alpha_k}{2}\right)A_k^*(s) + \frac{\alpha_k}{2}(-1)^k A_k^*(-s)\right] \quad (3.16)$$

has all zeros in the left half plane, it follows that $A_{k-1}(s)$ also has all zeros in the left half plane. Instead of analyzing (3.16) we will use imbedding, and consider the function

$$F(s, \alpha) = s^{-1}\left[\left(s - \frac{\alpha}{2}\right)A_k^*(s) + \frac{\alpha}{2}(-1)^k A_k^*(-s)\right] \quad (3.17)$$

Evaluation of Loss Functions for Continuous Time Systems

for arbitrary real α in the interval $0 \leqslant \alpha \leqslant \alpha_k$. Notice that

$$F(s, \alpha_k) = sA^*_{k-1}(s) \tag{3.18}$$

As A^*_k has no zeros in the right half plane it follows from Lemma 3.1 that

$$|A^*_k(s)| \geqslant |A^*_k(-s)| \quad \text{Re } s \geqslant 0$$

Now take s such that Re $s \geqslant 0$ and

$$\left| s - \frac{\alpha}{2} \right| > \left| \frac{\alpha}{2} \right|$$

It then follows from the triangle-inequality that

$$|F(s, \alpha)| = |s^{-1}| \left| \left(s - \frac{\alpha}{2} \right) A^*_k(s) + \frac{\alpha}{2} (-1)^k A^*_k(-s) \right|$$

$$\geqslant |s^{-1}| \left[\left| s - \frac{\alpha}{2} \right| |A^*_k(s)| - \left| \frac{\alpha}{2} \right| \cdot |A^*_k(-s)| \right] > 0$$

We thus find that the function F has no zeros in the set

$$S = \left\{ s; \text{ Re } s \geqslant 0 \quad \text{and} \quad \left| s - \frac{\alpha}{2} \right| > \left| \frac{\alpha}{2} \right| \right\} \tag{3.19}$$

Compare with Fig. 5.1.

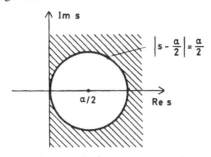

Fig. 5.1. Graph of the set $S = \{s; \text{ Re } s \geqslant 0 \text{ and } |s - \alpha/2| > |\alpha/2|\}$.

To show that this implies that F does not have any zeros in the right half plane, we will use a continuity argument. As the function F is continuous in α, its zeros will also be continuous in α. We have

$$F(s, 0) = A^*_k(s)$$

For $\alpha = 0$, we thus find that F has all zeros in the left half plane. As F does not have any zeros in the set (3.19), it follows that as α increases, no zeros will cross the imaginary axis for $s \neq 0$. The only possibility to obtain zeros in the right half plane is to have a zero enter the right half plane at the origin as α increases. But

$$F(0, \alpha) = a_0 - \alpha a_1 > 0, \qquad 0 \leqslant \alpha < \alpha_k$$

Hence as α increases $F(s, \alpha)$ will have a zero at the origin for $\alpha = a_0/a_1 = \alpha_k$. This is a single zero because $F'(0, \alpha) = a_1 > 0$. The function $F(s, \alpha_k)$ has one zero at the origin and all other zeros in the left half plane. Equation (3.18) then implies that $A_{k-1}^*(s)$ and thus also $A_{k-1}(s)$ has all zeros in the left half plane.

To prove that $2. \Longrightarrow 1.$ we assume that $A_{k-1}(s)$ has all zeros in the left half plane and that a_1^k and a_0^k are positive. Equations (3.6) and (3.9) give

$$A_{k-1}(s) = \left(1 - \frac{\alpha_k s}{2}\right) A_k(s) + (-1)^k \frac{\alpha_k s}{2} A_k(-s)$$

$$A_{k-1}(-s) = \left(1 + \frac{\alpha_k s}{2}\right) A_k(-s) - (-1)^k \frac{\alpha_k s}{2} A_k(s)$$

Elimination of $A_k(-s)$ between these equations gives

$$A_k(s) = \left(1 + \frac{\alpha_k s}{2}\right) A_{k-1}(s) - (-1)^k \frac{\alpha_k s}{2} A_{k-1}(-s)$$

As a_0^k and a_1^k are positive, α_k is positive. For Re $s \geqslant 0$ we have

$$\left| 1 + \frac{\alpha_k s}{2} \right| > \left| \frac{\alpha_k s}{2} \right|, \qquad \text{Re } s \geqslant 0$$

As $A_{k-1}(s)$ has all zeros in the left half plane, we can apply Lemma 3.1. Hence

$$|A_{k-1}(s)| \geqslant |A_{k-1}(-s)| \qquad \text{Re } s \geqslant 0$$

Combining the two inequalities given above, we find

$$\left| 1 + \frac{\alpha_k s}{2} \right| |A_{k-1}(s)| > \left| \frac{\alpha_k s}{2} \right| |A_{k-1}(-s)|, \qquad \text{Re } s \geqslant 0$$

Hence

$$|A_k(s)| = \left| \left(1 + \frac{\alpha_k s}{2}\right) A_{k-1}(s) - (-1)^k \frac{\alpha_k s}{2} A_{k-1}(-s) \right|$$

$$\geqslant \left| 1 + \frac{\alpha_k s}{2} \right| |A_{k-1}(s)| - \left| \frac{\alpha_k s}{2} \right| \cdot |A_{k-1}(-s)| > 0, \quad \text{Re } s \geqslant 0$$

The polynomial $A_k(s)$ thus does not have any zeros in the right half plane, and the proof of Theorem 3.1 is now completed.

By a repeated application of Theorem 3.1, we find that if the polynomial $A(s)$ has all roots in the left half plane, then all the polynomials $A_k(s)$, $k = n - 1, n - 2, \ldots, 0$ have also roots in the left half plane, and all the coefficients a_1^k are positive. Conversely, if all coefficients a_1^k are positive, we find that the polynomial $A(s)$ has all its roots in the left half plane. Hence we have Theorem 3.2.

THEOREM 3.2 (Routh)

Let $a_0^n > 0$, then the following conditions are equivalent
1. The polynomial $A(s)$ has all zeros in the left half plane.
2. All coefficients a_1^k are positive.

Main Result

We will now show how the integral (3.1) can be computed recursively. To do so we introduce

$$I_k = \frac{1}{2\pi i} \int_{-i\infty}^{i\infty} \frac{B_k(s) B_k(-s)}{A_k(s) A_k(-s)} ds$$

where the polynomials A_k and B_k are defined by (3.9) and (3.10). We observe that $I_n = I$. The main result is Theorem 3.3.

THEOREM 3.3

Assume that the polynomial A has all its roots in the left half plane. Then

$$I_k = I_{k-1} + \frac{\beta_k^2}{2\alpha_k} \qquad k = 1, 2, \cdots, n$$

$$I_0 = 0$$

Proof

The proof is based on elementary properties of analytic functions. As the coefficients α_k and β_k are continuous functions of the coefficients of the polynomial, it is sufficient to prove the theorem for the special case when all roots of $A_k(s)$ and $\tilde{A}_k(s)$ are distinct.

As the polynomial $A(s)$ has all zeros in the left half plane, it follows from Theorem 3.1 that all coefficients a_1^k are positive. The polynomials A_k and B_k can then be defined by (3.9) and (3.10). It also follows from Theorem 3.1 that all polynomials $A_k(s)$ have their zeros in the left half plane. We also observe that the polynomial $\tilde{A}_k(s)$ has all its zeros on the imaginary axis. It follows from (3.6) that

$$\tilde{A}_k(-s) = (-1)^{k-1} \tilde{A}_k(s)$$

and Lemma 3.1 gives

$$|\tilde{A}_k(s)| = \frac{1}{2} |A_k(s) - (-1)^k A_k(-s)| > \frac{1}{2}(|A_k(s)| - |A_k(s)|) = 0,$$

$$\text{Re } s > 0$$

It follows from (3.4), (3.5), and (3.6) that

$$A_k(-s) = \bar{A}_k(-s) + \tilde{A}_k(-s) = (-1)^k \bar{A}_k(s) + (-1)^{k-1} \tilde{A}_k(s)$$
$$= (-1)^k [\bar{A}_k(s) - \tilde{A}_k(s)] = (-1)^k A_k(s) + 2\tilde{A}_k(-s) \quad (3.20)$$

Now consider the functions

$$\frac{B_{k-1}(s) B_{k-1}(-s)}{A_{k-1}(s) A_{k-1}(-s)} \quad (3.21)$$

and

$$\frac{B_{k-1}(s) B_{k-1}(-s)}{A_{k-1}(s) 2\tilde{A}_k(-s)} \quad (3.22)$$

These functions have the same poles s_i in the left half plane which are given by $A_{k-1}(s_i) = 0$. The function (3.21) also has poles in the right half plane and the function (3.22) has poles on the imaginary axis.

At the poles which are strictly in the left half plane we have

$$A_{k-1}(s_i) = A_k(s_i) - \alpha_k s_i \tilde{A}_k(s_i) = 0 \quad (3.23)$$
$$A_{k-1}(-s_i) = A_k(-s_i) + \alpha_k s_i \tilde{A}_k(-s_i)$$
$$= A_k(-s_i) + \alpha_k s_i (-1)^{k-1} \tilde{A}_k(s_i)$$
$$= A_k(-s_i) + (-1)^{k-1} A_k(s_i) = 2\tilde{A}_k(-s_i)$$

where the first equality follows from (3.9), the second from (3.6), the third from (3.23), and the fourth from (3.6).

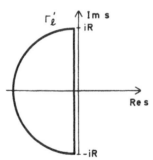

Fig. 5.2. The contour Γ_l is the limit of Γ_l' as $R \to \infty$.

As $A_{k-1}(s)$ has simple poles, the functions (3.21) and (3.22) have the same residues at the poles s_i. Integrating the functions (3.21) and (3.22) around a contour Γ_l (Fig. 5.2), which consists of a straight line slightly to the left of the imaginary axis and a semicircle with this line as a diameter, we find

$$I_{k-1} = \frac{1}{2\pi i} \int_{-i\infty-\varepsilon}^{i\infty-\varepsilon} \frac{B_{k-1}(s) B_{k-1}(-s)}{A_{k-1}(s) A_{k-1}(-s)} ds$$

$$= \frac{1}{2\pi i} \int_{-i\infty-\varepsilon}^{i\infty-\varepsilon} \frac{B_{k-1}(s)B_{k-1}(-s)}{A_{k-1}(s)2\tilde{A}_k(-s)} ds, \quad \varepsilon > 0 \quad (3.24)$$

because the integrands tend to zero as $|s|^{-2}$ for large s and the integrals along the semicircle thus vanish.

Now consider the functions

$$\frac{B_{k-1}(s)B_{k-1}(-s)}{A_{k-1}(s)2\tilde{A}_k(-s)} \quad (3.25)$$

and

$$\frac{B_{k-1}(s)B_{k-1}(-s)}{A_k(s)2\tilde{A}_k(-s)} \quad (3.26)$$

These functions have the same poles on the imaginary axis, namely at the zeros of $\tilde{A}_k$. They have no poles in the right half plane since the polynomials A_k and A_{k-1} have no zeros in the right half plane. Since $\tilde{A}_k(s_i) = 0$ we get

$$A_{k-1}(s_i) = A_k(s_i) - \alpha_k s_i \tilde{A}_k(s_i) = A_k(s_i)$$

If the poles s_i are distinct, we thus find that the functions (3.25) and (3.26) have the same residues at s_i. Integrating the functions (3.25) and (3.26) around a contour Γ_r which consists of a straight line slight to the left of the imaginary axis and a semicirle to the right, with the line as diameter (Fig. 5.3), we get

$$I_{k-1} = \frac{1}{2\pi i} \int_{-i\infty-\varepsilon}^{i\infty-\varepsilon} \frac{B_{k-1}(s)B_{k-1}(-s)}{A_{k-1}(s)2\tilde{A}_k(-s)} ds$$

$$= \frac{1}{2\pi i} \int_{-i\infty-\varepsilon}^{i\infty-\varepsilon} \frac{B_{k-1}(s)B_{k-1}(-s)}{A_k(s)2\tilde{A}_k(-s)} ds, \quad \varepsilon > 0 \quad (3.27)$$

because the integrands tend to zero as $|s|^{-2}$ for large $|s|$ and the integrals along the semicircles thus vanish.

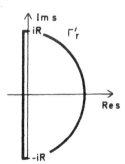

Fig. 5.3. The contour Γ_r is the limit of Γ_r' as $R \to \infty$.

Now consider the functions

$$\frac{B_{k-1}(s)B_{k-1}(-s)}{A_k(s)A_k(-s)} \tag{3.28}$$

and

$$\frac{B_{k-1}(s)B_{k-1}(-s)}{A_k(s)2\tilde{A}_k(-s)} \tag{3.29}$$

They have the same poles in the left half plane at the zeros of A_k. The function (3.28) also has poles in the right half plane and the function (3.29) has poles on the imaginary axis. At the poles in the left half plane we have

$$A_k(s_i) = 0 \tag{3.30}$$

It then follows from (3.7) and (3.9) that

$$A_k(-s_i) = (-1)^k A_k(s_i) + 2\tilde{A}_k(-s_i) = 2\tilde{A}_k(-s_i) \tag{3.31}$$

Since the zeros of A_k were assumed distinct, the functions (3.28) and (3.29) then have the same residues at their left half plane poles. Integrating (3.28) and (3.29) around the contour Γ_l (Fig. 5.2) we now find

$$\begin{aligned} I_{k-1} &= \frac{1}{2\pi i} \int_{-i\infty-\varepsilon}^{i\infty-\varepsilon} \frac{B_{k-1}(s)B_{k-1}(-s)}{A_k(s)2\tilde{A}_k(-s)} \, ds \\ &= \frac{1}{2\pi i} \int_{-i\infty-\varepsilon}^{i\infty-\varepsilon} \frac{B_{k-1}(s)B_{k-1}(-s)}{A_k(s)A_k(-s)} \, ds, \quad \varepsilon > 0 \end{aligned} \tag{3.32}$$

because the integrals along the semicircle with infinite radius will vanish because the integrand tends to zero as $|s|^{-2}$ for large $|s|$.

Equation (3.10) now gives

$$I_{k-1} = \frac{1}{2\pi i}\int_{-i\infty}^{i\infty}\frac{B_k(s)B_k(-s)}{A_k(s)A_k(-s)}\,ds - \frac{\beta_k}{2\pi i}\int_{-i\infty}^{i\infty}\frac{B_k(s)\tilde{A}_k(-s)}{A_k(s)A_k(-s)}\,ds \\ - \frac{\beta_k}{2\pi i}\int_{-i\infty}^{i\infty}\frac{\tilde{A}_k(s)B_k(-s)}{A_k(s)A_k(-s)}\,ds + \frac{\beta_k^2}{2\pi i}\int_{-i\infty}^{i\infty}\frac{\tilde{A}_k(s)\tilde{A}_k(-s)}{A_k(s)A_k(-s)}\,ds \tag{3.33}$$

The functions

$$\frac{B_k(s)\tilde{A}_k(-s)}{A_k(s)A_k(-s)} \tag{3.34}$$

and

$$\frac{B_k(s)\tilde{A}_k(-s)}{A_k(s)2\tilde{A}_k(-s)} \tag{3.35}$$

Evaluation of Loss Functions for Continuous Time Systems

have the same poles in the left half plane. The poles s_i are equal to the zeros of A_k. Since these zeros were assumed distinct, it follows from (3.30) and (3.31) that the functions (3.34) and (3.35) have the same residues at these poles. Integrating (3.34) and (3.35) around the contour Γ_l (Fig. 5.2) we now find

$$\frac{1}{2\pi i}\int_{-i\infty-\varepsilon}^{i\infty-\varepsilon}\frac{B_k(s)\tilde{A}_k(-s)}{A_k(s)A_k(-s)}ds = \frac{1}{2\pi i}\int_{\Gamma_l}\frac{B_k(s)\tilde{A}_k(-s)}{A_k(s)A_k(-s)}ds$$
$$= \frac{1}{2\pi i}\int_{\Gamma_l}\frac{B_k(s)\tilde{A}_k(-s)}{A_k(s)2\tilde{A}_k(-s)}ds = \frac{1}{2\pi i}\int_{\Gamma_l}\frac{B_k(s)}{2A_k(s)}ds = \frac{1}{2}\frac{b_1^k}{a_0^k}$$

(3.36)

where the first equality follows from the fact that the integral along the semicircle vanishes because the integrand tends to zero as $|s|^{-2}$ for large $|s|$. The second equality follows from the fact that both integrands have the same poles inside Γ_l and the same residues at these poles. The third equality is just an identity. Since A_k has all its zeros in the left half plane, the contour Γ_l can be changed to a circle around the origin without changing the value of the last integral. Observing that the integrand $B_k(s)/(2A_k(s))$ has a pole at infinity with the residue $b_1^k/(2a_0^k)$ we finally get the last equality.

We find similarly

$$\frac{1}{2\pi i}\int_{-i\infty}^{i\infty}\frac{\tilde{A}_k(s)A_k(-s)}{A_k(s)A_k(-s)}ds = \frac{1}{2\pi i}\int_{\Gamma_l}\frac{\tilde{A}_k(s)\tilde{A}_k(-s)}{A_k(s)2\tilde{A}_k(-s)}ds$$
$$= \frac{1}{2\pi i}\int_{\Gamma_l}\frac{\tilde{A}_k(s)}{2A_k(s)}ds = \frac{a_1^k}{2a_0^k}$$

(3.37)

Equations (3.33), (3.36), and (3.37) now give

$$I_{k-1} = I_k - \frac{\beta_k}{2}\frac{b_1^k}{a_0^k} - \frac{\beta_k}{2}\frac{b_1^k}{a_0^k} + \frac{\beta_k^2}{2}\frac{a_1^k}{a_0^k} = I_k - \frac{\beta_k^2}{2\alpha_k}$$

For $k = 1$ we have

$$I_1 = \frac{1}{2\pi i}\int_{-i\infty}^{i\infty}\frac{b_1^1}{a_0^1 s + a_1^1}\cdot\frac{b_1^1}{-a_0^1 s + a_1^1}ds = \frac{(b_1^1)^2}{2a_0^1 a_1^1} = \frac{\beta_1^2}{2\alpha_1}$$

The proof of the theorem is now completed.

Computational Aspects

Having established the recursive formula given by Theorem 3.3 we will now turn to the practical aspects of the computations. To obtain the value of the integral, we must first compute the coefficients of the polynomials $A_k(s)$ and $B_k(s)$. This is conveniently done using the following tables.

a_0^n	a_1^n	a_2^n	a_3^n	a_4^n	...	b_1^n	b_2^n	b_3^n	b_4^n	b_5^n	...
a_1^n	0	a_3^n	0	a_5^n	...	a_1^n	0	a_3^n	0	a_5^n	...
a_0^{n-1}	a_1^{n-1}	a_2^{n-1}	a_3^{n-1}	...		b_1^{n-1}	b_2^{n-1}	b_3^{n-1}	b_4^{n-1}	...	
a_0^{n-1}	0	a_3^{n-1}	0	...		a_1^{n-1}	0	a_3^{n-1}	0	...	
⋮						⋮					
a_0^2	a_1^2	a_2^2				b_1^2	b_2^2				
a_1^2	0					a_1^2	0				
a_0^1	a_1^1					b_1^1					
a_1^1	0										
a_0^0											

Each even row in the table of a_i^k coefficients is formed by shifting the elements of the preceding row one step to the left and putting zero in the every other position. The even rows of the table on the right are identical to those of the table on the left. The elements of the odd rows of both tables are formed from the two elements immediately above, using the formulas

$$a_i^{k-1} = \begin{cases} a_{i+1}^k & i \text{ even} \\ a_{i+1}^k - \alpha_k a_{i+2}^k & i \text{ odd}, \end{cases} \quad \alpha_k = a_0^k / a_1^k \quad i = 0, \cdots, k-1$$

$$b_i^{k-1} = \begin{cases} b_{i+1}^k & i \text{ even} \\ b_{i+1}^k - \beta_k a_{i+1}^k & i \text{ odd}, \end{cases} \quad \beta_k = b_1^k / a_1^k \quad i = 1, \cdots, k-1$$

These are obtained by identifying coefficients of powers of s in (3.9) and (3.10).

It follows from Routh's stability test (Theorem 3.2) that the polynomial A has all its zeros in the left-half plane if all the coefficients a_1^k are positive. The coefficients a_1^k are boldfaced in the table above.

Having obtained the values α_k and β_k, the value of the integral is then given by Theorem 3.3.

$$I = \sum_{k=1}^{n} \beta_k^2 / (2\alpha_k) = \sum_{k=1}^{n} (b_1^k)^2 / (2 a_0^k a_1^k)$$

As the computations are defined recursively, it is now an easy matter to obtain a computer algorithm. See the FORTRAN program (p. 139) for the computation.

Exercises

1. Evaluate the integral (3.1) for

```
       SUBROUTINE COLOSS (A, B, N, IERR, V, IN)
C
C      PROGRAM FOR EVALUATING THE INTEGRAL OF THE RATIONAL
C .    FUNCTION
C         1/(2*PI*I)*B(S)*B(-S)/(A(S)*A(-S))
C      ALONG THE IMAGINARY AXIS
C
C      A—VECTOR WITH THE COEFFICIENTS OF THE POLYNOMIAL
C         A(1)*S**N + A(2)*S**(N-1) + ··· + A(N + 1)
C      IT IS ASSUMED THAT A(1) IS POSITIVE
C      B—VECTOR WITH THE COEFFICIENTS OF THE POLYNOMIAL
C         B(1)*S**(N - 1) + B(2)*S**(N - 2) + ··· + B(N)
C
C      THE VECTORS A AND B ARE DESTROYED
C
C      N—ORDER OF THE POLYNOMIALS A AND B
C      IERR—WHEN RETURNING IERR = 1 IF ALL ZEROS OF A ARE IN LEFT
C         HALF PLANE IERR = 0 IF THE POLYNOMIAL A DOES NOT HAVE
C         ALL ZEROS IN LEFT HALF PLANE OR IF A(1) IS NOT POSITIVE
C      V—THE RETURNED LOSS
C      IN—DIMENSION OF A AND B IN MAIN PROGRAM
C
C      SUBROUTINE REQUIRED
C         NONE
C
       DIMENSION A(IN), B(IN)
C
       IERR = 1
       V = 0.
       IF (A(1)) 70, 70, 10
10     DO 20 K = 1, N
       IF (A(K + 1)) 70, 70, 30
30     ALFA = A(K)/A(K + 1)
       BETA = B(K)/A(K + 1)
       V = V + BETA**2/ALFA
       K1 = K + 2
       IF(K1 - N) 50, 50, 20
50     DO 60 I = K1, N, 2
       A(I) = A(I) - ALFA*A(I + 1)
60     B(I) = B(I) - BETA*A(I + 1)
20     CONTINUE
       V = V/2.
       RETURN
70     IERR = 0
       RETURN
       END
```

$$A(s) = s^6 + 3s^5 + 5s^4 + 12s^3 + 6s^2 + 9s + 1$$
$$B(s) = 3s^5 + s^4 + 12s^3 + 3s^2 + 9s + 1$$

2. Consider the feedback system whose block diagram is shown in Fig. 5.4, where the input signal u is a Wiener process with unit variance parameter. Determine the variance of the tracking error e as a function of K, and calculate the K-value which minimizes the variance of the tracking error.

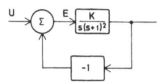

Fig. 5.4. Block diagram of the system of Exercise 2.

3. Consider the feedback system whose block diagram is shown in Fig. 5.5. The input u is a stationary process with the spectral density

$$\phi_u(\omega) = \frac{1}{\omega^2 + a^2}$$

and the measurement noise is white with the spectral density

$$\phi_n(\omega) = b^2$$

Determine the mean square deviation between the input and the output and the value of the gain parameter for which the mean square error is as small as possible.

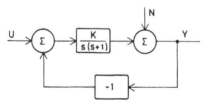

Fig. 5.5. Block diagram of the system of Exercise 3.

4. Show that for $n = 4$, the integral I defined by (3.1) can be computed as the first component x_1 of the following linear system

$$\begin{bmatrix} a_1 & a_0 & 0 & 0 \\ a_3 & a_2 & a_1 & a_0 \\ 0 & a_4 & a_3 & a_2 \\ 0 & 0 & 0 & a_4 \end{bmatrix} \begin{bmatrix} x_1 \\ x_2 \\ x_3 \\ x_4 \end{bmatrix} = \frac{(-1)^{n+1}}{2a_0} \begin{bmatrix} b_1^2 + b_2^2 + b_3^2 + b_4^2 \\ b_1 b_2 + b_2 b_3 + b_3 b_4 \\ b_1 b_3 + b_2 b_4 \\ b_1 b_4 \end{bmatrix}$$

Evaluation of Loss Functions for Continuous Time Systems

Notice that the matrix on the left is equal to the Hurwitz matrix for the polynomial $A(s)$.

5. Generalize the formula in Exercise 4 to arbitrary n.

6. When the function A_{k-1} is determined from A_k using (3.9), the highest power of A_k is eliminated. Show that it is possible to obtain results which are analogous to Theorem 3.3 through the following reductions

$$A_{k-1}(s) = \frac{1}{s}\left[A_k(s) - \frac{a_k^{\ k}}{a_{k-1}^k s}\tilde{A}_k(s)\right]$$

$$B_{k-1}(s) = \frac{1}{s}\left[B_k(s) - \frac{b_k^{\ k}}{a_{k-1}^k s}\tilde{A}_k(s)\right]$$

$$\tilde{A}_k(s) = \frac{1}{2}[A_k(s) - A_k(-s)]$$

Hint:

$$I_k = I_{k-1} + \frac{(b_k^{\ k})^2}{2a_k^{\ k} a_{k-1}^k}$$

7. Derive recursive formulas for evaluating the integral

$$\frac{1}{2\pi i}\int_{-i\infty}^{i\infty} \frac{s^k B(-s)}{A(s) A(-s)}\, ds$$

where A and B are polynomials with real coefficients

$$A(s) = a_0 s^n + a_1 s^{n-1} + \cdots + a_n$$
$$B(s) = b_0 s^m + b_1 s^{m-1} + \cdots + b_m$$

and $k + m \leqslant 2(n-1)$. The polynomial A has all its roots in the left half plane.

8. Derive recursive formulas for evaluating the integral

$$\frac{1}{2\pi i}\int_{-i\infty}^{i\infty} \frac{B(s)}{A(s) C(-s)}\, ds$$

where

$$A(s) = a_0 s^n + a_1 s^{n-1} + \cdots + a_n$$
$$B(s) = b_0 s^m + b_1 s^{m-1} + \cdots + b_m$$
$$C(s) = c_0 s^k + c_1 s^{k-1} + \cdots + c_k$$

and $m < n + k - 2$. The polynomials A and C have all their zeros in the left half plane.

9. Show that

$$\frac{1}{2\pi i} \int_{-i\infty}^{i\infty} \frac{1}{A(s)A(-s)} \, ds = \frac{1}{2a_1{}^1 a_0{}^1}$$

10. Show that the polynomial

$$\tilde{A}(s) = \frac{1}{2}[A(s) - (-1)^n A(-s)]$$

has all its zeros on the imaginary axis.

11. Consider two stationary stochastic processes with the spectral densities

$$\phi_x(\omega) = G_1(i\omega)G_1(-i\omega)$$
$$\phi_y(\omega) = G_2(i\omega)G_2(-i\omega)$$
$$\phi_{xy}(\omega) = G_1(i\omega)G_2(-i\omega)$$

where

$$G_1(s) = \frac{\omega^2}{s^2 + 2\zeta\omega s + \omega^2}$$

$$G_2(s) = \frac{\omega s}{s^2 + 2\zeta\omega s + \omega^2}$$

Determine Ex^2, Ey^2, and Exy using Theorem 3.3. Also solve the same problem by first applying the representation theorem of Chapter 4 (Theorem 5.2) and then using Theorem 6.1 of Chapter 3. Compare the computational efforts in the two cases.

12. Verify that the given FORTRAN program gives the desired result.

4. RECONSTRUCTION OF STATE VARIABLES FOR DISCRETE TIME SYSTEMS

Introduction

There are in practice many situations when only a few state variables of a dynamical system can be measured directly. Consider for example the following discrete time dynamical system

$$x(t + 1) = \Phi x(t) + \Gamma u(t) \quad (4.1)$$
$$y(t) = \theta x(t) \quad (4.2)$$

where x is an n-dimensional state vector, u an r-dimensional vector of inputs, and y a p-dimensional vector of output signals. The matrix Φ is $n \times n$, Γ is $n \times r$ and θ is $p \times n$. The elements of Φ, Γ, and θ may depend on t.

If the system (4.1), (4.2) is completely observable in Kalman's sense, the state vector can be reconstructed from at most n measurements of the

Reconstruction of State Variables for Discrete Time Systems

output signal. The state variables can, however, also be reconstructed from a mathematical model of the system. Consider for example the following model

$$\hat{x}(t + 1) = \Phi \hat{x}(t) + \Gamma u(t) \tag{4.3}$$

which has the same input as the original system (4.1).

If (4.3) is a perfect model, i.e., if the model parameters are identical to the system parameters, and if the initial conditions of (4.1) and (4.3) are identical, then the state $\hat{x}$ of the model will be identical to the true state variable x. If the initial conditions of (4.1) and (4.3) differ, the reconstruction x will converge to the true state variable x only if the system (4.1) is asymptotically stable. Notice, however, that the reconstruction (4.3) does not make use of the measurements of the state variables. By comparing y with $\theta \hat{x}$, it is possible to get an indication of how well the reconstruction (4.3) works. The difference $y - \theta \hat{x}$ can be interpreted physically as the difference between the actual measurements and the predictions of the measurements based on the reconstructed state variables. By exploiting the difference $y - \theta \hat{x}$, we can adjust the estimates $\hat{x}$ given by (4.3), for example, by using the reconstruction

$$\hat{x}(t + 1) = \Phi \hat{x}(t) + \Gamma u(t) + K[y - \theta \hat{x}] \tag{4.4}$$

where K is a suitably chosen matrix. If the reconstructed state vector $\hat{x}$ is identical to the true state vector, the reconstructions (4.3) and (4.4) are identical and both will give the correct result. In a practical case, we might also expect (4.4) to give better results than (4.3) because in (4.4) we use the measurements as well as the input signals for the reconstruction. To get some insight into the proper choice of K, we will consider the reconstruction error $\tilde{x} = x - \hat{x}$. By subtracting (4.4) from (4.1), and using (4.2), we get

$$\tilde{x}(t + 1) = \Phi \tilde{x}(t) - K[y(t) - \theta \hat{x}(t)] = [\Phi - K\theta] \tilde{x}(t) \tag{4.5}$$

If K is chosen in such a way that the system (4.5) is asymptotically stable, the reconstruction error $\tilde{x}$ will always converge to zero. Hence by introducing a feedback in the model it is possible to reconstruct state variables also in the case when the system itself is unstable. By a proper choice of K, the reconstruction error will always converge to zero for arbitrary states of (4.4).

A Parametric Optimization Problem

We have thus found that the state variables of a dynamical system can be reconstructed through the use of a mathematical model. The reconstruction contains a matrix K which can be chosen arbitrarily, subject to the constraint that the matrix Φ-$K\theta$ has all its zeros inside the unit circle. The problem now arises if there is an optimal choice of K. To pose such a problem we must introduce more structure into the problem. To do so we assume that the system is actually governed by a stochastic difference equation

$$x(t+1) = \Phi x(t) + \Gamma u(t) + v(t) \tag{4.6}$$

where $\{v(t), t \in T\}$ is a sequence of independent random n-vectors. The vector $v(t)$ has zero mean and covariance R_1. We also assume that the initial value $x(t_0)$ is normal with mean m and covariance R_0 and that there are measurement errors

$$y(t) = \theta x(t) + e(t) \tag{4.7}$$

where $\{e(t), t \in T\}$ is a sequence of independent random p-vectors. The vector $e(t)$ has zero mean and covariance R_2. The measurement errors e are assumed to be independent of v. The parameters $\Phi, \Gamma, \theta, R_1$, and R_2 may depend on time. Notice that even if the disturbances acting on the system are not white noise, they can often be described by a model of type (4.6) by enlarging the state space as was described in Chapter 4. A block diagram representation of the system described by the (4.6) and (4.7) is given in Fig. 5.6.

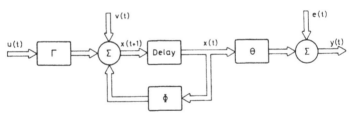

Fig. 5.6. Block diagram representation of the system described by (4.1) and (4.2).

In order to reconstruct the state variables we use the mathematical model

$$\hat{x}(t+1) = \Phi \hat{x}(t) + \Gamma u(t) + K[y(t) - \theta \hat{x}(t)] \tag{4.8}$$

A block diagram of the system (4.6), (4.7), and the reconstructor (4.8) is given in Fig. 5.7. We can now formulate a parametric optimization problem.

PROBLEM 4.1

Given an arbitrary constant vector a. Find a sequence of matrices $K(t)$ such that the error of the reconstruction of the scalar product $a^T x$ is as small as possible in the sense of mean square.

Reconstruction of State Variables for Discrete Time Systems

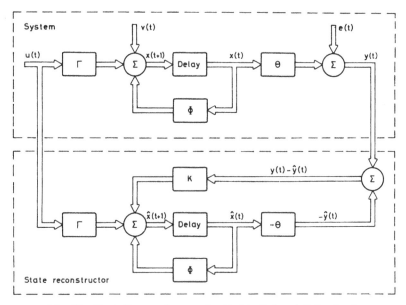

Fig. 5.7. Block diagram of system described by (4.6) and (4.7), of and state reconstructor given by (4.8).

Solution

The solution of the problem is straightforward. We first evaluate the mean and the variance of the reconstruction error, and we will then carry out the minimization. We will first derive an equation for the reconstruction error. Subtracting (4.8) from (4.6) we get

$$\tilde{x}(t+1) = x(t+1) - \hat{x}(t+1) = \Phi\tilde{x}(t) + v(t) - K[y(t) - \theta\hat{x}(t)]$$

Using (4.7) we find

$$\tilde{x}(t+1) = (\Phi - K\theta)\tilde{x}(t) + v(t) - Ke(t) \qquad (4.9)$$

The reconstruction error is thus governed by a linear stochastic difference equation. The analysis of such equations was discussed in Chapter 3 (Theorem 3.1). The mean value is given by

$$E\tilde{x}(t+1) = [\Phi - K\theta]E\tilde{x}(t) \qquad (4.10)$$

Hence if we choose the initial condition such that $\hat{x}(t_0) = m$, we find that $E\tilde{x}(t_0) = E(\hat{x}(t_0) - m) = 0$, and the reconstruction error has thus zero mean, irrespective of how K is chosen. The variance of the reconstruction error

$$P(t) = E[\tilde{x}(t) - E\tilde{x}(t)][\tilde{x}(t) - E\tilde{x}(t)]^T \qquad (4.11)$$

is thus given by

$$P(t + 1) = [\Phi - K\theta]P(t)[\Phi - K\theta]^T + R_1 + KR_2K^T$$
$$P(t_0) = R_0 \qquad (4.12)$$

This equation follows from Theorem 3.1 of Chapter 3. It can also be derived directly by multiplying (4.9) with its transpose and taking mathematical expectation.

Having obtained an equation for the variance of the reconstruction error, we will now determine the gain-matrix K in such a way that the variance of the scalar product $a^T \tilde{x}$ is as small as possible. We have

$$E(a^T \tilde{x})^2 = Ea^T \tilde{x} \tilde{x}^T a = a^T(E\tilde{x}\tilde{x}^T)a = a^T P(t)a$$

Using (4.12) we get

$$a^T P(t + 1)a = a^T \{\Phi P(t)\Phi^T + R_1 - K\theta P(t)\Phi^T - \Phi P(t)\theta^T K^T$$
$$+ K[R_2 + \theta P(t)\theta^T]K^T\}a \qquad (4.13)$$

We can now determine the time-varying gain recursively. Starting with $t = t_0$ we find that the right member of (4.13) is a quadratic function of K. By the proper choice of K we can then achieve that $P(t_0 + 1)$ is as small as possible. Having done this, we then put $t = t_0 + 1$ and we can then determine $K = K(t_0 + 1)$ in such a way that $P(t_0 + 2)$ is as small as possible. To carry out the details we rewrite (4.13) by completing the squares. We get

$$P(t + 1) = \Phi P(t)\Phi^T + R_1 - \Phi P(t)\theta^T[R_2 + \theta P(t)\theta^T]^{-1}\theta P(t)\Phi^T$$
$$+ \{K - \Phi P(t)\theta^T[R_2 + \theta P(t)\theta^T]^{-1}\}[R_2 + \theta P(t)\theta^T]$$
$$\times \{K - \Phi P(t)\theta^T[R_2 + \theta P(t)\theta^T]^{-1}\}^T \qquad (4.14)$$

Now consider the scalar

$$a^T P(t + 1)a = a^T \{\Phi P(t)\Phi^T + R_1 - \Phi P(t)\theta^T[R_2 + \theta P(t)\theta^T]^{-1}\theta P(t)\Phi^T\}a$$
$$+ a^T \{K - \Phi P(t)\theta^T[R_2 + \theta P(t)\theta^T]^{-1}\}[R_2 + \theta P(t)\theta^T]$$
$$\times \{K - \Phi P(t)\theta^T[R_2 + \theta P(t)\theta^T]^{-1}\}^T a \qquad (4.15)$$

The right member is thus a function of two terms; the first term is independent of K and the second term is nonnegative because the matrix $R_2 + \theta P(t)\theta^T$ is nonnegative. The smallest value of the left member is thus obtained by choosing K in such a way that the second term of the right member of (4.14) vanishes. Doing so we find

$$K = K(t) = \Phi P(t)\theta^T[R_2 + \theta P(t)\theta^T]^{-1} \qquad (4.16)$$
$$P(t + 1) = \Phi P(t)\Phi^T + R_1 - \Phi P(t)\theta^T[R_2 + \theta P(t)\theta^T]^{-1}\theta P(t)\Phi^T \qquad (4.17)$$

Notice that the result does not depend on a. Hence if we choose K in order to minimize the mean square reconstruction error of one linear combination of the state variables, we will at the same time minimize the mean square reconstruction error for all linear combinations. Also notice that (4.17) gives the variance of the reconstruction error for the optimal reconstruction.

The first term of the right member $\Phi P(t)\Phi^T$ shows how the reconstruction error at stage t will propagate to stage $t + 1$ through the system dynamics. The term R_1 represents the increase of the variance of the reconstruction error due to the disturbance v which acts on the system, and the third term of (4.17) shows how the reconstruction error decreases due to the information obtained from the measurements.

It follows from (4.16) and (4.17) that

$$P(t + 1) = \Phi P(t)\Phi^T + R_1 - K(t)\theta P(t)\Phi^T = [\Phi - K(t)\theta]P(t)\Phi^T + R_1$$
$$K(t)R_2 + K(t)\theta P(t)\theta^T = \Phi P(t)\theta^T$$

Postmultiplying the last equation by $K^T(t)$ and subtracting we find

$$\begin{aligned}P(t + 1) &= \Phi P(t)\Phi^T + R_1 - K(t)\theta P(t)\Phi^T - \Phi P(t)\theta^T K^T(t) \\ &\quad + K(t)R_2 K^T(t) + K(t)\theta P(t)\theta^T K^T(t) \\ &= [\Phi - K(t)\theta]P(t)[\Phi - K(t)\theta]^T + R_1 + K(t)R_2 K^T(t)\end{aligned}$$

From this equation we can immediately deduce that in a pure algebraical way, if $P(t)$ is nonnegative definite, then $P(t + 1)$ is also nonnegative definite.

Summarizing the results we find that the solution to the reconstruction problem is given by Theorem 4.1.

THEOREM 4.1

Consider the dynamical system (4.6) with the output signal (4.7). A reconstruction of the state variables of the system using the mathematical model (4.8) is optimal in the sense of mean square if the gain parameter K is chosen as

$$K(t) = \Phi P(t)\theta^T[R_2 + \theta P(t)\theta^T]^{-1} \qquad (4.16)$$

where $P(t)$ is the variance of the optimal reconstruction which is given by

$$\begin{aligned}P(t + 1) &= \Phi P(t)\Phi^T + R_1 - \Phi P(t)\theta^T[R_2 + \theta P(t)\theta^T]^{-1}\theta P(t)\Phi^T \\ &= [\Phi - K(t)\theta]P(t)\Phi^T + R_1 \\ &= [\Phi - K(t)\theta]P(t)[\Phi - K(t)\theta]^T + R_1 + K(t)R_2 K^T(t) \qquad (4.18)\end{aligned}$$

with

$$P(t_0) = R_0$$

Remark 1

Notice that we have solved a parametric optimization problem, i. e., we have determined the reconstruction with the structure (4.8) which gives the smallest mean square error. Hence there might be reconstructions with a different structure which give still smaller errors. Chapter 7 will show that the chosen structure is in fact optimal.

Remark 2

It follows from the derivation that Theorem 4.1 holds when the matrices Φ, Γ, θ, R_1 and R_2 are time dependent. Writing the time dependence explicitly the model given by (4.6) and (4.7) becomes

$$x(t+1) = \Phi(t+1; t)x(t) + \Gamma(t)u(t) + v(t)$$
$$y(t) = \theta(t)x(t) + e(t)$$

where the covariances of $v(t)$ and $e(t)$ are $R_1(t)$ and $R_2(t)$. The optimal reconstructor is then given by

$$\hat{x}(t+1) = \Phi(t+1; t)\hat{x}(t) + \Gamma(t)u(t) + K(t)[y(t) - \theta(t)\hat{x}(t)]$$

where $K(t)$ is given by (4.16) and (4.18) with $\Phi = \Phi(t+1; t)$,

$$\Gamma = \Gamma(t), \quad \theta = \theta(t), \quad R_1 = R_1(t) \quad \text{and} \quad R_2 = R_2(t).$$

Exercises

1. Consider a system described by

$$x(t+1) = \begin{bmatrix} 1 & h \\ 0 & 1 \end{bmatrix} x(t) + \begin{bmatrix} 0 \\ 1 \end{bmatrix} e(t)$$

$$y(t) = [1 \quad 0]x(t)$$

where $\{e(t), t \in T\}$ is a sequence of independent normal $(0, 1)$ random variables. Assume that the initial state $x(0)$ is normal with mean value

$$Ex(0) = \begin{bmatrix} 1 \\ 1 \end{bmatrix}$$

and covariance

$$\text{cov}\,[x(0), x(0)] = \begin{bmatrix} \sigma_1^2 & 0 \\ 0 & \sigma_2^2 \end{bmatrix}$$

Determine the gain vector in a minimal reconstructor of the form (4.8). Also determine the covariance matrix of the reconstruction error.

2. Consider the dynamical system

$$x(t+1) = \Phi x(t) + \Gamma u(t) + v(t)$$

whose output is given by

$$y(t) = \theta x(t) + e(t)$$

where $\{e(t)\}$ and $\{v(t)\}$ are discrete time white noise with zero mean values and the covariances

$$Ev(t)v^T(s) = \delta_{s,t}R_1$$
$$Ev(t)e^T(s) = \delta_{s,t}R_{12}$$
$$Ee(t)e^T(s) = \delta_{s,t}R_2$$

Show that the state variables can be reconstructed with the mathematical model

$$\hat{x}(t+1) = \Phi\hat{x}(t) + \Gamma u(t) + K[y(t) - \theta\hat{x}(t)]$$

where the best value of K is given by

$$K = K(t) = [\Phi P(t)\theta^T + R_{12}][\theta P(t)\theta^T + R_2]^{-1}$$

where

$$P(t+1) = \Phi P(t)\Phi^T + R_1 - K(t)[R_2 + \theta P(t)\theta^T]K^T(t)$$

3. The reconstructor given by (4.8) has the property that the value of the state vector at time t is reconstructed from observations $y(t-1)$, $y(t-2)$, ... It is possible to find reconstructors which also make use of the correct observation $y(t)$ to reconstruct $x(t)$. This can be achieved by the following equation

$$\hat{x}(t+1) = \Phi\hat{x}(t) + \Gamma u(t)$$
$$+ K(t+1)\{y(t+1) - \theta[\Phi\hat{x}(t) + \Gamma u(t)]\} \quad (*)$$

Show that if the system is governed by Eqs. (4.6) and (4.7) the optimal choice of K is given by

$$K(t) = P(t)\theta^T[R_2 + \theta P(t)\theta^T]^{-1}$$
$$P(t) = \Phi S(t-1)\Phi^T + R_1$$
$$S(t) = P(t) - K(t)\theta P(t)$$
$$P(t_0) = R_0$$

Also give a physical interpretation of the matrices P and S.

4. Consider the system of Exercise 1. Find the gain matrix of an optimal reconstructor of type (*). Determine the covariance matrix of the reconstruction error. Compare with the results of Exercise 1.

5. Equation (4.18) of Theorem 4.1 suggests three different ways of computing the matrix $P(t)$ recursively. Discuss the computational aspects of the different schemes.

5. RECONSTRUCTION OF STATE VARIABLES FOR CONTINUOUS TIME SYSTEMS

Introduction

We will now discuss the continuous time version of the problem of Section 4. Consider a system described by the stochastic differential equation

$$dx = Ax\,dt + Bu\,dt + dv \tag{5.1}$$

where x is an n-dimensional state vector, u an r-dimensional input vector, and v a Wiener process with incremental covariance $R_1 dt$. Let the output of the system be observed through a noisy channel described by

$$dy = Cx\,dt + de \tag{5.2}$$

where the output y is a p-vector and e is a Wiener process with incremental covariance $R_2\,dt$. The matrices A, B, C, R_1 and R_2 may be time varying. The elements are assumed to be continuous functions of time. R_2 is positive definite and R_1 positive semidefinite.

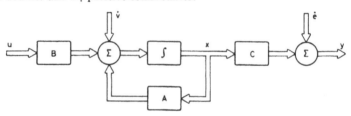

Fig. 5.8. Block diagram of a physical system which can be modeled by (5.1) and (5.2). The stochastic processes $\dot{v}$ and $\dot{e}$ have finite variances and spectral densities which are essentially constant over the frequency range $(-\omega_0, \omega_0)$ where ω_0 is considerably larger than the largest eigenvalue of A.

In Fig. 5.8 we give a block diagram of a physical system which can be modeled approximatively by (5.1) and (5.2). Using heuristic arguments analogous to those presented in Section, we find that it looks reasonable to try a reconstructor of the form

$$d\hat{x} = A\hat{x}\,dt + Bu\,dt + K[dy - C\hat{x}\,dt] \tag{5.3}$$

where K is an $n \times p$ matrix with time varying elements. To investigate if the model (5.3) can give a reasonable reconstruction we introduce the reconstruction error defined by

$$\tilde{x} = x - \hat{x}$$

Using (5.1) and (5.2) we find that the reconstruction error is thus a Gauss-

Reconstruction of State Variables for Continuous Time Systems 151

Markov stochastic process governed by the following stochastic differential equation

$$d\tilde{x} = (A - KC)\tilde{x}\,dt + dv - K\,de \qquad (5.4)$$

To investigate the properties of the reconstructor (5.3) we thus have to analyze the stochastic differential equation (5.4). Using Theorem 6.1 of Chapter 3, we find that the mean value of the reconstruction error is given by

$$\frac{d}{dt}(E\tilde{x}) = (A - KC)(E\tilde{x}) \qquad (5.5)$$

$$E\tilde{x}(t_0) = Ex(t_0) - \hat{x}(t_0) \qquad (5.6)$$

and that the covariance of the reconstruction error

$$P(t) = E\{\tilde{x}(t) - E\tilde{x}(t)\}\{\tilde{x}(t) - E\tilde{x}(t)\}^T \qquad (5.7)$$

is given by

$$\frac{dP}{dt} = (A - KC)P + P(A - KC)^T + R_1 + KR_2K^T \qquad (5.8)$$

$$P(t_0) = E\{x(t_0) - \hat{x}(t_0)\}\{x(t_0) - \hat{x}(t_0)\}^T = R_0 \qquad (5.9)$$

Hence if K is chosen in such a way that (5.5) is stable, the mean value of the reconstruction error will converge to zero. It also follows from (5.5) that if the reconstruction error has zero mean value at any time t_0, it will have zero mean value for all t.

It is instructive to give the physical interpretation of the different terms of (5.8). For this purpose we rewrite it as

$$\frac{dP}{dt} = AP + PA^T + R_1 - (KCP + PC^TK^T - KR_2K^T) \qquad (5.10)$$

The first two terms represent the propagation of the covariance of the reconstruction error through the system dynamics. The term R_1 represents the increase in the covariance of the reconstruction error due to the disturbance v which acts on the system, and the last term represents the reduction of the covariance of the reconstruction error due to the measurements. The last term naturally depends on the choice of the gain matrix.

A Parametric Optimization Problem

Having found the equations which characterize the reconstruction error we will now investigate if there is a choice of K which is optimal. It is assumed that $E\tilde{x}(t_0) = 0$ which means that the reconstruction error always has a zero mean value. The mean square error of the reconstruction of the linear combination of state variables a^Tx is chosen as the criterion. Hence

$$E(a^T\tilde{x})^2 = E(a^T\tilde{x})(\tilde{x}^Ta) = a^TE\tilde{x}\tilde{x}^Ta = a^TP(t)a \qquad (5.11)$$

where the last equality holds because the mean value of the reconstruction error is zero. Using the differential equation (5.8) we find

$$\frac{d}{dt} a^T P(t) a = a^T (AP + PA^T + R_1) a + a^T (KR_2K^T - KCP - PC^TK^T) a \tag{5.12}$$

To proceed we need the following result.

LEMMA 5.1

Let P and Q be solutions of the Riccati equations

$$\frac{dP}{dt} = AP + PA^T + R_1 + KR_2K^T - KCP - PC^TK^T \tag{5.13}$$

$$\frac{dQ}{dt} = AQ + QA^T + R_1 - QC^TR_2^{-1}CQ \tag{5.14}$$

where R_2 is assumed positive definite and the initial conditions are

$$P(t_0) = Q(t_0) = R_0 \tag{5.15}$$

where R_0 is symmetric. The matrix $P(t) - Q(t)$ is then positive semidefinite and

$$P(t) = Q(t) \tag{5.16}$$

for

$$K = PC^TR_2^{-1} \tag{5.17}$$

Proof

It follows from (5.13) and (5.14) that

$$\frac{d}{dt}(P - Q) = A(P - Q) + (P - Q)A^T + KR_2K^T - KCP - PC^TK^T$$
$$+ QC^TR_2^{-1}CQ = (A - KC)(P - Q) + (P - Q)(A - KC)^T$$
$$+ (K - QC^TR_2^{-1})R_2(K - QC^TR_2^{-1})^T \tag{5.18}$$

Let $\Psi(t; s)$ be the solution of the differential equation

$$\frac{d\Psi(t;s)}{dt} = [A(t) - K(t)C(t)] \Psi(t; s)$$
$$\Psi(t; t) = I \tag{5.19}$$

The solution of (5.18) can then be written as

$$P(t) - Q(t) = \int_{t_0}^{t} \Psi(t; s)[K(s) - Q(s)C^T(s)R_2^{-1}(s)]R_2(s)$$
$$\times [K(s) - Q(s)C^T(s)R_2^{-1}(s)]^T \Psi^T(t; s) ds \tag{5.20}$$

But the matrix of the right member is always nonnegative definite for all K. Furthermore for

$$K = QC^T R_2^{-1}$$

we get $P(t) = Q(t)$ which then implies (5.16) and completes the proof of the lemma.

We thus find that there is a universal choice of K which makes $a^T P(t) a$ as small as possible for all a. The optimal value of the gain matrix is given by (5.17). The optimal value of K will thus give the smallest error for reconstructing any linear combination of the state variables. Summing up we get Theorem 5.1.

THEOREM 5.1

Let a dynamical system subject to disturbances and measurement errors be described by (5.1) and (5.2). A reconstructor of the form (5.3) is optimal in the sense of mean squares if the initial value is chosen as

$$\hat{x}(t_0) = Ex(t_0) \tag{5.21}$$

and the gain parameter K is chosen as

$$K(t) = P(t) C^T R_2^{-1} \tag{5.17}$$

where $P(t)$ is the covariance of the optimal reconstruction errors. The matrix $P(t)$ is given by the Riccati equation

$$\frac{dP(t)}{dt} = AP + PA^T + R_1 - PC^T R_2^{-1} C P \tag{5.22}$$

with the initial condition

$$P(t_0) = R_0 \tag{5.23}$$

Remark

Notice that we have solved a parametric optimization problem for a state reconstructor with the structure (5.3). Chapter 7 will show that the structure (5.3) is in fact optimal.

Exercises

1. Consider the motion of a particle on a straight line. Assume that the acceleration of the particle can be described as a white noise process with the spectral density $1/(2\pi)$ and that the position of the particle is observed with a white noise measurement error with the spectral density $r/(2\pi)$. Determine a minimal variance reconstructor of the form (5.3) for the position and velocity of the particle. Also find the covariance matrix of the reconstruction error.

2. Consider a dynamical system described by (5.1) and (5.2) where e and v are correlated Wiener processes with zero mean values and the incremental covariance

$$E\begin{bmatrix} dv \\ de \end{bmatrix}[dv^T\ de^T] = \begin{bmatrix} R_1 & R_{12} \\ R_{12}^T & R_2 \end{bmatrix}dt$$

Show that a minimum variance reconstructor with the structure (5.3) is obtained when the initial value is chosen as $\hat{x}(t_0) = Ex(t_0)$ and the gain matrix is given by

$$K(t) = [P(t)C^T + R_{12}]R_2^{-1}$$

where the covariance matrix of the reconstruction error satisfies the Riccati equation

$$\frac{dP}{dt} = [A - R_{12}R_2^{-1}C]P + P[A - R_{12}R_2^{-1}C]^T$$
$$\quad + R_1 - R_{12}R_2^{-1}R_{12}^T - PC^TR_2^{-1}CP$$
$$P(t_0) = R_0$$

3. A simple vertical alignment system which consists of a platform servo and an accelerometer is shown in Fig. 5.9. The accelerometer signal provides information about the vertical indication error. The signal is used to drive the platform normal to the vertical position. For small deviations the accelerometer signal is

$$y = \theta + n$$

where θ is the vertical indication error and n a disturbance due to horizontal accelerations. The platform can be described by the equation

$$\frac{d\theta}{dt} = u$$

where u is the control signal. The servo loop can be described by

$$u = -ky$$

where k is the gain. Assuming that the noise can be modeled by white noise, we find that the system can be described by the stochastic differential equation

$$d\theta = -k\theta\,dt + k\,dv$$

where $\{v(t)\}$ is a Wiener process with variance parameter r. Assume that the initial state is normal $(0, \sigma)$; find a time varying gain $k = k(t)$ such that the variance of the vertical alignment error is as small as possible. Compare the results with those obtained for a constant gain when the alignment period is kept constant T; see Fig. 5.10.

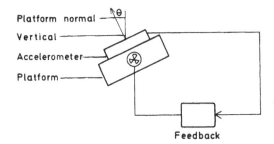

Fig. 5.9. Schematic diagram of a vertical alignment system.

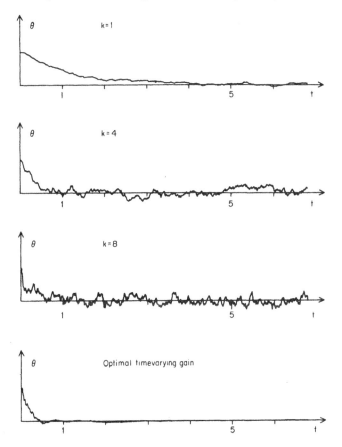

Fig. 5.10. Vertical alignment error θ versus time t for analog simulation of systems with different fixed gains and with optimal time varying gain.

4. Let P and Q be solutions to the Riccati equations

$$\begin{cases} \dfrac{dP}{dt} = AP + PA^T + R_1 - PC^T R_2^{-1} CP \\ P(t_0) = P_0 \end{cases}$$

$$\begin{cases} \dfrac{dQ}{dt} = AQ + QA^T + R_1 - QC^T R_2^{-1} CQ \\ Q(t_0) = Q_0 \end{cases}$$

Show that $P_0 > Q_0$ implies $P(t) > Q(t)$ for all t for which the Riccati equations have solutions.

5. Consider the Riccati equations

$$\begin{cases} \dfrac{dP}{dt} = AP + PA^T + R_1 - PC^T R_2^{-1} CP \\ P(t_0) = P_0 \end{cases}$$

$$\begin{cases} \dfrac{dQ}{dt} = AQ + QA^T + R_3 - QC^T R_2^{-1} CQ \\ Q(t_0) = P_0 \end{cases}$$

Show that $R_1 \geqslant R_3$ implies $P(t) \geqslant Q(t)$ for all t for which the Riccati equations have solutions.

6. Consider the Riccati equations

$$\begin{cases} \dfrac{dP}{dt} = AP + PA^T + R_1 - PC^T R_2^{-1} CP \\ P(t_0) = P_0 \end{cases}$$

$$\begin{cases} \dfrac{dQ}{dt} = AQ + QA^T + R_1 - QC^T R_3^{-1} CQ \\ Q(t_0) = P_0 \end{cases}$$

Show that $R_2 \geqslant R_3$ implies $P(t) \geqslant Q(t)$ for all t for which the Riccati equations have solutions.

6. BIBLIOGRAPHY AND COMMENTS

The idea of parametric optimization of stochastic systems is discussed in

James, H. M., Nichols, N. B., and Phillips, R. S., *Theory of Servomechanisms*, McGraw-Hill, New York, 1947

Newton, G. C., Jr., Gould, L. A., and Kaiser, J. F., *Analytical Design of Linear Feedback Controls*, Wiley, New York, 1957.

Bibliography and Comments

In these textbooks the integrals for the variance of the signals are evaluated using straightforward residue calculus. Tables for the integrals in the continuous time case are also given in these references.

Analogous results for discrete time systems are given in

Jury, E. I., *Theory and Application of the z-Transform Method*, Wiley, New York, 1964.

Theorem 2.2 is essentially the Schur-Cohn theorem for stability of a discrete time linear system. The proof presented is Section 2 is due to

Růžička, J., "Algebraická Kritéria Stability Impulsních Soustav," (Algebraic criteria for stability of sampled data systems, in Czech), *Strojnícky Časopis* XIII, č 5, 395–403 (1962).

The proof is also given in

Strejc, V., *Syntese von Regelungssystemen mit Prozessrechner*, Akademie-Verlag, Berlin, 1967.

Closely related theorems are also given in

Jury, E. I., "On the Roots of Real Polynomial Inside the Unit Circle and a Stability Criterion for Linear Discrete Systems," *Proc. of Sec. Congr. of IFAC*, Butterworths, London, 1964.

Toma, M., "Ein Einfaches Verfahren zur Stabilitätsprüfung von linearen Abtastsystemen," *Regelungstechnik* 10, 302–306 (1962).

Theorem 3.2 is in essence the Routh-Hurwitz theorem. The proof given in Section 3 which is patterned after the discrete time case is believed to be new.

The idea of evaluating the integrals recursively is due to Nekolný. See, e.g.,

Nekolný, J., *Nova Jednoduchá Methodika Testu Jakosti Regulace*, Prague, 1957

Nekolný, J. and Beneš, J., "Simultaneous Control of Stability and Quality of Adjustment-Application of Statistical Dynamics," in Coales *et al.* "Automatic and Remote Control," *Proc. of First IFAC Congr. Moscow 1960*, Vol. 2, Butterworths, London, 1961.

The algorithm for the discrete time case Theorem 2.3 is given in

Åström, K. J., "Recursive Formulas for the Evaluation of Certain Complex Integrals," Rep. 6804, Lund Institute of Technology, Division of Automatic Control. September 1968.

The corollary of Theorem 2.3 and a computer algorithm is given in

Åström, K. J., Jury, E. I., and Agniel, R. G., "A Numerical Method for the Evaluation of Complex Integrals," *IEEE Trans.* AC (1970).

Algol programs for the continuous time algorithms are found in

Peterka, V. and Vidinčev, P., "Rational-Fraction Approximation of Transfer Functions," *Proc. IFAC Symp. on Identification in Autom. Control Sys.* Prague, 1967.

The idea of reconstructing the state of a dynamical system using a mathematical model discussed in Sections 4 and 5 is probably very old. It was, e.g., encountered in discussions with John Bertram in 1961. An early explicit reference to a model reconstructor without feedback from the measurements is given in

Kalman, R. E. and Bertram, J. E., "General Synthesis Procedure for Computer Control of Single and Multiloop Linear Systems," *Trans. AIEE* **77** (1958).

A detailed discussion with several refinements is given in

Luenberger, D. G., "Observing the State of a Linear System," *IEEE Trans. on Military Electron.* **8**, 74-80 (1964).

Luenberger, D. G., "Observers for Multivariable Systems," *IEEE Trans. AC* **11**, 190-191 (1966).

The recursive equations given by Theorem 4.1 and Theorem 5.1 are identical to the Kalman Bucy filtering algorithms. See

Kalman, R. E., "A New Approach to Linear Filtering and Prediction Problems," *ASME J. Basic Eng.* **82**, 35-45 (1960)

Kalman, R. E. and Bucy, R. S., "New Results in Linear Filtering and Prediction Theory," *ASME J. of Basic Eng.* **83**, 95-107, (1961).

Also compare with the results of Chapter 7.

CHAPTER 6

MINIMAL VARIANCE CONTROL STRATEGIES

1. INTRODUCTION

Parametric optimization of systems where the regulator has a known structure were discussed in Chapter 5. The present chapter will consider more general optimization problems where the structure of the regulator is unknown. The purpose of this chapter is to present a problem in a very simple setting, so that the essential ideas of stochastic optimal control can be developed with very little mathematical sophistication. We will thus analyze a simple regulation problem for a linear time-invariant system with one input and one output. It is assumed that the disturbance acting on the system can be described as a realization of a normal stationary stochastic process with a rational spectral density, and that the purpose of control is to minimize the variance of the output. To keep the analysis simple, we will also specialize to discrete time systems. Throughout this section the index set T will thus be the set of integers.

A first order system is treated in Section 2. For this particular example, it is very easy to prove the separation theorem. We thus find that the optimal regulator can be thought of as consisting of two parts: one predictor which predicts the effect of the disturbance on the output, and one dead-beat regulator which computes the control signal required to make the predicted output equal to the desired value.

Thus having established the connection between prediction theory and stochastic optimal control, we have a motivation to study the prediction problem. Section 3 solves the problems of predicting a stationary discrete time stochastic process with a rational spectral density. The central result is Theorem 3.1 which gives a recursive formula for the predictor as well as an expression for the prediction error. It is also shown that the coefficients of the recursive formula are easily obtained by identifying the coefficients of a polynomial identity.

Section 4 returns to the control problem. Using the results of the prediction problem, we derive a formula for the minimal variance control strategy. It is also shown that the control error for the optimal strategy equals the prediction error. The sensitivity of the optimal regulator to variations in the model parameters is discussed in Section 5. It is shown that under certain circumstances the optimal system can be extremely sensitive to parameter variations. A method for deriving suboptimal strategies which are less sensitive to parameter variations is also given.

Section 6 gives an example of an industrial application of the minimal variance control strategies. The problem of basis weight control on a paper machine is discussed. Particular attention is given to the formulation of the problem and a discussion of the applicability of the theory. The identification problem, i.e., how to obtain the mathematical models of process dynamics and disturbances from experimental data, is discussed. Results from actual operations using the minimal variance strategy are given.

2. A SIMPLE EXAMPLE

We will first consider a special case which illustrates the basic idea with very little analysis. Let the process to be regulated be described by the first order equation

$$y(t) + ay(t-1) = u(t-1) + e(t) + ce(t-1) \qquad (2.1)$$

where u is the control variable, y the output, and $\{e(t)\}$ a sequence of independent, equally distributed normal $(0, 1)$ random variables. It is assumed that $|c| < 1$. Let the purpose of the control be to minimize the variance of the output. Furthermore, let the admissible control strategies be such that the value of the control variable at time t is a function of the measured output up to time t, that is, $y(t), y(t-1), \ldots,$ and the past values of the control variable, that is, $u(t-1), u(t-2), \ldots$.

To determine the minimum variance control strategy, we will first argue intuitively. Consider the situation at time t. We have

$$y(t+1) = -ay(t) + u(t) + e(t+1) + ce(t) \qquad (2.2)$$

We observe that $y(t+1)$ can be changed arbitrarily by a proper choice of $u(t)$. The problem is to choose $u(t)$ so as to minimize $Ey^2(t+1)$. We first notice that $e(t+1)$ is independent of $y(t), u(t),$ and $e(t)$. Hence

A Simple Example

$$Ey^2(t+1) \geq Ee^2(t+1) = 1$$

The best control law will thus make the variance of the output greater than or equal to 1.

The information available for determining $u(t)$ is all the past outputs $y(t), y(t-1), \ldots,$ and all the past inputs $u(t-1), u(t-2), \ldots$. When this information is available, we can compute $e(t)$ exactly from (2.1) which describes the system. The first and fourth terms of the right member of (2.2) are thus known. If we choose the control law

$$u(t) = ay(t) - ce(t) \qquad (2.3)$$

we thus find that

$$y(t+1) = e(t+1) \qquad (2.4)$$

which gives the smallest possible variance of $y(t+1)$. If the control law (2.3) is used in every step, (2.4) will thus hold for all t. The computation of $e(t)$ from the available data then reduces to (2.4) and the control law (2.3) becomes

$$u(t) = (a-c)y(t) \qquad (2.5)$$

So far we have argued heuristically. We will now analyze the properties of the control law (2.5). Put (2.5) into (2.1) and we get

$$y(t+1) + cy(t) = e(t+1) + ce(t)$$

Hence

$$[y(t+1) - e(t+1)] + c[y(t) - e(t)] = 0$$

Solving this difference equation with initial value $y(t_0) - e(t_0) = K$, we get

$$y(t) = e(t) + K \cdot (-c)^{(t-t_0)}$$

As $|c| < 1$ we find $y(t) = e(t)$ as $t_0 \to -\infty$. Hence if the control law (2.5) is used, the output will in the steady state have the smallest possible variance. The optimal control law is thus given by (2.5).

Having obtained the result, we will now discuss some of its implications. The quantity $-ay(t) + bu(t) + ce(t)$ can be interpreted as the best estimate (in the sense of mean square) of the output at time $t+1$ based on the data available at time t. The prediction error is $e(t+1)$. By choosing $u(t)$ according to the control law, we thus make the predicted value of $y(t+1)$ equal to the desired value 0. The control error then equals the prediction error. We thus find that stochastic control is closely connected with prediction. Also notice that the dynamics of the optimal system are uniquely given by the parameter c and that the output of the optimal system equals white noise.

It is also of interest to observe that we would get the same result if we relax the normality assumption on the $e:s$ and restrict the admissible control laws to be linear functions of the observations. We would then assume that $e(t)$ has zero mean value and unit variance for all t, that $e(t)$ and $e(s)$ are uncorrelated for $t \neq s$, and that $u(t)$ is a linear function of $y(t), y(t-1), \ldots, u(t-1), \ldots$.

3. OPTIMAL PREDICTION OF DISCRETE TIME STATIONARY PROCESSES

In the example of Section 2 we found that the solution of the stochastic control problem was closely related to a prediction problem. This section will discuss the prediction problem in some detail. The next section will then generalize the problem of Section 2.

Prediction theory can be stated in many different ways which differ in the assumptions made on the process, the criterion, and the admissible predictors. This section will use the following set of assumptions:

- The process to be predicted is stationary, Gaussian with a rational spectral density.
- The best predictor is the one which minimizes the variance of the prediction error.
- An admissible predictor for $y(t+k)$ is any function of all past observations. Hence to predict $y(t+k)$ we can use all observations $y(t)$, $y(t-1), y(t-2), \ldots$.

Notice that we will obtain the same results if we restrict the admissible predictors to be linear functions of the observations and relax the first assumption from normal processes to second order processes. In the analysis which follows we will first consider a simple example using straightforward analysis. Guided by the results, we will then streamline the analysis by introducing proper notations. Having done this, we will then solve the problem in full generality.

Example

Consider a first order normal process with the spectral density

$$\phi(\omega) = \frac{1 + c^2 + 2c \cos \omega}{1 + a^2 + 2a \cos \omega}, \qquad |a| < 1, \quad |c| < 1 \tag{3.1}$$

The spectral density is a rational function of $\exp(i\omega)$ because

$$\phi(\omega) = \frac{1 + c^2 + c(e^{i\omega} + e^{-i\omega})}{1 + a^2 + a(e^{i\omega} + e^{-i\omega})} = \frac{(1 + ce^{i\omega})(1 + ce^{-i\omega})}{(1 + ae^{i\omega})(1 + ae^{-i\omega})} \tag{3.2}$$

It now follows from Theorem 2.2 of Chapter 4 that the stochastic process $\{y(t), t = 0, \pm 1, \pm 2, \ldots\}$ can be represented by the stochastic difference equation

Optimal Prediction of Discrete Time Stationary Processes

$$y(t+1) + ay(t) = e(t+1) + ce(t) \tag{3.3}$$

where $\{e(t), t = 0, \pm 1, \pm 2, \ldots\}$ is a sequence of equally distributed normal random variables with zero mean. To construct predictors for the stochastic process, we will use the representation (3.3) of the process. To simplify we will first discuss one-step-ahead prediction. Consider the situation at time t. We have thus observed $y(t), y(t-1), \ldots$. Based on this information, we would like to make the best possible estimate of $y(t+1)$. Equation (3.3) gives

$$y(t+1) = -ay(t) + ce(t) + e(t+1) \tag{3.4}$$

where $e(t)$ and $e(t+1)$ are normal $(0, 1)$ independent random variables. The random variable $y(t+1)$ can thus be expressed as the sum of three terms. The first $-ay(t)$ is known directly from the observations. The second term $ce(t)$ can be calculated exactly from the observed data using (3.3) recursively. The third term $e(t+1)$ is independent of all observations $y(t), y(t-1), \ldots$. The best prediction of $y(t+1)$ is thus given by the first two terms of the right member of (3.4). We thus have the outline of the solution of the problem, and we will now carry out the details, i.e., we will show how to calculate $e(t)$ from the observations $y(t), y(t-1), \ldots$ using (3.3). This computation obviously requires an initial condition. As $|c| < 1$ and we have infinitely many $y:s$ available, the initial condition is immaterial. To show this we will first assume that $e(t_0)$ is known and that $y(t_0), y(t_0+1), \ldots y(t)$ have been observed. To compute $e(t)$ we have to solve the first order difference (3.3) which we rewrite as

$$[e(\tau+1) - y(\tau+1)] + c[e(\tau) - y(\tau)] = (a - c)y(\tau)$$

The solution is given by

$$e(t) = [e(t_0) - y(t_0)](-c)^{t-t_0} + y(t) + (a - c) \sum_{n=t_0}^{t-1} (-c)^{t-1-n} y(n)$$

As $|c| < 1$, the first term converges to zero for all initial conditions as $t_0 \to -\infty$. We thus find that $e(t)$ can be computed exactly from the available information using the formula

$$e(t) = y(t) + (a - c) \sum_{n=-\infty}^{t-1} (-c)^{t-n-1} y(n)$$

The best prediction of $y(t+1)$ based on $y(t), y(t-1), y(t-2), \ldots$ is thus given by

$$\hat{y}(t+1 \mid t) = -ay(t) + cy(t) + c(a - c) \sum_{n=-\infty}^{t-1} (-c)^{t-n-1} y(n)$$

$$= (c - a) \sum_{n=-\infty}^{t} (-c)^{t-n} y(n) \tag{3.5}$$

The prediction error is given by
$$\tilde{y}(t+1|t) = y(t+1) - \hat{y}(t+1|t) = e(t+1)$$
We have thus solved the one step prediction problem. The formula (3.5) is, however, not very convenient to use. For example, if we want to implement the predictor using a digital computer we have to store infinitely many observations. This storage requirement can be reduced drastically if the result is rewritten as a recursive equation. We get from (3.5)

$$\begin{aligned}\hat{y}(t+1|t) &= (c-a)\left[y(t) + \sum_{n=-\infty}^{t-1}(-c)^{t-n}y(n)\right]\\ &= (c-a)\left[y(t) + (-c)\sum_{n=-\infty}^{t-1}(-c)^{t-n-1}y(n)\right]\\ &= -c\hat{y}(t|t-1) + (c-a)y(t)\end{aligned} \quad (3.6)$$

The one step predictor can thus be described by a first order difference equation. Using (3.6) to compute the one step prediction, it is only necessary to store one number $\hat{y}(t|t-1)$. Notice that the dynamics of the predictor is determined by the number c.

Streamlined Notations

We will now introduce a formalism which will simplify the analysis. For this purpose we introduce the forward shift operator q defined by
$$qx(t) = x(t+1)$$
Equation (3.3) can then be written as
$$y(t+1) = \frac{1+cq^{-1}}{1+aq^{-1}}e(t+1) = e(t+1) + \frac{c-a}{1+aq^{-1}}e(t) \quad (3.7)$$
The last term is a linear combination of $e(t), e(t-1), \ldots$ which can be computed from $y(t), y(t-1), \ldots$ using (3.3) which can be written as
$$e(t) = \frac{1+aq^{-1}}{1+cq^{-1}}y(t) \quad (3.8)$$
Eliminating $e(t)$ between (3.7) and (3.8) we get
$$y(t+1) = e(t+1) + \frac{c-a}{1+cq^{-1}}y(t) \quad (3.9)$$
Now let $\hat{y}$ be any function of the available observations $y(t), y(t-1), \ldots$. As $e(t+1)$ is independent of the observations, we have
$$E[y(t+1) - \hat{y}]^2 = Ee^2(t+1) + E\left[\frac{c-a}{1+cq^{-1}}y(t) - \hat{y}\right]^2$$
Hence

Optimal Prediction of Discrete Time Stationary Processes

$$E[y(t+1) - \hat{y}]^2 \geq Ee^2(t+1) = 1$$

where equality is obtained for

$$\hat{y} = \hat{y}(t+1 \mid t) = \frac{c-a}{1+cq^{-1}} y(t)$$

We thus find that the optimal one step predictor is given by the difference equation

$$\hat{y}(t+1 \mid t) + c\hat{y}(t \mid t-1) = (c-a)y(t) \tag{3.10}$$

Compare with Eq. (3.6).

Two Step Predictor

Having obtained a formalism, we will now proceed to determine the two step predictor. Consider the situation at time t when $y(t), y(t-1), \ldots$ are observed and we want to determine $y(t+2)$. Equation (3.3) gives

$$y(t+2) = \frac{1+cq^{-1}}{1+aq^{-1}} e(t+2) \tag{3.11}$$

The right member is a linear function of $e(t+2), e(t+1), e(t), e(t-1)$. The stochastic variables $e(t), e(t-1), \ldots$ can be determined exactly from the available observations $y(t), y(t-1), \ldots$. The random variables $e(t+2)$ and $e(t+1)$ are independent of the observations. Rewriting (3.11) we get

$$y(t+2) = e(t+2) + \frac{(c-a)}{1+aq^{-1}} e(t+1)$$

$$= e(t+2) + \frac{(c-a)(1+aq^{-1}) - a(c-a)q^{-1}}{1+aq^{-1}} e(t+1)$$

$$= e(t+2) + (c-a)e(t+1) - \frac{a(c-a)}{1+aq^{-1}} e(t)$$

where the last term can be calculated exactly from the available observations. We get from (3.3)

$$e(t) = \frac{1+aq^{-1}}{1+cq^{-1}} y(t)$$

Hence

$$y(t+2) = e(t+2) + (c-a)e(t+1) - \frac{a(c-a)}{1+cq^{-1}} y(t)$$

Now let $\hat{y}$ be a function of the available observations $y(t), y(t-1), \ldots$. As $e(t+1)$ and $e(t+2)$ are independent of the observations we have

$$E[y(t+2) - \hat{y}]^2 = E[e^2(t+2)] + (c-a)^2 E[e^2(t+1)]$$
$$+ E\left[\hat{y} + \frac{a(c-a)}{1+cq^{-1}} y(t)\right]^2$$

Hence
$$E[y(t+2) - \hat{y}]^2 \geqslant 1 + (c-a)^2$$
where equality holds for
$$\hat{y} = \hat{y}(t+2 \mid t) = -\frac{a(c-a)}{1+cq^{-1}} y(t)$$
The two step predictor thus satisfies the following difference equation
$$\hat{y}(t+2 \mid t) = -c\hat{y}(t+1 \mid t-1) - a(c-a)y(t)$$
The two step prediction error is a moving average of second order
$$\tilde{y}(t+2 \mid t) = y(t+2) - \hat{y}(t+2 \mid t) = e(t+2) + (c-a)e(t+1)$$

The General Problem

We will now consider the problem of determining the k step predictor for a stationary normal process with rational spectral density. Let $\{y(t), t = 0, t = \pm 1, t = \pm 2, \ldots\}$ be such a process. According to Theorem 3.1 of Chapter 4, we can for such a process always find two polynomials, $A(z)$ and $C(z)$, such that the spectral density $\phi(\omega)$ of the process can be written as

$$\phi(\omega) = \lambda^2 \frac{C(e^{i\omega})C(e^{-i\omega})}{A(e^{i\omega})A(e^{-i\omega})} = \lambda^2 \frac{C(e^{i\omega})C^*(e^{i\omega})}{A(e^{i\omega})A^*(e^{i\omega})} \quad (3.12)$$

where A^* is the reciprocal polynomial defined by $A^*(z) = z^n A(z^{-1})$. In the representation (3.12) the polynomials A and C can be chosen so that A has all zeros inside the unit circle and C has all zeros inside or on the unit circle. Throughout this chapter it will be assumed that the polynomial C does not have zeros on the unit circle. The coefficients of A and C are denoted as follows

$$A(z) = z^n + a_1 z^{n-1} + \cdots + a_n \quad (3.13)$$
$$C(z) = z^n + c_1 z^{n-1} + \cdots + c_n \quad (3.14)$$

It follows from Theorem 3.2 of Chapter 4 that the stochastic process $\{y(t)\}$ can be represented as

$$A(q)y(t) = \lambda C(q)e(t) \quad (3.15')$$
or
$$A^*(q^{-1})y(t) = \lambda C^*(q^{-1})e(t) \quad (3.15)$$

where $\{e(t), t \in T\}$ is a sequence of independent normal $N(0, 1)$ random variables. We will now determine the best k step predictor, i.e., we will construct a function $\hat{y}(t+k \mid t)$ of $y(t), y(t-1), \ldots$ such that

$$E[y(t+k) - \hat{y}(t+k \mid t)]^2$$

is as small as possible.

Optimal Prediction of Discrete Time Stationary Processes

To derive the predictor we consider the situation at time t. The outputs $y(t), y(t-1), \ldots$ have been observed and we want to predict $y(t+k)$. Equation (3.15) gives

$$y(t+k) = \lambda \frac{C^*(q^{-1})}{A^*(q^{-1})} e(t+k) \quad (3.16)$$

The right member is a linear combination of $e(t+k), e(t+k-1), \ldots, e(t+1), e(t), e(t-1), \ldots$. The random variables $e(t), e(t-1), \ldots$ can be computed exactly from the observed data. The random variables $e(t+1), \ldots, e(t+k)$ are independent of the observations. The right member thus consists of terms which can be computed exactly from the observations and terms which are independent of the observations. To separate these groups of terms we write the right member of (3.16) as

$$\lambda \frac{C^*(q^{-1})}{A^*(q^{-1})} e(t+k) = \lambda \left[F^*(q^{-1}) e(t+k) + q^{-k} \frac{G^*(q^{-1})}{A^*(q^{-1})} e(t+k) \right] \quad (3.17)$$

where F^* and G^* are polynomials of order $k-1$ and $n-1$, respectively.

$$F^*(z) = 1 + f_1 z + \cdots + f_{k-1} z^{k-1} \quad (3.18)$$
$$G^*(z) = g_0 + g_1 z + \cdots + g_{n-1} z^{n-1} \quad (3.19)$$

Equation (3.16) can then be written as

$$y(t+k) = \lambda F^*(q^{-1}) e(t+k) + \lambda q^{-k} \frac{G^*(q^{-1})}{A^*(q^{-1})} e(t+k)$$
$$= \lambda F^*(q^{-1}) e(t+k) + \lambda \frac{G^*(q^{-1})}{A^*(q^{-1})} e(t) \quad (3.20)$$

The second term of the right member is a function of $e(t), e(t-1), \ldots$ and can thus be computed from the available measurements $y(t), y(t-1), \ldots$. We get

$$\lambda e(t) = \frac{A^*(q^{-1})}{C^*(q^{-1})} y(t)$$

and (3.20) can be reduced to

$$y(t+k) = \lambda F^*(q^{-1}) e(t+k) + \frac{G^*(q^{-1})}{C^*(q^{-1})} y(t) \quad (3.21)$$

The first term of the right member is a linear function of $e(t+1), e(t+2), \ldots, e(t+k)$, which are all independent of the available measurements. The second term is a linear function of the available measurements. Let $\hat{y}$ be an arbitrary function of $y(t), y(t-1), \ldots$. Then

$$E[y(t+k) - \hat{y}]^2 = E[\lambda F^*(q^{-1}) e(t+k)]^2 + E\left[\hat{y} - \frac{G^*(q^{-1})}{C^*(q^{-1})} y(t) \right]^2$$
$$+ 2E[\lambda F^*(q^{-1}) e(t+k)] \left[\hat{y} - \frac{G^*(q^{-1})}{C^*(q^{-1})} y(t) \right]$$

The last term will vanish because $e(t + 1), e(t + 2), \ldots, e(t + k)$ are independent of $y(t), y(t - 1), \ldots,$ and $e(t)$ has zero mean value for all t. We thus get

$$E[y(t+k) - \hat{y}]^2 = E[\lambda F^*(q^{-1})e(t+k)]^2 + E\left[\hat{y} - \frac{G^*(q^{-1})}{C^*(q^{-1})}y(t)\right]^2 \quad (3.22)$$

$$E[y(t+k) - \hat{y}]^2 \geq \lambda^2[1 + f_1^2 + \cdots + f_{k-1}^2]$$

where equality is obtained for

$$\hat{y} = \hat{y}(t+k \mid t) = \frac{G^*(q^{-1})}{C^*(q^{-1})}y(t) \quad (3.23)$$

The optimal k-step predictor is thus given by the difference equation

$$\hat{y}(t+k \mid t) + c_1\hat{y}(t+k-1 \mid t-1) + \cdots + c_n\hat{y}(t+k-n \mid t-n)$$
$$= g_0 y(t) + g_1 y(t-1) + \cdots + g_{n-1} y(t-n+1) \quad (3.24)$$

The prediction error is given by

$$\tilde{y}(t+k \mid t) = y(t+k) - \hat{y}(t+k \mid t) = \lambda F^*(q^{-1})e(t+k)$$
$$= \lambda[e(t+k) + f_1 e(t+k-1) + \cdots + f_{k-1} e(t+1)] \quad (3.25)$$

To obtain the predictor we thus have to determine the coefficients of the polynomials $F(z)$ and $G(z)$ defined by (3.18) and (3.19). Equation (3.17) gives the following identity

$$C^*(q^{-1}) = A^*(q^{-1})F^*(q^{-1}) + q^{-k}G^*(q^{-1}) \quad (3.26)$$

If A^* and C^* are arbitrary polynomials in q^{-1}, there exists two unique polynomials F^* and G^* which satisfy (3.26). These polynomials can be determined using long division. The polynomial F^* is thus the quotient when dividing C^* by A^* and $q^{-k}G^*(q^{-1})$ is the remainder.

The polynomials F^* and G^* can also be determined by equating the coefficients of different powers of q^{-1}. Hence

$$\begin{aligned}
c_1 &= a_1 + f_1 \\
c_2 &= a_2 + a_1 f_1 + f_2 \\
&\vdots \\
c_{k-1} &= a_{k-1} + a_{k-2} f_1 + a_{k-3} f_2 + \cdots + a_1 f_{k-2} + f_{k-1} \\
c_k &= a_k + a_{k-1} f_1 + a_{k-2} f_2 + \cdots + a_1 f_{k-1} + g_0 \\
c_{k+1} &= a_{k+1} + a_k f_1 + a_{k-1} f_2 + \cdots + a_2 f_{k-1} + g_1 \\
&\vdots \\
c_n &= a_n + a_{n-1} f_1 + a_{n-2} f_2 + \cdots + a_{n-k+1} f_{k-1} + g_{n-k} \\
0 &= a_n f_1 + a_{n-1} f_2 + \cdots + a_{n-k+2} f_{k-1} + g_{n-k+1} \\
&\vdots \\
0 &= a_n f_{k-1} + g_{n-1}
\end{aligned} \quad (3.27)$$

The coefficients of the polynomials F^* and G^* can thus be determined recursively. We summarize the result as Theorem 3.1.

THEOREM 3.1

Let $\{y(t), t \in T\}$ be a normal discrete time stochastic process which has the representation

$$A^*(q^{-1})y(t) = \lambda C^*(q^{-1})e(t) \qquad (3.15)$$

where the polynomials A and C have all zeros inside the unit circle, and $\{e(t), t \in T\}$ is a sequence of normal $(0, 1)$ random variables. The k-step predictor, which minimizes the variance of the prediction error, is given by the difference equation

$$C^*(q^{-1})\hat{y}(t+k \mid t) = G^*(q^{-1})y(t) \qquad (3.23)$$

where the polynomial $G(z)$ of order $n-1$ is defined by the identity

$$C^*(q^{-1}) = A^*(q^{-1})F^*(q^{-1}) + q^{-k}G^*(q^{-1}) \qquad (3.26)$$

The prediction error is a moving average of order k

$$\tilde{y}(t+k \mid t) = \lambda F^*(q^{-1})e(t+k)$$
$$= \lambda[e(t+k) + f_1 e(t+k-1) + \cdots + f_{k-1}e(t+1)] \qquad (3.25)$$

and has the variance

$$\text{var}\,[\tilde{y}(t+k \mid t)] = \lambda^2(1 + f_1^2 + f_2^2 + \cdots + f_{k-1}^2) \qquad (3.28)$$

Remark 1

Notice that the best predictor is linear and that the result does not depend critically on the criterion. As the distribution of y is normal, we find, for example, that the result would be the same if the criterion was to minimize

$$Eh(y(t+k) - \hat{y}) \qquad (3.29)$$

where h is an arbitrary symmetrical function.

Remark 2

Notice that the assumption, $e(t)$ and $e(s)$ being independent for $t \neq s$, is crucial for (3.22) to hold. If the stochastic variables $e(t)$ and $e(s)$ are not independent, it is not true that the mathematical expectation of a product of $e(t + \tau)$ with an arbitrary function of $y(t), y(t-1)$ is zero for $\tau > 0$. However, if we restrict the admissible predictors $\hat{y}$ to linear functions of the observations, the expression for $E[y(t+k) - \hat{y}]^2$ will only contain quadratic terms, and (3.22) will then also hold when $e(t+1)$, $e(t+2), \ldots, e(t+k)$ are uncorrelated with $y(t), y(t-1), \ldots$.

Remark 3

Notice that the predictor is a dynamical system of n:th order whose characteristic polynomial is $C(z)$.

Remark 4

Notice that it follows from (3.25) that

$$y(t) - \hat{y}(t \mid t - 1) = \lambda e(t)$$

The stochastic variables λe are thus the innovations of the process $\{y(t), t \in T\}$. Compare Section 3 of Chapter 4.

Remark 5

Notice that the crucial properties used in the proof are the following:

- The stochastic variables $e(t)$ and $e(s)$ are independent for $t \neq s$ and $e(t)$ is independent of $y(t-1), y(t-2), \ldots$.
- The representation (3.15) of the stochastic process is invertible in the sense that $y(t)$ can be computed in terms of $e(t), e(t-1), \ldots$, and vice versa using *stable* difference equations. (To carry out the computation we actually need initial conditions for the difference equations.)
- The assumptions on stability of the A and C polynomials and of infinite durations of the processes make, however, the result independent of initial conditions.

Exercises

1. Consider the stochastic process $\{y(t)\}$ defined by

$$y(t) - 1.5y(t-1) + 0.5y(t-2)$$
$$= 2[e(t) - 1.2e(t-1) + 0.6e(t-2)]$$

   where $\{e\{(t)\}$ is a sequence of independent normal $(0, 1)$ random variables. Determine the k-step predictor which minimizes the mean square prediction error.

2. Consider the stochastic process $\{y(t)\}$ defined by

$$y(t) + ay(t-1) = \lambda[e(t) + ce(t-1)]$$

   where $|a| < 1, |c| < 1$ and $\{e(t)\}$ is a sequence of independent normal $(0, 1)$ random variables. Determine the k-step predictor which minimizes the mean square prediction error.

3. Find the two step predictor which minimizes the mean square error and determine the prediction error for the stochastic process defined by

$$y(t) = \frac{1}{1 + aq^{-1}} e(t) + \sigma v(t)$$

where $\{e(t), t \in T\}$ and $\{v(t), t \in T\}$ are sequences of independent normal (0, 1) random variables and $|a| < 1$.

4. A stochastic process $\{y(t), t \in T\}$ has the representation
$$y(t) + 0.7y(t-1) = e(t) + 2e(t-1)$$
where $\{e(t), t \in T\}$ is a sequence of independent normal (0, 1) random variables. Determine the best one-step-ahead predictor and the variance of the prediction error. The best predictor is the one which minimizes the mean square prediction error.

5. Consider the stochastic process $\{y(t), t = t_0, t_0 + 1, \ldots\}$ defined by
$$y(t) = \sum_{k=t_0}^{t} g(t; k)e(k) \quad (*)$$
where $\{e(t), t = t_0, t_0 + 1, \ldots\}$ is a sequence of independent normal (0, 1) random variables. If $g(t; t) \neq 0$ the relation (*) always has an inverse. Let this be denoted by
$$e(t) = \sum_{k=t_0}^{t} h(t; k)y(k)$$
Determine the k-step predictor of the process $\{y(t), t = t_0, \ldots\}$ which minimizes the mean square prediction error.

Hint: Use the method of the proof of Theorem 3.1.

6. Consider the stochastic process $\{y(t), t \in T\}$ defined by
$$y(t) = \lambda \frac{C^*(q^{-1})}{A^*(q^{-1})} e(t)$$
where $\{e(t)\}$ is a sequence of independent normal (0, 1) random variables and
$$A(z) = a_0 z^n + a_1 z^{n-1} + \cdots + a_n$$
$$C(z) = c_0 z^m + c_1 z^{m-1} + \cdots + c_m$$
Determine the best mean square k-step predictor. The best predictor is a dynamical system. What is the order?

7. Consider the stochastic process $\{y(t), t \in T\}$ with the representation
$$y(t) - 2.6y(t-1) + 2.85y(t-2) - 1.4y(t-3)$$
$$+ 0.25y(t-4) = e(t) - 0.7e(t-1)$$
where $\{e(t), t \in T\}$ is a sequence of independent normal random variables. Determine the best one step predictor for the process $\{y(t), t \in T\}$. What is the order of the dynamical system which represents the predictor?

8. Find a state space representation of the system described by (3.15).

4. MINIMAL VARIANCE CONTROL STRATEGIES

Having obtained a solution to the prediction problem we will now generalize the stochastic control problem discussed in Section 2. To formulate an optimal control problem it is necessary to specify the process dynamics, the environment, the criterion, and the restrictions on the control law.

The Process Dynamics

It is assumed that the process to be regulated is a sampled, time invariant linear dynamical system of m:th order with one input u and one output y. The input-output relation can thus be described by an m:th order difference equation

$$y(t) + a_1^1 y(t-1) + \cdots + a_m^1 y(t-m)$$
$$= b_0^1 u(t-k) + b_1^1 u(t-k-1) + \cdots + b_m^1 u(t-k-m) \quad (4.1)$$

where the sampling interval is chosen as the time unit. Introduce the forward shift operator q, the polynomials A_1 and B_1 defined by

$$A_1(z) = z^m + a_1^1 z^{m-1} + \cdots + a_m^1 \quad (4.2)$$
$$B_1(z) = b_0^1 z^m + b_1^1 z^{m-1} + \cdots + b_m^1 \quad (4.3)$$

and their reciprocal polynomials A_1^* and B_1^*. The input-output relation (4.1) can then be written as

$$y(t) = \frac{B_1^*(q^{-1})}{A_1^*(q^{-1})} u(t-k) = \frac{B_1(q)}{A_1(q)} u(t-k) \quad (4.4)$$

The Environment

It is assumed that the influence of the environment on the process can be characterized by disturbances which are stochastic processes. As the system is linear, we can use the principle of superposition and represent all disturbances as a single disturbance acting on the output. The process and its environment can thus be described by the model

$$y(t) = \frac{B_1^*(q^{-1})}{A_1^*(q^{-1})} u(t-k) + v(t) \quad (4.5)$$

If it is furthermore assumed that the disturbance $v(t)$ is a stationary Gaussian process with rational spectral density, it can be represented as

$$v(t) = \lambda \frac{C_1^*(q^{-1})}{A_2^*(q^{-1})} e(t) \quad (4.6)$$

where $\{e(t), t = 0, \pm 1, \pm 2, \ldots\}$ is a sequence of equally distributed independent normal $(0, 1)$ random variables and C_1 and A_2 are polynomials.

Minimal Variance Control Strategies

We thus find that the system and its environment can be represented by the equation

$$y(t) = \frac{B_1^*(q^{-1})}{A_1^*(q^{-1})} u(t-k) + \lambda \frac{C_1^*(q^{-1})}{A_2^*(q^{-1})} e(t) \qquad (4.7)$$

A block diagram representation of the system is shown in Fig. 6.1.

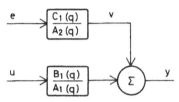

Fig. 6.1. Block diagram representation of the system described by (4.7).

Equation (4.7) is thus a canonical form for a sampled time invariant dynamical system with one input and one output with a time delay that is an integer multiple of the sampling interval, subject to disturbances which are stationary with rational spectral densities. The polynomials $A_2(z)$ and $C_1(z)$ can always be chosen to have their zeros inside or on the unit circle. As the disturbance v was assumed to be stationary, A_2 cannot have any zeros on the unit circle.†

To simplify the analysis we will also rewrite the model (4.7) to a slightly different form, namely

$$A^*(q^{-1})y(t) = B^*(q^{-1})u(t-k) + \lambda C^*(q^{-1})e(t) \qquad (4.8)$$

where A, B and C are polynomials defined by $A = A_1 A_2$, $B = B_1 A_2$ and $C = C_1 A_1$. Put

$$A(z) = z^n + a_1 z^{n-1} + \cdots + a_n \qquad (4.9)$$
$$B(z) = b_0 z^n + b_1 z^{n-1} + \cdots + b_n \qquad (4.10)$$
$$C(z) = z^n + a_1 z^{n-1} + \cdots + c_n \qquad (4.11)$$

There is no loss in generality to assume that all polynomials are of order n, because we can always put trailing coefficients equal to zero. The polynomial $C(z)$ can always be chosen to have all zeros inside or on the unit circle. We will make the additional assumption that there are no zeros on the unit circle.

The Criterion

The criterion for the control problem is chosen so as to minimize the variance of the output y.

---

† If we can postulate a representation of the form (4.6) for v directly, it is not necessary to assume that A_2 has all zeros inside the unit circle.

Admissible Control Laws

It is assumed that the control law should have the property that the value of u at time t is a function of observed outputs up to and including time t, that is, $y(t), y(t-1), y(t-2), \ldots$ and of all past control signals $u(t-1), u(t-2), \ldots$.

Problem Statement

Consider the dynamical system described by (4.8). Find an admissible control law such that the variance of the output is as small as possible. The optimal control law is called the *minimal variance strategy*.

Solution

To solve the problem we consider the situation at time t. We have thus obtained the measurements $y(t), y(t-1), \ldots$, and we know all past control actions $u(t-1), u(t-2), \ldots$. The problem is to determine $u(t)$ in such a way that the variance of the output is as small as possible. It follows from Eq. (4.8) that the control signal $u(t)$ will influence $y(t+k)$ but not any earlier outputs. Consider

$$y(t+k) = \frac{B^*(q^{-1})}{A^*(q^{-1})} u(t) + \lambda \frac{C^*(q^{-1})}{A^*(q^{-1})} e(t+k) \qquad (4.12)$$

The last term is a linear function of $e(t+k), e(t+k-1), \ldots, e(t+1)$, $e(t), e(t-1), \ldots$. It follows from (4.8) that $e(t), e(t-1), \ldots$ can be computed from the information available at time t. To do this explicitly we rewrite (4.12) using the identity

$$C^*(q^{-1}) = A^*(q^{-1}) F^*(q^{-1}) + q^{-k} G^*(q^{-1}) \qquad (4.13)$$

where F and G are polynomials of degrees $k-1$ and $n-1$ defined by (3.18) and (3.19).

Compare with (3.20) in the solution of the prediction problem. Hence

$$y(t+k) = \lambda F^*(q^{-1}) e(t+k) + \frac{B^*(q^{-1})}{A^*(q^{-1})} u(t) + \lambda \frac{G^*(q^{-1})}{A^*(q^{-1})} e(t) \qquad (4.14)$$

Solving (4.8) for $\lambda e(t)$, we get

$$\lambda e(t) = \frac{A^*(q^{-1})}{C^*(q^{-1})} y(t) - \frac{B^*(q^{-1})}{C^*(q^{-1})} q^{-k} u(t) \qquad (4.15)$$

Eliminating $e(t)$ between (4.14) and (4.15) we find

$$y(t+k) = \lambda F^*(q^{-1}) e(t+k)$$
$$+ \left[\frac{B^*(q^{-1})}{A^*(q^{-1})} - q^{-k} \frac{B^*(q^{-1}) G^*(q^{-1})}{A^*(q^{-1}) C^*(q^{-1})} \right] u(t) + \frac{G^*(q^{-1})}{C^*(q^{-1})} y(t)$$

Minimal Variance Control Strategies

Using the identity (4.13), the second term of the right member can be reduced and we get

$$y(t+k) = \lambda F^*(q^{-1})e(t+k) + \frac{G^*(q^{-1})}{C^*(q^{-1})} y(t) + \frac{B^*(q^{-1})F^*(q^{-1})}{C^*(q^{-1})} u(t) \quad (4.16)$$

Now let $u(t)$ be an arbitrary function of $y(t), y(t-1), \ldots$ and $u(t-1)$, $u(t-2), \ldots$. Then

$$Ey^2(t+k) = E[\lambda F^*(q^{-1})e(t+k)]^2$$
$$+ E\left[\frac{G^*(q^{-1})}{C^*(q^{-1})} y(t) + \frac{B^*(q^{-1})F^*(q^{-1})}{C^*(q^{-1})} u(t)\right]^2 \quad (4.17)$$

The mixed terms will vanish because $e(t+1), e(t+2), \ldots, e(t+k)$ are independent of $y(t), y(t-1), \ldots$ and $u(t-1), u(t-2), \ldots$. Hence

$$Ey^2(t+k) \geqslant \lambda^2[1 + f_1^2 + f_2^2 + \cdots + f_{k-1}^2] \quad (4.18)$$

where equality holds for

$$B^*(q^{-1})F^*(q^{-1})u(t) + G^*(q^{-1})y(t) = 0 \quad (4.19)$$

which gives the desired control law. We summarize the result as Theorem 4.1.

THEOREM 4.1

Consider the process described by

$$A^*(q^{-1})y(t) = B^*(q^{-1})u(t-k) + \lambda C^*(q^{-1})e(t) \quad (4.8)$$

where $\{e(t), t \in T\}$ is a sequence of independent normal $(0, 1)$ random variables. Let the polynomial $C(z)$ have all its zeros inside the unit circle. The minimum variance control law is then given by

$$B^*(q^{-1})F^*(q^{-1})u(t) = -G^*(q^{-1})y(t) \quad (4.19)$$

where the polynomials F and G of order $k-1$ and $n-1$ respectively are defined by the identity

$$C^*(q^{-1}) = A^*(q^{-1})F^*(q^{-1}) + q^{-k}G^*(q^{-1}) \quad (4.13)$$

The regulation error of the optimal system is a moving average of order k

$$y(t) = \lambda F^*(q^{-1})e(t)$$
$$= \lambda[e(t) + f_1 e(t-1) + \cdots + f_{k-1}e(t-k+1)] \quad (4.20)$$

Remark 1

Notice that the theorem still holds if it is only assumed that $e(t)$ and $e(s)$ are uncorrelated for $t \neq s$ but a linear control law is postulated.

Remark 2

A comparison with the solution of the prediction problem in Section 3 shows that the term

$$\frac{G^*(q^{-1})}{C^*(q^{-1})} y(t) + \frac{B^*(q^{-1})F^*(q^{-1})}{C^*(q^{-1})} u(t)$$

of (4.16) can be interpreted as the k-step prediction of $y(t + k)$ based on $y(t)$, $y(t - 1)$, ... and that the control error equals the k-step prediction error. Theorem 4.1 thus implies that the minimum variance control law is obtained by determining the k-step predictor and then choosing the control variable such that the predicted output coincides with the desired output. The stochastic control problem can thus be separated into two problems, one prediction problem and one control problem. Theorem 4.1 is therefore referred to as the *separation theorem*.

Remark 3

Notice that the control error is a moving average of order k when the minimum variance strategy is used. The covariance function of the control error will thus vanish for arguments greater than k. This observation is very convenient to use if we want to test a system in operation to find out if the control strategy in use is optimal.

Remark 4

Notice that the poles of the closed loop system equal the zeros of the polynomial $C^*(z)$.

We will now give an example of the calculation of minimum variance control strategies.

EXAMPLE 4.1

Consider a system described by

$$y(t) = \frac{B^*(q^{-1})}{A^*(q^{-1})} u(t - k) + \frac{C^*(q^{-1})}{A^*(q^{-1})} e(t)$$

with

$$A^*(q^{-1}) = 1 - 1.7q^{-1} + 0.7q^{-2}$$
$$B^*(q^{-1}) = 1 + 0.5q^{-1}$$
$$C^*(q^{-1}) = 1 + 1.5q^{-1} + 0.9q^{-2}$$

We will first consider the case $k = 1$. To compute the control strategy, we use the identity (4.13)

$$C^*(q^{-1}) = A^*(q^{-1})F^*(q^{-1}) + q^{-1}G^*(q^{-1})$$

We get

$$(1 + 1.5q^{-1} + 0.9q^{-2}) = (1 - 1.7q^{-1} + 0.7q^{-2}) + q^{-1}(g_0 + g_1 q^{-1})$$

Identification of the coefficients give

$$1.5 = -1.7 + g_0$$
$$0.9 = 0.7 + g_1$$

Hence $g_0 = 3.2$, $g_1 = 0.2$. The minimal variance strategy is thus given by

$$u(t) = -\frac{G^*(q^{-1})}{B^*(q^{-1})F^*(q^{-1})} y(t) = -\frac{3.2 + 0.2q^{-1}}{1 + 0.5q^{-1}} y(t)$$

or

$$u(t) = -0.5u(t-1) - 3.2y(t) - 0.2y(t-1)$$

Using the optimal strategy we find from Theorem 4.1 (4.20) that the control error is given by

$$y(t) = e(t)$$

We will now investigate how much the performance of the optimal system is degraded if there is an additional time delay in the system. For $k = 2$ the identity (4.13) becomes

$$(1 + 1.5q^{-1} + 0.9q^{-2})$$
$$= (1 - 1.7q^{-1} + 0.7q^{-2})(1 + f_1q^{-1}) + q^{-2}(g_0 + g_1q^{-1})$$

Identification of the coefficients give

$$1.5 = -1.7 + f_1$$
$$0.9 = 0.7 - 1.7f_1 + g_0$$
$$0 = 0.7f_1 + g_1$$

Solving these equations we get $f_1 = 3.2$, $g_0 = 5.64$, $g_1 = -2.24$. The minimal variance control strategy becomes

$$u(t) = -\frac{G^*(q^{-1})}{B^*(q^{-1})F^*(q^{-1})} y(t) = -\frac{5.64 - 2.24q^{-1}}{1 + 3.7q^{-1} + 1.6q^{-2}} y(t)$$

Hence

$$u(t) = -5.64y(t) + 2.24y(t-1) - 3.7u(t-1) - 1.6u(t-2)$$

The control error is

$$y(t) = e(t) + f_1 e(t-1) = e(t) + 3.2e(t-1)$$

The variance of the control error is

$$\text{var } y(t) = 1 + f_1^2 = 11.24$$

In this case we thus find that the control error increases considerably when there is an additional time delay in the system.

In Figs. 6.2, 6.3, and 6.4 we show simulations of the output without control, and control signals, and output signals for the system with minimum variance control strategies for $k = 1$ and $k = 2$.

178   *Minimal Variance Control Strategies*

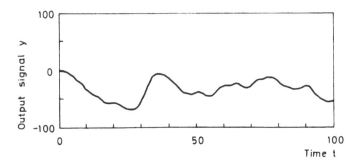

Fig. 6.2. Simulation of the output of the system in Example 4.1 when the control signal is zero.

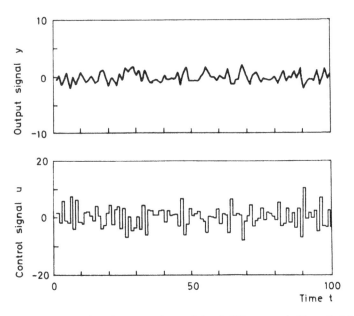

Fig. 6.3. Simulation of output and control signal of the system in Example 4.1 with $k = 1$ and the minimal variance control strategy. A comparison with Fig. 6.2 shows that the control strategy reduces the variations of the output considerably.

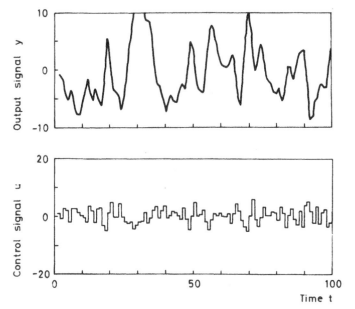

Fig. 6.4. Simulation of output signal and control signal of the system in Example 4.1 with $k = 2$ and the minimal variance control strategy. A comparison with Fig. 6.3 shows the deterioration of the performance due to the additional time-delay.

Exercises

1. Consider the system described by

$$y(t) = \frac{1}{1 + 0.5q^{-1}} u(t-1) + \frac{1 + 0.7q^{-1}}{1 - 0.2q^{-1}} e(t)$$

   where $\{e(t)\}$ is a sequence of independent normal $(0, 1)$ random variables. Determine the minimal variance control strategy.

2. Consider a system described by

$$y(t) + ay(t-1) = bu(t-k) + \lambda[e(t) + ce(t-1)]$$

   where $\{e(t), t \in T\}$ is a sequence of independent normal $(0, 1)$ random variables. Determine the minimal variance control strategies and the control errors for $k = 1, 2,$ and 3.

3. Consider a deterministic system described by

$$A^*(q^{-1})y(t) = B^*(q^{-1})u(t-k)$$

   Show that the control strategy

$$u(t) = -\frac{G^*(q^{-1})}{B^*(q^{-1})F^*(q^{-1})} y(t) \qquad (*)$$

where the polynomials F and G are determined by the identity

$$1 = A^*(q^{-1})F^*(q^{-1}) + q^{-k}G^*(q^{-1})$$

has the property that it brings the output of the system to zero in k steps. The strategy (*) is called a dead-beat strategy.

4. Consider the system

$$A^*(q^{-1})y(t) = B^*(q^{-1})u(t-k) + \lambda C^*(q^{-1})e(t)$$

Determine the variance of the output when the system is regulated using the dead-beat strategy (*) given in Exercise 3. Show that the control error is a moving average and compare the variance with the minimal variance.

5. Consider a system described by (4.7). Show that the minimal variance control strategy will always contain an integration if the polynomial A_2 has a simple zero at $z = 1$ and if $A_1(1) \ne 0$.

6. Show that the numbers f_i defined by the identity (4.13) can be interpreted as the impulse response of a discrete dynamical system with the pulse transfer function $C(z)/A(z)$.

7. Determine the minimum variance control strategy and the variance of the control error for a system described by

$$y(t) + 0.5y(t-1) = e(t) + 2e(t-1) + u(t-1)$$

8. Consider the system described by

$$y(t) = \frac{1}{1+2q^{-1}} u(t-1) + \frac{1+0.7q^{-1}}{1-0.2q^{-1}} e(t)$$

where $\{e(t)\}$ is a sequence of independent normal $(0, 1)$ random variables. Determine the minimal variance control strategy for the system. (Compare with Exercise 1.)

9. A digital version of a PI-regulator can be described by the equation

$$u(t) = u(t-1) + K\left[\left(1 + \frac{h}{T}\right)y(t) - y(t-1)\right] \qquad (**)$$

where K, h, T are positive numbers. Give the most general system of the form (4.8) for which the algorithm (**) can be a minimal variance control law.

10. Consider the polynomial identity

$$C(z) = A(z)F(z) + B(z)G(z)$$

where A, B and C are given polynomials of degree n_A, n_B and n_C respectively. Assume that $n_C < n_A + n_B - 1$. Show that there exist unique polynomials F and G of degree $n_F = n_B - 1$ and $n_G = n_A - 1$ which satisfy the identity. Also show that there exist many polynomials of degree $n_F > n_B - 1$ and $n_G > n_A - 1$ which satisfy the identity.

11. Consider the system (4.8). Show that there exist a control strategy such that the regulation error equals the error in predicting the output l steps ahead when $l > k$. Hint.: Use the identity

$$C^*(z^{-1}) = A^*(z^{-1})F_1^*(z^{-1}) + z^{-l}G_1^*(z^{-1}).$$

12. Determine the minimum variance control strategy for the system (4.8) when the admissible control strategies are such that $u(t)$ is a function of $y(t-1), y(t-2), \ldots, u(t-1), u(t-2), \ldots$.

13. Find a state space representation of the system described by (4.12).

5. SENSITIVITY OF THE OPTIMAL SYSTEM

It is well-known that optimal solutions under special circumstances may be very sensitive to parameter variations. We shall therefore investigate this matter in our particular case. To do so we shall assume that the system is governed by (4.8) which we write as

$$A^0(q)y(t) = B^0(q)u(t-k) + \lambda^0 C^0(q)e(t) \quad (5.1)$$

but that the control law is calculated under the assumption that the system model is

$$A(q)y(t) = B(q)u(t-k) + \lambda C(q)e(t) \quad (5.2)$$

where the coefficients of A, B, and C differ slightly from those of A^0, B^0, and C^0.

Notice that the orders n of the models (5.1) and (5.2) are the same. The minimal variance control strategy for the model (5.2) is

$$u(t) = -\frac{G^*(q^{-1})}{B^*(q^{-1})F^*(q^{-1})} y(t) = -\frac{q^k G(q)}{B(q)F(q)} y(t) \quad (5.3)$$

where F and G are polynomials of degree $k-1$ and $n-1$ defined by the identity (4.13).

We shall now investigate what happens if the system (5.1) is controlled with the control law (5.3). Introducing (5.3) into (5.1) we get

$$\left[A^0(q) + \frac{B^0(q)G(q)}{B(q)F(q)}\right]y(t) = \lambda^0 C^0(q)e(t) \quad (5.4)$$

Let q^{n+k-1} operate on the identity (4.13) and use the definition of reciprocal polynomial. We find

$$q^{k-1}C(q) = A(q)F(q) + G(q) \tag{5.5}$$

Equations (5.4) and (5.5) now give

$$[q^{k-1}B^0(q)C(q) + (A^0(q)B(q) - A(q)B^0(q))F(q)]y(t)$$
$$= \lambda^0 B(q)C^0(q)F(q)e(t) \tag{5.6}$$

The characteristic equation of the system is thus

$$z^{k-1}B^0(z)C(z) + [A^0(z)B(z) - A(z)B^0(z)]F(z) = 0 \tag{5.7}$$

If $A = A^0$, $B = B^0$ and $C = C^0$, the characteristic polynomial reduces to $z^{k-1}B^0(z)C^0(z)$. For small perturbations in the parameters, the modes of the system (5.6) are thus close to the modes associated with $z^{k-1}B^0(z)C^0(z)$, that is, $k-1$ modes with poles at the origin, n modes with poles at the zeros of B^0, and n modes with poles at the zeros of C^0. Furthermore when the design parameters equal the true parameters, the factor B^0C^0 cancels in (5.6). This implies that the modes associated with B^0C^0 are uncoupled to the input e if $A = A^0$, $B = B^0$ and $C = C^0$ or that the corresponding state variables under the same conditions are not controllable from the input e. Hence if the control law is calculated from a model which deviates from the true model, the input might excite all the modes associated with the solutions of the characteristic (5.7). This is not a serious matter if the modes are stable. However, if some modes are unstable, it is possible to get infinitely large errors if the model used for designing the control law deviates from the actual model by an arbitrarily small amount. This situation will occur if the polynomial B^0C^0 has zeros outside or on the unit circle. It follows from the representation theorems for stationary random processes (Theorem 3.2 of Chapter 4) that C^0 can always be chosen to have zeros inside or on the unit circle. As far as C^0 is concerned the only critical case would be if C^0 had a zero on the unit circle. The polynomial B^0 will have zeros outside the unit circle if the dynamics of the sampled system is nonminimum phase. Hence, if either the dynamical system to be controlled is nonminimum phase, or if the numerator of the spectral density of the disturbances has a zero on the unit circle, the minimum variance control law will be extremely sensitive to variations in the model parameters. In these situations it is of great practical interest to derive control laws which are insensitive to parameter variations and whose variances are close to the minimal variances.

Suboptimal Strategies

There are many ways to obtain control laws which are not sensitive to parameter variations. One possibility is to proceed as follows. To fix

Sensitivity of the Optimal System

the ideas, we will assume that $B(z)$ can be factored as

$$B(z) = B_1(z)B_2(z) \tag{5.8}$$

where B_1 is of degree n_1 and has all zeros inside the unit circle and B_2 is of degree n_2 and has all zeros outside the unit circle.

When resolving the identity (5.5) we impose the additional requirement that $G(z)$ contain $B_2(z)$ as a factor, i.e., we use the identity

$$q^{n_2+k-1}C(q) = A(q)F'(q) + B_2(q)G'(q) \tag{5.9}$$

instead of (5.5). Going through the arguments used when deriving Theorem 4.1, we find the control law

$$u(t) = -\frac{q^k G'(q)}{B_1(q)F'(q)} y(t) \tag{5.10}$$

which gives the control error

$$y(t) = \lambda\{e(t) + f_1 e(t-1) + \cdots + f_{k-1} e(t-k+1) \\ + f'_k e(t-k) + \cdots + f'_{k+n_2-1} e(t-k-n_2+1)\} \tag{5.11}$$

The control law (5.10), which is not optimal, gives an error with the variance

$$\text{var } y = \min(\text{var } y) + \lambda^2 \{f'^2_k + \cdots + f'^2_{k+n_2-1}\} \tag{5.12}$$

The control law (5.10) is not extremely sensitive to variations in system parameters. To realize this we assume again that the system is governed by the model $(A^0, B^0, C^0, \lambda^0)$ but that the control law is calculated from the model (A, B, C, λ) with slightly different parameters. The equation describing the controlled system then becomes

$$[A^0(q)B_1(q)F'(q) + B^0(q)G'(q)]y(t) = \lambda^0 C^0(q)B_1(q)F'(q)e(t) \tag{5.13}$$

When the parameters equal the true parameters the characteristic equation of the system becomes

$$z^{n_2+k-1} B_1^0(z) C^0(z) = 0 \tag{5.14}$$

and it now follows from the definition of B_1^0 and the assumption made on C^0, that all modes are stable when the design parameters equal the actual parameters. The stability for small perturbations of the parameters now follows by continuity.

A comparison of (5.7) and (5.14) shows that the control law (5.10) brings the zeros of B_2 to the origin. There are other suboptimal control strategies which instead move the zeros of B_2 to arbitrary positions inside the unit circle. Hence, whenever the polynomial B^0 has zeros outside or on the unit circle, the optimal system will be so sensitive to parameter variations that it is useless. Infinitely small parameter variations can make the variance infinite. Even if B^0 or C^0 have zeros inside the unit circle, the optimal system might still be sensitive to parameter variations. To illustrate this we will analyse a specific case.

EXAMPLE

Consider a system governed by

$$y(t) + a^0 y(t-1) = u(t-1) + b^0 u(t-2) + e(t) + c^0 e(t-1) \quad (5.15)$$

with $a^0 = -0.7$, $b^0 = 0.99$ and $c^0 = 0.95$. To obtain the minimum variance control strategy for this system, we use the identity (4.13)

$$(1 + c^0 q^{-1}) = (1 + a^0 q^{-1}) + q^{-1} g_0$$

Hence $g^0 = c^0 - a^0$. The minimal variance strategy is

$$u(t) = -\frac{c^0 - a^0}{1 + b^0 q^{-1}} y(t) = -\frac{1.65}{1 + 0.99 q^{-1}} y(t) \quad (5.16)$$

Now assume that the control law is computed from a system with a different value of the coefficient of $u(t-2)$ and we get the control law

$$u(t) = -\frac{1.65}{1 + b q^{-1}} y(t)$$

Using this control law we find that the output is given by

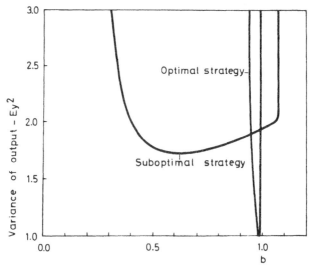

Fig. 6.5. Shows the value of the variance of the output signal when the control strategy is computed from a model with a parameter b which deviates from the true parameter value $b^o = 0.99$. Notice the extreme sensitivity of the optimal system as compared to the suboptimal system.

Sensitivity of the Optimal System

$$y(t) = \frac{q^2 + q(b + c^0) + bc^0}{q^2 + q(b + c^0) + (b - b^0)a^0 + c^0 b^0} e(t) = H(q)e(t)$$

The variance of the output signal is then given by

$$V(b) = Ey^2 = \frac{1}{2\pi i} \oint H(z)H(z^{-1}) \frac{dz}{z}$$

In Fig. 6.5 we show how the variance of the output depends on the parameter b. It is clear from the figure that the optimal system is very sensitive to variations in the b parameter.

To obtain a control law which is *not optimal* but which is less sensitive for variations in the b-parameter, we can use the control law (5.10). The identity (5.9) gives

$$(1 + c^0 q^{-1}) = (1 + a^0 q^{-1})(1 + fq^{-1}) + q^{-1}(1 + b^0 q^{-1})g$$

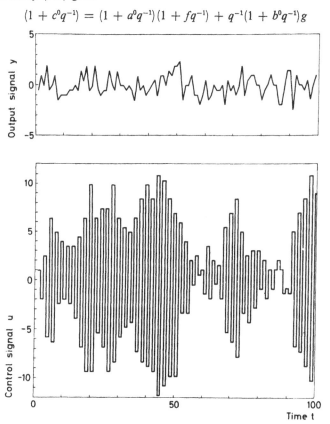

Fig. 6.6. Simulation of the output signal and the control signal of the system (5.15) with the minimal variance control strategy (5.16).

Hence

$$f = b^0(c^0 - a^0)/(b^0 - a^0) = 0.966$$
$$g = -a^0(c^0 - a^0)/(b^0 - a^0) = 0.684$$

and the control law (5.10) becomes

$$u(t) = -\frac{g}{1 + fq^{-1}} y(t) = -\frac{0.684}{1 + 0.966q^{-1}} y(t) \quad (5.17)$$

It follows from (5.11) that the control strategy (5.17) gives the control error

$$y(t) = e(t) + fe(t - 1) = e(t) + 0.966e(t - 1)$$

The control strategy (5.17) thus gives the variance

$$\text{var } y = 1.93$$

Compare with the minimal variance 1.

Assuming that the control strategy is evaluated from a model whose parameter b differs from the value b^0 of the system, we find that the control error is

$$y(t) = \lambda \frac{q^2 + q(c_0 + f) + c^0 f}{q^2 + q(a^0 + f + g) + a^0 f + b^0 g} e(t)$$

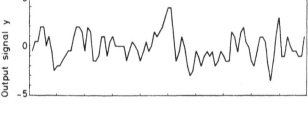

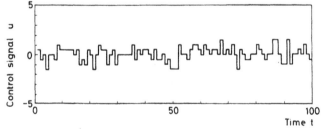

Fig. 6.7. Simulation of the output signal and the control signal of the system (5.15) with the suboptimal control strategy (5.17). Compare with the simulation of the optimal strategy shown in Fig. 6.6.

Sensitivity of the Optimal System

Evaluating the variance of the output for different values of the parameter b, we get the result shown in Fig. 6.5. We thus find that the control law (5.17) is much less sensitive to variations in the parameter b than the optimal control law (5.16).

In Figs. 6.6 and 6.7 we also show simulations of the outputs and the control signals when the nominal system is controlled using the optimal (5.16) and the suboptimal control law (5.17).

Exercises

1. Consider the system
$$y(t) + 0.64y(t-1) + 0.22y(t-2) = 6.4u(t-3)$$
$$+ 19.2u(t-4) + \lambda[e(t) - 0.82e(t-1) + 0.21e(t-2)]$$

   Determine the minimal variance control law and discuss its sensitivity. Construct a control law which is less sensitive to parameter variations.

2. Compare the variance of the control error using the control law (5.10) with the variance of the error of the $k + n_2 - 1$ step predictor.

3. Replace the identity (5.9) by
$$H(q)C(q) = A(q)F'(q) + B_2(q)G'(q)$$
   where H is an arbitrary polynomial of degree $n_2 + k - 1$. Show that the control law
$$u(t) = -\frac{G'(q)}{B_1(q)F'(q)} y(t)$$
   gives the control error
$$H(q)y(t) = F'(q)e(t)$$
   and that the characteristic equation of the closed loop system is given by
$$H(z)B_1(z)C(z) = 0$$

4. Consider the system given by (4.8). Show that the control strategy
$$u(t) = -\frac{G_1^*(q^{-1})}{F_1^*(q^{-1})} y(t)$$
   where the polynomials F_1 and G_1, of order $n + k - 1$ and $n - 1$, respectively, are defined by the identity
$$C^*(q^{-1}) = A^*(q^{-1})F_1^*(q^{-1}) + q^{-k}B^*(q^{-1})G_1^*(q^{-1})$$
   gives a closed loop system with the characteristic equation
$$z^{n+k-1}C(z) = 0$$

and that the closed loop system is not particularly sensitive to parameter variations. Also evaluate the variance of the control error and compare with the minimum variance.

5. Consider the system
$$y(t) + ay(t-1) = u(t-1) + 2.5u(t-1)$$
$$+ u(t-2) + e(t) + ce(t-1).$$

Determine the minimal variance control strategy and the minimal regulation error. Show that the optimal system is extremely sensitive to parameter variations and derive suboptimal control strategies which are not as sensitive to parameter variations using the techniques discussed in Section 5 and in Exercise 4. Determine the regulation error for the two suboptimal control strategies.

6. AN INDUSTRIAL APPLICATION

Introduction

This section will give an example of an industrial application of the theory presented in Sections 4 and 5. The material is based on work carried out at the IBM Nordic Laboratory in Stockholm, in connection with the installation of a process computer in the Billerud Kraft Paper Mill at Gruvön, Sweden. Much of the material is taken from the paper "Computer Control of a Paper Machine—an Application of Linear Stochastic Control Theory" which appeared in IBM Journal of Research and Development, July 1967. The permission from IBM to use this material is gratefully acknowledged.

Stochastic control theory was used to solve several problems both in quality control and process control in connection with the Billerud study. In this section we will describe the application to one typical problem, namely basis weight control. We discuss the applicability of linear stochastic control theory and give a mathematical formulation of the problem. The identification problem, i.e., how to obtain mathematical models of process dynamics and disturbances from experimental data, is also discussed briefly. This is naturally a very important problem in practical applications. Finally we give some practical experiences with on-line control using optimal strategies.

Basis weight fluctuations were investigated during a feasibility study before the system was under computer control. The fluctuations had a standard deviation of 1.3 g/m^2. In the feasibility study, the target value for standard deviation of basis weight under computer control was set to 0.7 g/m^2.

An Industrial Application

The control computer was installed late in December, 1964. Two experiments for determination of process dynamics were performed in March, 1965 and the first operation of on-line basis weight control was done on April 28, 1965 covering a period of 10 hours. Since that date a large number of tests have been performed and the basis weight loop has been in continuous operation since the beginning of 1966. In actual operation we can now consistently achieve standard deviations of 0.5 g/m² wet basis weight and 0.3 g/m² dry basis weight.

Applicability of Stochastic Control Theory

The theory presented in Section 4 is based on the following assumptions:

- Process dynamics are characterized by linear differential equations with constant coefficients. There is one input and one output.
- Disturbances are characterized as stationary Gaussian processes with rational spectral densities.
- The criterion is to minimize the variance of the output signal.

During the initial phase of the project we performed experiments which indicated that the disturbances occurring during normal operation were so small that the system could be described by linear equations. The indications obtained in these early experiments were subsequently verified by experiments on the computer-controlled system.

Using conventional spectral analysis we verified, during a feasibility study, that the normal fluctuations could be described reasonably well as stationary processes. The probability distributions were also shown to be approximately normal.

Admittedly there are sometimes "upsets" which give rise to large deviations. These upsets may originate in many different ways, for example, from equipment or instrument malfunctions, which all require particular corrective actions. It is questionable whether these types of disturbances can be described by probabilistic models. In the following we disregard these upsets.

Basis weight is an important quality variable for kraft paper. When paper is sold limits of the quality variables are usually specified. The customers test procedures are usually set up in such a way that a batch is accepted if the test sample shows that the quality variables are within the test limits with a specified probability. Since there are always fluctuations in quality during normal operation, the paper manufacturer chooses the set points for the basis weight regulator well above the lower test limit in order to make sure that the product satisfies the customers specifications. If variations in basis weight can be reduced, it is then possible to move the set point closer to the acceptance limit without changing the probability for acceptance. See Fig. 6.8. By doing this there is a capital gain which can be expressed as a gain of raw material or as an increase in production. In the particular application, the gain obtained from this was an essential contribution to the net capital gain obtained through computer control.

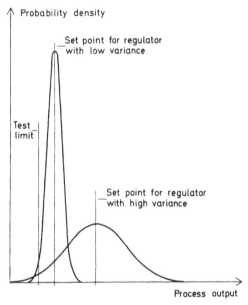

Fig. 6.8. Illustrates that a decrease of the variance of the output signal makes it possible to move the set point closer to the test limit.

The possibility to produce paper with smaller fluctuations in quality variables also has other advantages. It may, for example, simplify the subsequent processing of the paper by the customer. These advantages are, however, often difficult to evaluate objectively.

Since it was found that basis weight fluctuations were normally distributed, the problem of operating as close as possible to the tolerance limits is equivalent to controlling the process in such a way that the variance of the process output is as small as possible.

To summarize, we thus found reasonable evidence to believe that the theory of Section 4 can be applied to the basis weight control problem. This is not a strange coincidence since the theory of Section 4 was actually developed for the basis weight control problem. The belief was also verified in the sense that the results predicted, using stochastic control theory, could also be achieved in practice.

An Industrial Application 191

Mathematical Formulation of the Basis Weight Regulation Problem

Figure 6.9 shows a simplified diagram of the parts of the paper machine that are of interest for basis weight control. Thick stock, i.e., a water fiber mixture with a fiber concentration of about 3% comes from the machine chest. The thick stock is diluted with white water so that the headbox concentration is reduced to 0.2 to 0.5%. On the wire, the fibers are separated from the water and a web is formed. Water is pressed out of the paper web in the presses, and the paper is then dried on steam-heated cylinders in the dryer section.

In this particular case it is possible to influence the basis weight by varying the thick stock flow and/or the thick stock consistency (fiber concentration in thick stock). Both these variables will directly influence the amount of fibers flowing out of the headbox and thus also the basis weight. The control variables are manipulated via set points of analog regulators which control the thick stock flow valve and the thick stock dilution valve shown in Fig. 6.9.

Basis weight is measured by a beta-ray gauge set at a fixed position at the dry end. The output of this instrument will be proportional to the mass of fibers and water per unit area, i.e., the *wet basis weight*. This is because the coefficient of absorption of beta-rays in fibers and water is approximately the same. In order to obtain *dry basis weight*, i.e., the mass of fibers per unit area, the beta-ray gauge reading has to be compensated for the moisture in the paper sheet. Moisture is measured by a capacitance gauge. In our particular case, this gauge can traverse the paper web although it is normally set at a fixed position.

There is also a beta-ray gauge before the drying section. Basis weight is also measured by the machine tender and in the test-laboratories. When a reel of paper is produced, its weight and size are determined giving a very accurate value of the average basis weight of the reel. This information is used to calibrate the other gauges. An analysis of the information sources for basis weight has shown that:

- The reel weight and dimension information can be used to compensate the drift in the beta-ray gauge.
- The high frequency fluctuations in the moisture gauge and the beta-ray gauge signal have similar characteristics and a good estimate of dry basis weight is

$$y = WSP(1 - MSP) \qquad (6.1)$$

where *WSP* is the calibrated beta-ray gauge signal and *MSP* is the signal from the moisture gauge. The difference between dry basis weight and the estimate y of (6.1) is essentially a stationary random process which contains many high frequencies.

192 *Minimal Variance Control Strategies*

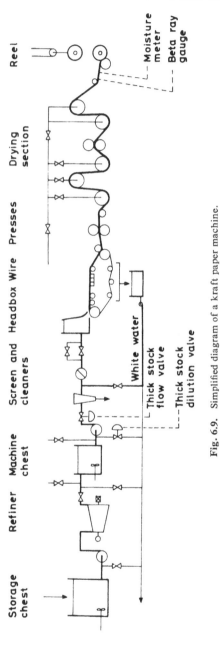

Fig. 6.9. Simplified diagram of a kraft paper machine.

An Industrial Application

- The estimate of dry basis weight given by (6.1) is improved very slightly when laboratory measurements are taken into account.

Fluctuations in basis weight during normal operation have been investigated. There are variations of weight in both the machine direction and the cross-direction. In our case, it was found that the cross-direction profile is stable if certain precautions are taken. The fluctuations observed can be described as normal random processes. There is a considerable amount of low-frequency variation. Data in the records have been divided into samples covering about five hours for analysis. Before carrying out a time-series analysis, the trend is removed. Figure 6.10 shows the covariance fluction of basis weight variations in a typical case. In all cases studied, we found that the variations in basis weight had a standard deviation equal to or greater than 1.3 g/m^2.

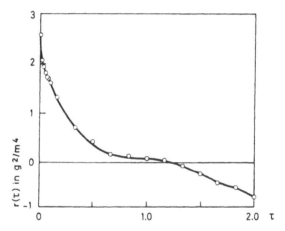

Fig. 6.10. Covariance function of fluctuations in wet basis weight obtained during the feasibility study, before control computer was installed.

We investigated the possibilities of controlling the basis weight by careful regulation of machine speed, thick stock flow, and consistency. Experiments have also been performed to establish the correlation of fluctuations in basis weight with fluctuations in thick stock flow and consistency. The results of these investigations have shown that, in this particular application, it is not possible to keep the basis weight constant by careful regulation of machine speed and fiber flow. It was therefore decided to control basis weight by feedback from the measurements at the dry end of the machine to thick stock flow or thick stock consistency.

The dynamics of the paper machine are such that there is a coupling between dry basis weight and moisture content. An increase of thick stock flow or thick stock consistency results in an increase in moisture content as well as dry basis weight. A change of steam pressure in the drying section will, however, influence the moisture content of the paper but not the dry basis weight. This coupling was not known to us initially, but the first identification experiment showed the effect clearly. The special character of the coupling implies that the loop for controlling dry basis weight can be considered as a single-input, single-output system. The control actions of the basis weight loop (thick stock flow) will, however, introduce disturbances in moisture content. These disturbances can be eliminated by using thick stock flow as an input to the moisture control loop. Control of *dry* basis weight can thus be considered as a system with one input, thick stock flow, and one output, the variable y as given by (6.1). Notice, however, that the moisture control system must be considered as a system with two inputs and one output.

We have found previously that the variance of dry basis weight is a good measure of the performance of the basis weight control loop. Hence if the estimate of dry basis weight y given by (6.1) differs from dry basis weight by a quantity which is essentially high frequency noise, minimizing the variance of dry basis weight is equivalent to minimizing the variance of y. Notice that this reasoning fails if the deviation contains low frequencies!

The criterion is thus to control the system in such a way that the variance of the output signal y is minimal. To complete the formulation of the control problem, we now need a description of the process dynamic and the characteristics of the disturbances.

The corrections that are required to control the process during normal operation are so small that the system can be described by linear differential equations with a delayed input. The time-delay T_d depends on the time it takes to transport the fibers along the paper machine. The equations determining the dynamics can be partly determined by continuity equations for the flow. The degree of mixing in the tanks is uncertain, however. There is also a rather complicated mechanism that determines the amount of fibers that pass through the wire. A direct derivation will thus give a very uncertain model for process dynamics.

Since a digital computer is to be used to implement the control law, we will consider a discrete time model directly. If it is assumed that the sampling interval T_s is chosen so that T_d is an integral multiple of T_s and if we also assume that the control signal is constant over the sampling interval, the process dynamics can be expressed by the general linear model given by (4.1).

Since the process is linear, we can use superposition and reduce the disturbances to an equivalent disturbance in the output. Compare (4.5). If the disturbances moreover are stationary with rational spectral densities,

An Industrial Application

they can always be represented by (4.6) and we thus find that the process and the disturbances can be represented by (4.7), which can be reduced to (4.8).

Summing up we thus have the problem of controlling a system described by (4.8) in such a way that the variance of the output is as small as possible.

If the coefficients of the polynomials A, B, and C, and the integers n and k are known, the solution to the problem follows immediately from Theorem 4.1. In a practical problem we have, however, also the additional problem to determine the parameters of the model.

Process Identification

As stated previously, it is very difficult to derive the mathematical model from first principles. Instead we have determined the model given by (4.8) directly from measurements on the process. When making the measurements, the control variable of the process is perturbed and the resulting variations in the output are observed. On the basis of recorded input-output pairs $\{u(t), y(t), t = 1, 2, \ldots, N\}$ we then determine a model (4.8) of the process and the disturbances. Since a complete presentation of identification techniques is outside the scope of the book, we will not go into any details. References are given in Section 7.

Let it suffice to say that the problem is solved by determining the maximum likelihood estimate of the parameters

$$\theta = (a_1, a_2, \ldots, a_n, b_0, b_1, \ldots, b_{n-1}, c_1, c_2, \ldots, c_n)^\dagger$$

of the model based on a sequence of input-output pairs $\{u(t), y(t), t = 1, 2, \ldots, N\}$. It has been shown that maximizing the likelihood function is equivalent to minimizing the loss function

$$V(\theta) = \frac{1}{2} \sum_{t=1}^{N} \varepsilon^2(t) \qquad (6.2)$$

where the numbers $\varepsilon(t)$ called *residuals* are related to the input-output signal by the equation

$$C^*(q^{-1})\varepsilon(t) = A^*(q^{-1})y(t) - B^*(q^{-1})u(t - k) \qquad (6.3)$$

The numbers $\varepsilon(t)$ can be interpreted as being one-step-ahead prediction errors. Compare Theorem 4.1.

When we have found a $\theta = \hat{\theta}$ such that $V(\hat{\theta})$ is minimal, we get the maximum likelihood estimate of λ from

$$\hat{\lambda}^2 = \frac{2V(\hat{\theta})}{N} \qquad (6.4)$$

† Throughout this section the coefficient b_n of the polynomial B is assumed to be zero.

The identification problem is thus reduced to a proplem of finding the minimum of a function of several variables.

The function V is minimized recursively by a gradient routine which involves computation of the gradient V_θ of V with respect to the parameters as well as the matrix of second partial derivatives $V_{\theta\theta}$. Due to the particular choice of model structure, the computation of the derivatives of the loss function can be done very economically. In fact for large N, the computations increase only linearly with the order of the model.

To obtain a starting value for the maximizing algorithm, we set $c_i = 0$. The function V is then quadratic in a_i and b_i, and the algorithm converges in one step, giving the least squares estimate. This is then taken as the starting point for the gradient routine. To investigate whether $V(\theta)$ has local minima, we also choose several other starting points.

It is shown in the paper by Åström, Bohlin, and Wensmark (quoted in Section 7) that the maximum likelihood estimate is consistent, asymptotically normal, and efficient under mild conditions. The conditions are closely related to the information matrix. An estimate of this matrix is provided by

$$\hat{I} = \lambda^{-2} V_{\theta\theta} \qquad (6.5)$$

The matrix $V_{\theta\theta}$, which was computed in order to get a fast convergence for the gradient routine, will thus have a physical interpretation.

Practical Experiments to Determine Process Dynamics

The identification procedure just described briefly has been applied extensively in connection with the Billerud project for quality control, basis weight control, moisture content control, and refiner control. We will now present some of the practical results obtained when applied to basis weight control.

The control computer is used to perform the experiments. The input signal used in the experiment is represented as a sequence of numbers stored in the control computer. The numbers of the sequence are read periodically and converted to analog signals by the D/A converter and the regular D/A conversion subroutines. The output signals from the process are converted to digital numbers using the control computer's A/D converter. In this way, we represent both the input and output signals by numbers that appear in the control computer in precisely the way they occur when the computer is controlling the process. The dynamics of signal transducers, transmission lines, and A/D and D/A converters are thus included in the model. Disturbances in transducers and signal converters, as well as round-off errors, are thus also included in the disturbances of the model. The whole experiment is executed by a program. The result of a typical identification experiment is illustrated in Fig. 6.11.

An Industrial Application

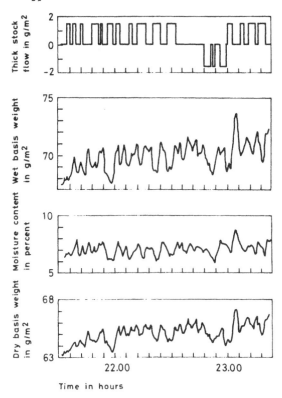

Fig. 6.11. Results of an experiment for determination of process dynamics showing response of wet basis weight, moisture content, and dry basis weight to perturbation in thick stock flow.

Choosing the Input Signal

The choice of the input signal involves certain considerations. That is, it is desirable to have large signal amplitudes in order to get good estimates. Large input signals may, however, drive the system outside the linear region and may also cause unacceptably large variations in the process variables. In our particular case, we had to make all experiments during normal production. This was one major reason for using a fairly sophisticated identification procedure. Notice that in order to obtain a specified accuracy, it is possible to compromise between signal amplitude and length of the sample. In the identification of models required for the design of basis weight control laws, we usually used samples one to five hours in length. The amplitudes of the signals shown in Fig. 6.11 corresponding to 1.7 g/m² peak-to-peak are typical. This number was a suitable compromise. Notice that the standard deviation during normal operation with no control is typically 1.3 g/m².

Pseudo-random binary signals have been used successfully as input signals. We have found, however, that if some knowledge of the process is available, it is desirable to tailor the test signals to the specific purpose.

Examples of Numerical Identification

We will now present some examples which illustrate the numerical identification procedure. These examples are based on the data shown in Fig. 6.11. Mathematical models which relate changes in dry basis weight (*WSPO*) and wet basis weight (*WSP*) to changes in (the set point of the) thick stock flow (regulator) will be discussed. Figure 6.11 shows that the output is drifting. The drift is even more pronounced in test experiments of longer duration. To take care of this drift we have used models which relate changes in the output to changes in the input

$$\nabla y(t) = \frac{b_0 + b_1 q^{-1} + \cdots + b_{n-1} q^{n-1}}{1 + a_1 q^{-1} + \cdots + a_n q^{-n}} \nabla u(t - k)$$
$$+ \lambda \frac{1 + c_1 q^{-1} + \cdots + c_n q^{-n}}{1 + a_1 q^{-1} + \cdots + a_n q^{-n}} e(t) \quad (6.6)$$

where ∇ is the backward difference operator

$$\nabla y(t) = y(t) - y(t - 1)$$

Rewriting the equation we find

$$y(t) = \frac{b_0 + b_1 q^{-1} + \cdots + b_{n-1} q^{-n+1}}{1 + a_1 q^{-1} + \cdots + a_n q^{-n}} u(t - k)$$
$$+ \lambda \frac{1 + c_1 q^{-1} + \cdots + c_n q^{-n}}{(1 - q^{-1})(1 + a_1 q^{-1} + \cdots + a_n q^{-n})} e(t) \quad (6.7)$$

The time interval in all cases has been 0.01 hour. All examples are based on data of Fig. 6.11 in the time interval 21.53 to 22.58 hours.

As stated previously, the identification procedure is carried out recursively, starting with a first-order system, continuing with a second order system, etc. To obtain the value of k for a fixed order, the identification is also repeated with the input signal shifted.

EXAMPLE 1—*Model Relating Dry Basis Weight to Thick Stock Flow*

The first numerical example will be a model relating dry basis weight to thick stock flow. First we shall identify a first order model having the structure (6.6). Applying our numerical identification algorithm we get the results shown in Table I.

An Industrial Application

TABLE I—Successive parameter iterates for a first-order model relating dry basis weight to thick stock flow; $k = 4$, $N = 101$.

Step	a_1	b_0	c_1	V	$\frac{\partial V}{\partial a} \times 10^5$	$\frac{\partial V}{\partial b} \times 10^5$	$\frac{\partial V}{\partial c} \times 10^5$
0	0	0	0	6.7350	91683	39509	−91683
1	−0.0122	13.0054	0	4.1603	0	0	193777
2	−0.3924	13.9356	−0.6320	3.3764	−78727	1190	51707
3	−0.3492	14.6689	−0.6542	3.3360	1339	−69	2575
4	−0.3502	14.6468	−0.6572	3.3360	106	−3	−165
5	−0.3500	14.6468	−0.6569				

Starting with the initial parameter estimate $\theta = 0$, the first step of the identification algorithm gives the least squares estimate of the parameters, and this estimate is then successively improved until the loss function $V(\theta)$ of (6.2) is minimized and the maximum likelihood estimate obtained. Notice, in particular, the significant difference between the least squares estimate (Step 1) and the maximum likelihood estimate. Also notice the convergence rate.

The value of the matrix of second partial derivatives at the last step of the iteration is

$$V_{\theta\theta} = \begin{bmatrix} 19.28 & -0.29 & -8.86 \\ -0.29 & 0.04 & 0.06 \\ -8.86 & 0.06 & 12.05 \end{bmatrix}$$

Repeating the identification for different values of the time-delay k we obtain the results given in Table II.

TABLE II—Results of identification of first-order models relating dry basis weight to thick stock flow for different time-delays.

k	a_1	b_0	c_1	λ	V
1	−0.807	9.846	−0.994	0.297	4.491
4	−0.350	14.647	−0.657	0.257	3.336
5	−0.749	1.286	−0.958	0.351	6.152

We thus find that the loss-function V has its smallest value for $k = 4$. To find the accuracy of the model parameters we proceed as follows:
An estimate of Fisher's information matrix is obtained from the matrix of second partial derivatives

$$\hat{I} = \lambda^{-2} V_{\theta\theta}$$

It can be shown that the estimate is asymptotically normal with mean θ_0 and covariance $V_{\theta\theta} I^{-1}$ if $V_{\theta\theta}$ is nonsingular.

We have the following estimate of the covariance of the asymptotic distribution:

$$\hat{I}^{-1} = \lambda^2 V_{\theta\theta}^{-1} = \begin{bmatrix} 0.006 & 0.042 & 0.004 \\ 0.042 & 2.202 & 0.020 \\ 0.004 & 0.020 & 0.008 \end{bmatrix}$$

Summarizing, we thus find the following numerical values for the best first-order model (the computations are based on 100 pairs of input-output data).

$k = 4$ $\quad\quad\quad c = -0.66 \pm 0.09$
$a = -0.35 \pm 0.08 \quad\quad \lambda = 0.257 \pm 0.017$
$b = 14.6 \pm 1.5 \quad\quad V = 3.34 \pm 0.44$

Proceeding to a second-order model, the identification algorithm gives the following results, based again on 100 pairs of input-output data:

$k = 3 \quad\quad\quad\quad c_1 = -0.73 \pm 0.18$
$a_1 = -0.46 \pm 0.14 \quad\quad c_2 = 0.12 \pm 0.16$
$a_2 = 0.04 \pm 0.12 \quad\quad \lambda = 0.249 \pm 0.017$
$b_0 = 3.4 \pm 1.6 \quad\quad V = 3.15 \pm 0.43$
$b_1 = 12.3 \pm 2.2$

The matrix of second partial derivatives at the minimum is:

$$V_{\theta\theta} = \begin{bmatrix} 22.47 & 13.82 & -0.08 & 0.36 & -7.61 & -1.87 \\ 13.83 & 22.47 & -0.17 & -0.08 & -4.94 & -7.59 \\ -0.08 & -0.17 & 0.04 & 0.02 & 0.05 & -0.05 \\ 0.36 & -0.08 & 0.02 & 0.04 & 0.06 & 0.06 \\ -7.61 & -4.94 & 0.05 & 0.06 & 11.06 & 6.60 \\ -1.87 & -7.59 & -0.05 & 0.06 & 6.60 & 10.56 \end{bmatrix}$$

Assuming that asymptotic theory can be applied, we can now solve various statistical problems. We will, for example, test the hypothesis that the model is of first order, i.e., our null hypothesis is

$$H_0: (a_2^0 = b_0^0 = c_2^0 = 0)$$

Using the asymptotic theory we find that the statistic

$$\xi = \frac{V_2 - V_1}{V_2} \cdot \frac{N - 6}{3}$$

has an $F(3, N - 6)$ distribution under the null hypothesis. The symbol V_2 denotes the minimal value of the loss function for the second-order model, V_1, the minimal value for the first-order model, and N, the number of input-output pairs. In this particular case, we have $\xi = 1.9$. At a risk

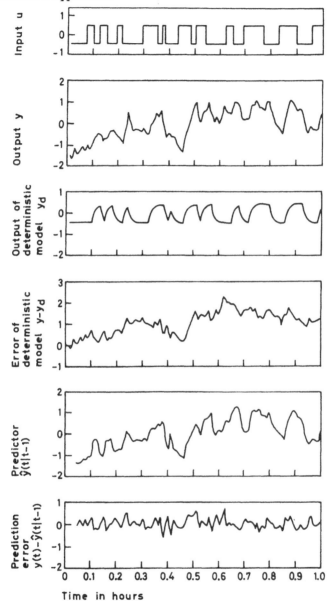

Fig. 6.12. Illustration of the results of the identification of a first-order model relating dry basis weight to thick stock flow. Notice, in particular, the relative magnitudes of the output of the deterministic model, the error of the deterministic model, and the error of the one-step-ahead predictor. Also notice the trend in the error of the deterministic model.

level of 10% we have $F(3, 96) = 2.7$, and the null hypothesis—that the system is of first-order—thus has to be accepted.

The results of the identification procedure are illustrated in Fig. 6.12. In this figure we show

- the input u
- the output y
- the deterministic output y_d defined by

$$y_d(t) = \frac{b_0 + b_1 q^{-1} + \cdots + b_{n-1} q^{-n+1}}{1 + a_1 q^{-1} + \cdots + a_n q^{-n}} u(t)$$

- the error of the deterministic model $e_d(t) = y(t) - y_d(t)$
- the one-step-ahead predictor $\hat{y}(t|t-1)$ of $y(t)$
- the one-step-ahead predictor error $y(t) - \hat{y}(t|t-1)$

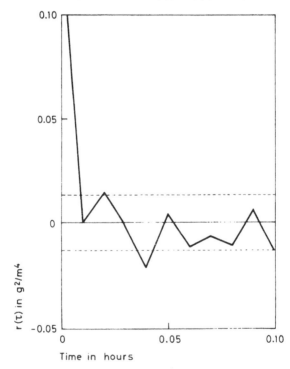

Fig. 6.13. Sample covariance function for the residual $\varepsilon(t)$ of the first-order model for basis weight. According to the assumption made in the identification theory, $r(\tau)$ should equal zero when $\tau \neq 0$. The dashed line gives the one sigma bound for $r(\tau)$, $\tau \neq 0$.

An Industrial Application

Figure 6.12 illustrates the properties of the identification procedure. The deterministic output $y_d(t)$ shows how much of the output $y(t)$ can be explained by the input $u(t)$. The error $e_d(t)$ thus represents the part of the output that is caused by the disturbances. Notice in particular the drifting character of the error e_d. The one-step-ahead predictor illustrates how well the output can be predicted one-step-ahead. Recall that the model was in fact constructed so as to minimize the sum of squares of the one-step-ahead prediction error.

The identification procedure was based on the assumption that the residuals were normal and uncorrelated. Having performed the identification and calculated the residuals or the one-step-ahead prediction errors $\varepsilon(t)$, we thus have the possibility of checking this assumption. Figure 6.13 shows the correlation function of the one-step-ahead prediction errors.

EXAMPLE 2–*Model Relating Wet Basis Weight to Thick Stock Flow*

As our second illustration of the numerical identification procedure, we will now use the data of Fig. 6.11 to find a model relating wet basis weight to thick stock flow. In this case, we find that the minimum value of the loss function for the first-order case occurs at $k = 4$, and the coefficients of the best first-order model are:

$$
\begin{aligned}
k &= 4 & c_1 &= -0.62 \pm 0.10 \\
a_1 &= -0.38 \pm 0.05 & \lambda &= 0.364 \pm 0.025 \\
b_0 &= 27.1 \pm 2 & V &= 6.60 \pm 0.94
\end{aligned}
$$

Similarly the best second-order model is given by the coefficients:

$$
\begin{aligned}
k &= 3 & c_1 &= -0.82 \pm 0.14 \\
a_1 &= -0.64 \pm 0.11 & c_2 &= -0.21 \pm 0.14 \\
a_2 &= 0.22 \pm 0.09 & \lambda &= 0.335 \pm 0.024 \\
b_0 &= 6.4 \pm 2.0 & V &= 5.73 \pm 0.80 \\
b_1 &= 20.2 \pm 3.0
\end{aligned}
$$

The matrix of second-order partial derivatives at the minimal point is

$$
V_{\theta\theta} = \begin{bmatrix}
79.24 & 53.37 & -0.13 & 0.76 & -12.68 & -0.13 \\
53.37 & 79.12 & -0.40 & -0.13 & -5.93 & -11.44 \\
-0.13 & -0.40 & 0.04 & 0.02 & 0.06 & -0.07 \\
0.76 & -0.13 & 0.02 & 0.04 & 0.10 & 0.10 \\
-12.68 & -5.93 & 0.06 & 0.10 & 17.64 & 7.83 \\
-0.13 & -11.44 & -0.07 & 0.10 & 7.83 & 15.12
\end{bmatrix}
$$

We now test the null hypothesis that the system is of first order

$$H: (a_2^o = b_0^o = c_2^o = 0).$$

Using the asymptotic results, we find $\xi = 4.8$, and the hypothesis thus has to be rejected. Increasing the order to three does not give any significant improvements in the loss function.

Hence, if we consider dry basis weight as the output of the system, we find that the model is of first order, but if we consider wet basis weight as the output, the model is of second order. This also shows up very clearly in Fig. 6.14 where we illustrate the results of the identification of the models for wet basis weight. There is a physical explanation for this difference in behavior. As mentioned previously, and as can be seen from

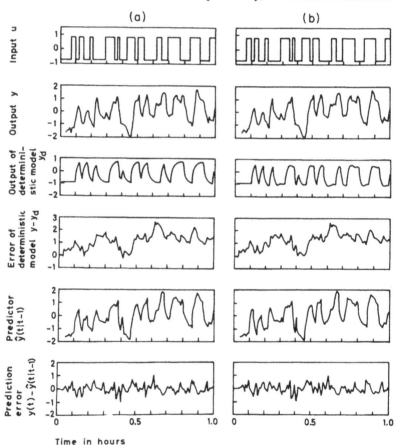

Fig. 6.14. Illustration of the results of the identification of models for wet basis weight. A first-order model is shown in (a) and a second-order model in (b). Notice in particular the differences between the outputs of the deterministic models for first- and second-order system.

Fig. 6.11, a change in thick stock flow will influence dry basis weight as well as moisture content. After an increase in thick stock flow, we find that both dry basis weight and moisture content will increase. The increase in moisture content will then be eliminated by the moisture control feedback loop which controls the set point of the fourth drying section by feedback from the moisture gauge.

These two effects will explain the overshoot in the response of the wet basis weight. It is also clear from this discussion that the response of the wet basis weight will be influenced by the settings of the moisture control loop. This fact is another argument for using dry basis weight as the control variable, when the basis weight loop is considered as a single-input, single-output system.

Practical Experiences with On-line Basis Weight Control

We shall now summarize some of the practical results achieved with on-line basis weight control. The experimental program that was carried out had a dual purpose: to arrive at control strategies for the particular application at hand and to test the applicability of stochastic control theory to practical control problems. This dual purpose led us to continue some experiments, even though these particular loops have been working satisfactorily. Several control schemes have been investigated. We have chosen thick stock flow as well as thick stock concentration as control variables. We have regulated both wet and dry basis weight. In the first experiments, the concentration of the thick stock was chosen as the control variable. This was later changed to thick stock flow mainly for two reasons. We found that the basis weight responds faster to changes in the set point of the thick stock flow regulator than to changes in the thick stock concentration. We also found that the dynamics of the concentration regulator changed with operating conditions, thereby introducing variations in the dynamics of the control loop.

In general it is very difficult to evaluate the performance of the control loops in practice, and in particular it is difficult to compare different control laws. The main reason for this is that there are variations in the disturbance level. This implies that in order to evaluate the different control loops we need test periods of considerable length.

It is also very difficult to judge the improvement unless reference values are available. In the case of basis weight, we had the results of the feasibility study. In all cases studied before the control computer was installed, standard deviation of basis weight was greater or equal to 1.3 g/m^2, and this value was therefore chosen as a conservative reference value. In the feasibility study, the target value for basis weight fluctuations was set to 0.7 g/m^2. In actual operation we can now consistently achieve standard deviations of 0.5 g/m^2 wet basis weight and 0.3 g/m^2 dry basis weight.

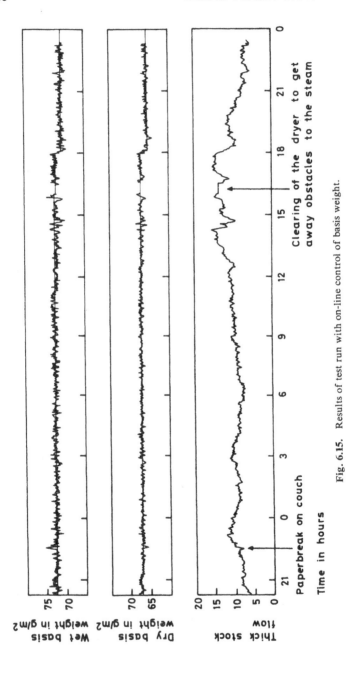

Fig. 6.15. Results of test run with on-line control of basis weight.

An Industrial Application

Basis weight was controlled successfully on-line on April 28, 1965 for a test period of 10 hours. The first experiments showed that it was indeed possible to obtain the variances predicted from the results of the process identification. We could also show that the deviations for the controlled system were moving averages of the appropriate order. The basis weight control loops has been subject to extensive investigations and has been in continuous operation since the beginning of 1966.

Two types of experiments have been performed. In one, the control loop is permitted to operate in the normal way for several weeks. Some data are collected at comparatively long sampling intervals (0.1 hour). The results are not analyzed extensively, and the performance of the control system is evaluated on the basis of the maximum deviations of test laboratory data, inspection of strip-chart recorders, and the judgement of machine tenders.

The other type of experiment is a controlled experiment extending over periods of 30 to 100 hours. Important process variables are logged at a sampling interval of 0.01 hour and analyzed. When analyzing the data, we compute covariance functions of the controlled variables and test whether they are moving averages of appropriate order (Theorem 4.1). Variances were checked against reference values. In some cases we also identify dynamic models, calculate minimum variance control strategies, and update the parameters of the control algorithms, if required.

In Fig. 6.15 we give a sample covering 24 hours of operation of the basis weight control loop. In the diagram we show wet basis weight, dry basis weight (the controlled output), and thick stock flow (the control signal). The scale for the control signal (thick stock flow) is chosen as dry basis weight. The magnitude of the control signal will thus directly indicate how much of the fluctuations in dry basis weight are removed by the control law. The control signal will thus approximately show the disturbances in the output of the system. Notice the different characteristics of the disturbances at different times. The large disturbances occurring at times 14.30 and 18.00 are due to large fluctuations in thick stock consistency.

Also notice that there are two interrupts in the operation of the system, one paper break and one interrupt to clear the drying section. In these instances the basis weight control loop is automatically switched off and the control signal is kept constant until the disturbances are cured, when the loops are automatically switched on again. Notice that a paper break does not introduce any serious disturbances. Also notice that there are some grade changes from which we can judge the response of the controlled system to step changes in the references values.

Moisture content was controlled by feedback from the moisture meter to the set point of the pressure regulator of the fourth drying section. The standard deviation of moisture content was 0.4%. Figure 6.16 shows the covariance function of dry basis weight in the time interval. As is to be expected from the Theorem 4.1, this is the covariance function of a moving average of fourth order.

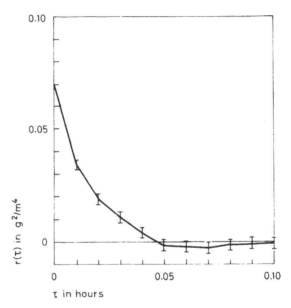

Fig. 6.16. Covariance function for fluctuations of dry basis weight in the time interval 23.00 to 12.00 of Fig. 6.15. Compare with Fig. 6.10 which shows the covariance function without computer control.

We have also made experiments to verify that the high frequency fluctuations in moisture content and basis weight have the same characteristics. This was one essential assumption made. If the assumption is true, the variance in dry basis weight would be independent of dry or wet basis weight control. In Table III we give standard deviations recorded during a 30 hour test, where alternatively wet and dry basis weight was controlled.

TABLE III—Standard deviations for wet basis weight (controlled) and dry basis weight (controlled) recorded over 30 hrs.

	Standard deviation	
	Wet basis weight	Dry basis weight
Wet basis weight controlled	0.50	0.38
Dry basis weight controlled	0.52	0.28

Exercise

1. Determine the minimum variance control strategies for the models of of Example 1 and Example 2 of Section 4. Discuss the sensitivity of the optimal strategies with respect to parameter variations.

7. BIBLIOGRAPHY AND COMMENTS

The results of this chapter are based on

Aström, K. J., "Notes on the Regulation problem," Report CT211, IBM Nordic Laboratory, August 30 1965.

The algorithm for the k step predictor given in Theorem 3.1 is believed to be new. The special case $k = 1$ was given in

Åström, K. J., and Bohlin, T., "Numerical Identification of Linear Dynamic System from Normal Operating Records," in *Theory of Self-Adaptive Control Systems*, P. H. Hammond (ed.) pp. 94-111, Plenum Press, New York, 1966.

Theorem 4.1 was published in

Aström, K. J., "Computer Control of a Paper Machine—an Application of Linear Stochastic Control Theory," *IBM J. Res. Develop.* 11, 389-405 (1967).

The practical example in Section 6 is also based upon this paper.

The idea of solving control and prediction problems by representing the stochastic processes as sequences of independent equally distributed stochastic variables is also given in

Whittle, P., *Prediction and Regulation*, English Univ. Press, London, 1963.

A related approach to the regulation problem has also been given by Box and Jenkins, e.g.,

Box, G.E.P. and Jenkins, G.M., *Time Series Analysis, Forecasting and Control*, Holden Day, San Fransisco, 1970.

Additional material on the control of the paper machine is found in

Åström, K. J., "Control Problems in Papermaking," *Proc. IBM Sci. Comput. Symp. Control Theory and Appl.* October 19-21, 1964.

Ekström, A., "Integrated Computer Control of a Paper Machine-System Summary," *Proc. Billerud, IBM Symp.* June, 1966.

The maximum likelihood method for process identification is described in full detail in

Åström, K. J., Bohlin, T. and Wensmark, S., "Automatic Construction of Linear Stochastic Dynamic Models for Stationary industrial Processes with Random Disturbances using Operating Records," Tech. Paper TP 18.150, IBM Nordic Laboratory (1965).

CHAPTER 7

PREDICTION AND FILTERING THEORY

1. INTRODUCTION

We found in Chapter 6 that the stochastic control problem was closely related to the statistical prediction problem. To be specific, we found for the simple regulator problem that the optimal control strategy was obtained simply by computing the k-step prediction and choosing a control signal such that the k-step prediction agrees with the desired output. In this chapter we will study the prediction problem in more detail. The purpose is to derive those results of filtering and prediction theory which are required in order to solve the linear quadratic control problem in the general case. Filtering and prediction theory is a very rich subject. We have no intention to cover the field but we will concentrate on those results which are needed for the stochastic control problem.

The formulation of filtering and prediction problems is treated in Section 2. Particular attention is given to the problem of estimating the state of a dynamical system. Relationships between different problem formulations are also given. It is shown that in many cases the best estimate is given by the conditional mean of the signal to be predicted, given all observed data.

In Section 3 we derive properties of conditional distributions for multivariate normal distributions. A geometric interpretation of the result is also given.

State estimation for discrete time systems is discussed in Section 4. This problem is closely related to the state reconstruction problem of Chapter 5. It is shown that the reconstructors obtained in Chapter 5 are in fact optimal. The main result is the formula for the Kalman filter for estimating the state. This result is derived using the properties of multivariate Gaussian distributions derived in Section 3. Section 5 shows that the filtering problem is the dual of a deterministic optimal control problem. The idea of duality is then exploited in Section 6 to obtain a solution to the continuous time version of the state estimation problem.

2. FORMULATION OF PREDICTION AND ESTIMATION PROBLEMS

The problem we will discuss can be formulated as follows: Consider two real stochastic processes $\{s(t), t \in T\}$ and $\{n(t), t \in T\}$ which are referred to as signal and noise respectively. Assume that the sum

$$y(t) = s(t) + n(t)$$

can be observed or measured. At time t we have thus obtained a realization $y(\tau)$, $t_0 \leqslant \tau \leqslant t$ of the measured variable. Based on this realization, we will determine the best estimate of the value of the signal at time t_1. If $t_1 < t$, the problem is called a *smoothing* problem or an *interpolation* problem. If $t_1 = t$, it is called a *filtering* problem, and if $t_1 > t$, a *prediction* problem.

In a slightly more general formulation, it is required to estimate a functional of the signal such as ds/dt or $\int s \, dt$ from observations of signal plus noise.

It is often convenient to have a notation for the set of observations which are available to estimate $s(t_1)$. This is called $\mathscr{Y}$. For discrete time processes, T is the set of integers and $\mathscr{Y}$ is simply a vector,

$$\mathscr{Y} = \left(y^T(t_0), y^T(t_0 + 1), \ldots, y^T(t) \right)^T.$$

For continuous time processes, $\mathscr{Y}$ is a function defined on the interval (t_0, t). Sometimes we would also like to indicate explicitly that $\mathscr{Y}$ depends on t. To do this we use $\mathscr{Y}_t$. Let $\mathscr{Y} \in Y$ and $s \in S$. An *estimator* (interpolator, filter, predictor) is a function which maps Y into S. The value of this function for a particular value of $\mathscr{Y}$ is called an estimate and is denoted by $\hat{s}$.

To describe any of these problems completely we must specify several things.

- The signal and noise processes
- The criterion which defines the best estimate
- The restrictions on the admissible estimators

The signal and noise processes can be specified in many different ways. Apart from the distinction of continuous time and discrete time, the processes can, for example, be characterized by covariance functions, spectral densities, stochastic difference or differential equations.

There are also many possibilties to define the "best" estimate. We can, for example, define a loss function l, which is a real function with the properties $l \geqslant 0$, $l(x) = l(-x)$, and l nondecreasing for $x > 0$. The loss is then the stochastic variable $l(s - \hat{s})$ and the best estimate $\hat{s}$ is the one which minimizes the average loss $E\, l(s - \hat{s})$. There are, of course, also other ways to define the best estimate. We have previously defined the estimator as a function which maps Y into S. This function can be required to be linear or to have other special properties.

State Estimation

There are apparently many possibilities to formulate the estimation problem. The special case when the signal and noise processes can be represented by stochastic difference or stochastic differential equations is of particular interest to stochastic control theory. In the discrete time case we thus have

$$x(t + 1) = \Phi x(t) + v(t) \tag{2.1}$$
$$y(t) = \theta x(t) + e(t) \tag{2.2}$$

where $\{v(t)\}$ and $\{e(t)\}$ are sequences of independent Gaussian random variables. In the continuous time case we have

$$dx = Ax\, dt + dv \tag{2.3}$$
$$dy = Cx\, dt + de \tag{2.4}$$

where $\{v(t)\}$ and $\{e(t)\}$ are Wiener processes. It will be assumed that a realization of the output $y(\tau)$, $t_0 \leqslant \tau \leqslant t$ has been observed, and that we want to estimate the state vector of (2.1) or (2.3). This particular problem is called the state-estimation problem.

Preliminaries

Before we go into specific details, we will make a few observations which will enable us to find the equivalence between different formulations of the estimation problem. We first notice that all relevant statistical information which the observations give about the stochastic variable $s(t_1)$ is contained in the conditional distribution

$$P\{s(t_1) \leqslant \sigma \,|\, y(\tau) = \eta(\tau),\, t_0 \leqslant \tau \leqslant t\} = F(\sigma\,|\,\eta) \tag{2.5}$$

The density of the distribution is denoted by $f(\sigma\,|\,\eta)$. Further, assume that the best estimate is defined as the one which minimizes the mean value of a loss function l.

To find the best estimator we thus have to find a function $\hat{s} = \hat{s}(\eta)$ such that the criterion

$$El(s - \hat{s}) \tag{2.6}$$

is minimal. To do this we rewrite the criterion in such a way that the dependence of $\hat{s}$ on η is explicit. We have

$$El(s - \hat{s}) = E_\eta[E\{l(s - \hat{s}) \mid \eta\}] \tag{2.7}$$

where $E\{\cdot \mid \eta\}$ denotes the conditional expectation given $y(\tau) = \eta(\tau)$, $t_0 \leqslant \tau \leqslant t$. To minimize (2.7) with respect to all functions, $\hat{s} = \hat{s}(\eta)$ is thus equivalent to minimize

$$E\{l(s - \hat{s}) \mid \eta\} = \int_{-\infty}^{\infty} l(\sigma - \hat{s}) f(\sigma \mid \eta) \, d\sigma \tag{2.8}$$

We have the following fundamental result.

THEOREM 2.1

Assume that the conditional distribution of $s(t_1)$ given $y = \eta$ has a density function which is symmetric around the conditional mean $m = \int \sigma f(\sigma \mid \eta) \, d\sigma$ and nonincreasing for $\sigma \geqslant m$. Let the loss function l be symmetric and nondecreasing for positive arguments. The best estimate is then given by the conditional mean

$$\hat{s} = \hat{s}(\eta) = E\{s \mid \eta\} = \int \sigma f(\sigma \mid \eta) \, d\sigma \tag{2.9}$$

Proof

The proof is based on an elementary lemma for real functions.

LEMMA

Let g and h be two integrable real functions with the properties

$$g(x) \geqslant 0, \quad h(x) \geqslant 0$$
$$g(x) = g(-x) \quad \text{and} \quad h(x) = h(-x)$$
$$g(x) \text{ nondecreasing for } x \geqslant 0$$
$$h(x) \text{ nonincreasing for } x \geqslant 0$$

Then

$$\int_{-\infty}^{\infty} g(x + a) h(x) \, dx \geqslant \int_{-\infty}^{\infty} g(x) h(x) \, dx$$

if the integrals exist.

Proof

Assume that the integrals exist and that $a \geqslant 0$. Then

$$\int_{-\infty}^{\infty} [g(x + a) h(x) - g(x) h(x)] \, dx$$
$$= \int_{-\infty}^{-a/2} [g(x + a) - g(x)] h(x) \, dx + \int_{-a/2}^{\infty} [g(x + a) - g(x)] h(x) \, dx$$

$$= \int_{a/2}^{\infty} [g(x-a) - g(x)] h(x)\, dx + \int_{a/2}^{\infty} [g(x) - g(x-a)] h(x-a)\, dx$$

$$= \int_{a/2}^{\infty} [g(x) - g(x-a)][h(x-a) - h(x)]\, dx$$

The first equality is obtained by splitting the integration interval. The second equality is obtained by substituting $x \to -x$ in the first integral and $x \to x - a$ in the second integral. Now consider the function $g(x) - g(x-a)$. We have

$$g(x) - g(x-a) = g(x) - g(a-x) \geqslant 0, \quad \text{for } a/2 \leqslant x \leqslant a$$

because g is nondecreasing for positive arguments. For the same reason we also find

$$g(x) - g(x-a) \geqslant 0 \quad \text{for } x \geqslant a$$

Hence

$$g(x) - g(x-a) \geqslant 0 \quad \text{for } x \geqslant a/2$$

In the same way we find that

$$h(x) - h(x-a) \leqslant 0 \quad \text{for } x \geqslant a/2$$

The integrand $[g(x) - g(x-a)][h(x-a) - h(x)]$ is thus nonnegative over the whole range of integration and the lemma is proven.

To prove Theorem 2.1 we now observe that the best estimate $\hat{s}$ is obtained by minimizing the function

$$E\{l(s - \hat{s}) \mid \eta\} = \int_{-\infty}^{\infty} l(\sigma - \hat{s}) f(\sigma \mid \eta)\, d\sigma = \int_{-\infty}^{\infty} l(t + m - \hat{s}) f(t + m \mid \eta)\, dt$$

The last equality is obtained by substituting $t = \sigma - m$. Put $g(x) = l(x)$ and $h(x) = f(t + m \mid \eta)$, and we find that the conditions of the lemma are fulfilled. We thus find

$$E[l(s - \hat{s}) \mid \eta] = \int_{-\infty}^{\infty} l(t + m - \hat{s}) f(t + m \mid \eta)\, dt \geqslant \int_{-\infty}^{\infty} l(t) f(t + m \mid \eta)\, dt$$

where equality is obtained for

$$\hat{s} = \hat{s}(\eta) = m = \int_{-\infty}^{\infty} \sigma f(\sigma \mid \eta)\, d\sigma$$

and the theorem is proven.

Remark 1

The theorem implies that if the conditional density satisfies the conditions of the theorem, the choice of loss function is immaterial as long as it is symmetric and nondecreasing for positive arguments.

Formulation of Prediction and Estimation Problems

Remark 2

Notice that the normal density satisfies the conditions of the theorem. For normal random variables, we also know that the conditional mean $E\{s|\eta\}$ is a linear function of η. For normal processes, the best estimate will thus be a linear function of the observations for all loss functions which satisfy the conditions of Theorem 2.1.

Remark 3

When the signal and noise processes are characterized as second order random processes, we can formulate an estimation problem using the criterion

$$E(s - \hat{s})^2 \quad \text{minimal}$$

if we also require that $\hat{s}$ is linear in η. It follows from Remark 2 that the best estimator will be the same as the one which is obtained, if the processes are assumed to be Gaussian, with the same first and second moments as the second order process, and no restrictions are given on the estimators.

Theorem 2.1 might lead us to belive that the conditional mean will always give the solution to prediction problems. The following example shows that this is indeed not the situation.

EXAMPLE 2.1

Consider the stochastic process defined by

$$dx = -x \, dw \qquad (2.10)$$

where $\{w(t), t \in T\}$ is a Wiener process with incremental variance $r \, dt$. We will determine the best predictor over the interval $(t, t + h)$ when the criterion is minimum mean square and the most probable value respectively.

As the equation is homogeneous in time, we can determine the predictor over $(0, h)$. The equation (2.10) has the solution

$$x(t) = e^{-[w(t) + \frac{1}{2}rt]} x(0)$$

The conditional distribution of $\log x(t)$ given $x(0)$ is thus normal

$$N(-\tfrac{1}{2}rt + \log x(0), \sqrt{rt})$$

and the conditional distribution of $x(t)$ given $x(0)$ is logarithmico normal. The density function is

$$f(\xi, t) = \frac{1}{\xi \sqrt{2\pi rt}} \exp\left\{ -\frac{[\log(\xi/x(0)) + \tfrac{1}{2}rt]^2}{2rt} \right\}$$

The mean value is

$$Ex(t) = \int_0^\infty \xi f(\xi, t) \, d\xi = x(0)$$

The mode of the distribution, that is, the ξ-value for which the density function $f(\xi, t)$ has its maximum, is given by

$$x_0(t) = e^{-\frac{3}{2}rt}x(0)$$

We thus find that the minimum mean square prediction of $x(t + h)$ based on observations of $x(t)$ is

$$\hat{x}(t + h|t) = x(t)$$

while the "most probable" prediction is

$$\hat{x}(t + h|t) = e^{-\frac{3}{2}rh}x(t)$$

Exercises

1. Let the signal $\{s(t), t \in T\}$ and the noise $\{n(t), t \in T\}$ be independent stationary stochastic processes with covariance functions $r_s(t)$ and $r_n(t)$ respectively. Asssume that the linear combination

   $$y = s + n$$

   is observed and that it is desirable to predict s over an interval of length h by the linear operation

   $$\hat{s}(t + h) = \int_{-\infty}^{t} g(t - \tau)y(\tau)\,d\tau = \int_{0}^{\infty} g(u)y(t - u)\,du$$

   Show that the mean square prediction error is given by

   $$E[s(t + h) - \hat{s}(t + h)]^2 = r_s(0) - 2\int_{0}^{\infty} g(u)r_s(h + u)\,du$$
   $$+ \int_{0}^{\infty} g(u)\,du \int_{0}^{\infty} g(v)[r_s(u - v) + r_n(u - v)]\,dv$$

   The prediction problem can thus be formulated as to minimize the mean square prediction error.

2. Consider the mean square prediction error obtained in Exercise 1 as a functional of the weighting function g of the predictor

   $$J[g] = r_s(0) - 2\int_{0}^{\infty} g(u)r_s(u + h)\,du$$
   $$+ \iint_{0}^{\infty} g(u)[r_s(u - v) + r_n(u - v)]g(v)\,du\,dv$$

   Show that the first and second variations of this functional are given by

   $$J[g + \delta g] = J[g] + J_1 + J_2$$

   where

Formulation of Prediction and Estimation Problems

$$J_1 = -2\int_0^\infty \delta g(u)\left\{r_s(u+h) - \int_0^\infty [r_s(u-v) + r_n(u-v)]g(v)\,dv\right\}du$$

$$J_2 = \int\int_0^\infty \delta g(u)\delta g(v)[r_s(u-v) + r_n(u-v)]\,du\,dv$$

Then show that a necessary and sufficient condition for g to be the weighting function of the optimal predictor is

$$r_s(t+h) - \int_0^\infty [r_s(t-v) + r_n(t-v)]g(v)\,dv = 0 \qquad t \geq 0 \qquad (*)$$

and $r_s + r_n$ positive definite. (The integral equation (*) is called the Wiener Hopf equation.)

3. Consider the scalar discrete time stochastic process $\{y(t), t \in T\}$ defined by

$$y(t) = \sum_{n=t_0}^{t} g(t,n)e(n)$$

where $\{e(t), t \in T\}$ is a sequence of equally distributed normal $(0, 1)$ random variables and $g(t, t) \neq 0$ for each t. Determine the best mean square k step predictor and the prediction error for the process.

Hint: $e(t_0), e(t_0+1), \ldots, e(t)$ can be computed exactly from $y(t_0)$, $y(t_0+1), \ldots, y(t)$. Compare Section 3 of Chapter 6.

4. Consider the scalar continuous time stochastic process $\{y(t), t \in T\}$ defined by

$$y(t) = \int_{t_0}^{t} g(t,s)\,dw(s)$$

where $\{w(t), t \in T\}$ is a Wiener process with unit variance parameter Determine the best mean square predictor over the interval $(t, t+h)$. Also determine the prediction error. Specialize to $g(t,s) = (t-s)e^{-(t-s)}$ and $t_0 = -\infty$.

Hint: The operator G defined by

$$Gu = \int_{-\infty}^{t} (t-s)e^{-(t-s)}u(s)\,ds$$

has the inverse

$$G^{-1}u = \frac{d^2u}{dt^2} + 2\frac{du}{dt} + u$$

The best predictor then becomes

$$\hat{y}(t+h) = \int_{-\infty}^{t} (t+h-s)e^{-(t+h-s)}\left[y(s)\,ds + 2\frac{dy}{ds}\,ds + d\left(\frac{dy}{ds}\right)\right]$$

Notice that dy/dt is continuous in the mean square. Partial integration then gives

$$\hat{y}(t + h) = (1 + h)e^{-h}y(t) + he^{-h}\frac{dy}{dt}$$

5. The optimal predictor of the problem of Exercise 4 contains a differentiator. Determine the best mean square predictor of the form

$$\hat{y}(t + h) = \alpha y(t)$$

for the problem of Exercise 4. Compare with the results of Exercise 4.

3. PRELIMINARIES

In this section we will develop some preliminary results which are required for the solution of the estimation problem. We will first derive some properties of conditional distributions of Gaussian random variables. We will then give a geometric interpretation of the results and we will finally give a brief discussion on the generalization to infinitely many variables.

Multivariate Gaussian Distribution

Let x be an n-dimensional normal vector with the mean value m and the covariance matrix R. The distribution is called *singular* if R is singular, and regular if R is regular. A singular distribution has all its mass in a hyperplane in n-dimensional space. Throughout this section we will assume that R is not singular. This is no serious restriction, because if the distribution is singular we can always make a projection on the hyperplane where the mass is concentrated and get a nonsingular distribution. Formally, the problem can be handled by introducing the pseudoinverse whenever the inverse of the covariance function is needed.

The probability density function of a normal variable with mean m and covariance R is given by

$$f(x) = (2\pi)^{-n/2}(\det R)^{-\frac{1}{2}} \exp -\tfrac{1}{2}(x - m)^T R^{-1}(x - m) \quad (3.1)$$

Compare with Chapter 2. It is often of interest to consider the properties of two vectors which are jointly Gaussian.

THEOREM 3.1

Let x and y be $n \times 1$ and $p \times 1$ vectors. Assume that the vector $\begin{bmatrix} x \\ y \end{bmatrix}$ is Gaussian with mean $\begin{bmatrix} m_x \\ m_y \end{bmatrix}$ and covariance $R = \begin{bmatrix} R_x & R_{xy} \\ R_{yx} & R_y \end{bmatrix}$

The vector

Preliminaries

$$z = x - m_x - R_{xy}R_y^{-1}(y - m_y) \tag{3.2}$$

is independent of y, has zero mean and the covariance

$$R_z = R_x - R_{xy}R_y^{-1}R_{yx} \tag{3.3}$$

Proof

We have

$$Ez = Ex - m_x - R_{xy}R_y^{-1}(Ey - m_y) = 0$$

Consider

$$Ez(y - m_y)^T = E(x - m_x)(y - m_y)^T - R_{xy}R_y^{-1}E(y - m_y)(y - m_y)^T$$
$$= R_{xy} - R_{xy}R_y^{-1}R_y = 0$$

The vectors z and y are thus uncorrelated. As they are Gaussian they are then also independent. As $Ez = 0$ we have

$$R_z = Ezz^T = E[x - m_x - R_{xy}R_y^{-1}(y - m_y)][x - m_x - R_{xy}R_y^{-1}(y - m_y)]^T$$
$$= E[x - m_x][x - m_x]^T - E(x - m_x)(y - m_y)^T R_y^{-1} R_{yx}$$
$$- E R_{xy} R_y^{-1}(y - m_y)(x - m_x)^T$$
$$+ E R_{xy} R_y^{-1}(y - m_y)(y - m_y)^T R_y^{-1} R_{yx} = R_x - R_{xy}R_y^{-1}R_{yx}$$

which completes the proof.

THEOREM 3.2

Let x and y be two vectors which are jointly Gaussian. The conditional distribution of x given y is normal with mean

$$E[x|y] = m_x + R_{xy}R_y^{-1}(y - m_y) \tag{3.4}$$

and covariance

$$E\{[x - E(x|y)][x - E(x|y)]^T | y\} = R_x - R_{xy}R_y^{-1}R_{yx} = R_z \tag{3.5}$$

The stochastic variables y and $x - E[x|y]$ are independent.

Proof

The theorem can be proven in a straightforward way using the definition of conditional density and the formula for the probability density of a Gaussian vector (3.1). The algebra is somewhat simpler if we use the variables z and y defined by

$$\begin{bmatrix} z \\ y - m_y \end{bmatrix} = \begin{bmatrix} I & -R_{xy}R_y^{-1} \\ 0 & I \end{bmatrix} \begin{bmatrix} x - m_x \\ y - m_y \end{bmatrix}$$

Hence

$$\begin{bmatrix} x - m_x \\ y - m_y \end{bmatrix} = \begin{bmatrix} I & R_{xy}R_y^{-1} \\ 0 & I \end{bmatrix} \begin{bmatrix} z \\ y - m_y \end{bmatrix}$$

As the Jacobian of the transformation is one, we find that the joint density of x and y can be written as

$$f(x, y) = (2\pi)^{-(n+p)/2}(\det R)^{-1/2} \exp - \tfrac{1}{2}[z^T R_z^{-1} z + (y - m_y)^T R_y^{-1}(y - m_y)]$$

where

$$R = \begin{bmatrix} R_x & R_{xy} \\ R_{yx} & R_y \end{bmatrix}$$

The density of y is

$$f(y) = (2\pi)^{-p/2}(\det R_y)^{-1/2} \exp - \tfrac{1}{2}(y - m_y)^T R_y^{-1}(y - m_y)$$

But

$$\det R = \det \begin{bmatrix} R_x & R_{xy} \\ R_{yx} & R_y \end{bmatrix} = \det \begin{bmatrix} R_x - R_{xy} R_y^{-1} R_{yx} & 0 \\ R_{yx} & R_y \end{bmatrix}$$
$$= \det(R_x - R_{xy} R_y^{-1} R_{yx}) \det R_y = \det R_z \det R_y$$

The second equality is obtained by subtracting the second row multiplied by $R_{xy} R_y^{-1}$ from the left, from the first row. The conditional density of x given y is then given by

$$f(x|y) = \frac{f(x, y)}{f(y)} = (2\pi)^{-n/2}(\det R_z)^{-1} \exp - \tfrac{1}{2} z^T R_z^{-1} z$$
$$= (2\pi)^{-n/2}(\det R_z)^{-1/2} \exp - \tfrac{1}{2}[x - m_x - R_{xy} R_y^{-1}(y - m_y)]^T$$
$$\times R_z^{-1}[x - m_x - R_{xy} R_y^{-1}(y - m_y)]$$

and we thus find that the conditional distribution is normal with mean (3.4) and covariance (3.5). The last statement of the theorem follows from Theorem 3.1.

THEOREM 3.3

Let x, u, and v be random vectors with a jointly Gaussian distribution, and let u and v be independent, then

$$E[x|u, v] = E[x|u] + E[x|v] - Ex \tag{3.6}$$

Proof

Put

$$y = \begin{bmatrix} u \\ v \end{bmatrix}$$

then

$$R_y = \begin{bmatrix} R_u & 0 \\ 0 & R_v \end{bmatrix}$$

and

Preliminaries

$$\text{cov}[x, y] = E[x - m_x][y - m_y]^T = E(x - m_x)\begin{bmatrix} u - m_u \\ v - m_v \end{bmatrix}^T = (R_{xu}, R_{xv})$$

Hence

$$R_{xy}R_y^{-1} = (R_{xu}, R_{xv})\begin{bmatrix} R_u^{-1} & 0 \\ 0 & R_v^{-1} \end{bmatrix} = (R_{xu}R_u^{-1}, R_{xv}R_v^{-1})$$

Using Theorem 3.2 we get

$$E[x|u, v] = m_x + (R_{xu}R_u^{-1}, R_{xv}R_v^{-1})\begin{bmatrix} u - m_u \\ v - m_v \end{bmatrix}$$
$$= m_x + R_{xu}R_u^{-1}(u - m_u) + R_{xv}R_v^{-1}(v - m_v)$$
$$= E[x|u] + E[x|v] - m_x$$

It is instructive to interpret these theorems as estimation problems. To do so we take two stochastic vectors x and y, which are jointly Gaussian. Consider the problem of estimating x from observations of y in such a way that the estimation error is as small as possible in the sense of mean square. Let $\hat{x}$ denote the best estimate and $\tilde{x}$ the estimation error. According to Theorem 2.1, the best estimate is then given by the conditional mean $E[x|y]$. Theorem 3.2 further implies that the best estimate is given by

$$\hat{x} = E[x|y] = m_x + R_{xy}R_y^{-1}(y - m_y)$$

and that the estimation error has the covariance

$$E[\tilde{x}\tilde{x}^T|y] = R_x - R_{xy}R_y^{-1}R_{yx}$$

Theorem 3.1 and Theorem 3.2 also imply that the estimation error

$$\tilde{x} = x - \hat{x} = x - E[x|y] = x - m_x - R_{xy}R_y^{-1}(y - m_y)$$

is independent of y.

To give an interpretation of Theorem 3.3, we will consider the problem of estimating x based on observations of u and v where u and v are independent. Let $\hat{x}(u)$, $\hat{x}(v)$ and $\hat{x}(u, v)$ denote the best estimates of x based on u, v and (u, v) respectively. Theorems 3.2 and 3.3 then imply that the best estimate of x based on u and v is given by

$$\hat{x}(u, v) = \hat{x}(u) + \hat{x}(v) - m_x$$

Geometric Interpretations

The theorems on multivariable Gaussian distributions, which have been derived previously, can be given geometric interpretations which have a strong intuitive appeal. To show this, we will first consider the simplest case of two stochastic variables x and y. For the sake of simplicity we will assume that both variables have zero mean. We represent the variables x

and y as elements in an Euclidean space whose scalar product is defined by

$$(x, y) = \text{cov}\,[x, y] \tag{3.7}$$

The norm is thus given by

$$\|x\|^2 = (x, x) = \text{cov}\,[x, x] \tag{3.8}$$

To be specific we thus take two lines l_1 and l_2 which intersect at the origin, where the angle θ between the lines is given by

$$\cos\theta = \frac{(x, y)}{\|x\| \cdot \|y\|} \tag{3.9}$$

The stochastic variable x is represented as a vector along l_1 with the length $\|x\| = \sqrt{Ex^2}$ and the stochastic variable y is represented by a vector along l_2 with the length $\|y\| = \sqrt{Ey^2}$. Compare Fig. 7.1.

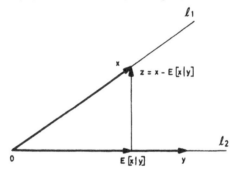

Fig. 7.1. Geometric illustration of the conditional mean values of normal random variables. The conditional mean $E[x|y]$ is represented by the projection of x on y.

Recalling the assumptions of zero mean values we find that Theorem 3.1 implies that the stochastic variable defined by

$$z = x - R_{xy}R_y^{-1}y \tag{3.10}$$

is independent of y. Hence

$$(z, y) = \text{cov}\,[z, y] = 0$$

Theorem 3.1 thus implies that z is orthogonal to y and that the norm of z is

$$\|z\|^2 = R_x - R_{xy}R_y^{-1}R_{xy} = \|x\|^2 - \frac{(x, y)^2}{\|y\|^2}$$

Notice that the projection of x on y is given by

$$\left(x, \frac{y}{\|y\|}\right)\frac{y}{\|y\|} = \frac{(x, y)y}{\|y\|^2} = R_{xy}R_y^{-1}y = E[x|y] \tag{3.11}$$

Preliminaries

where the second equality follows from (3.7) and (3.8), and the last equality follows from Theorem 3.2.

The variable $x - z = R_{xy}R_y^{-1}y = E[x|y]$, which equals the best mean square estimate of x based on y, can thus be interpreted geometrically as the projection of x on y. See Fig. 7.1.

To interpret Theorem 3.3 we introduce a three-dimensional Euclidean space. As u and v are independent we find that the corresponding vectors are orthogonal. Using the geometric interpretation of Theorem 3.2 we find that $E[x|u, v]$ can be interpreted as the orthogonal projection of x on the linear space spanned by u and v, that $E[x|u]$ can be interpreted as the orthogonal projection of x on u, and $E[x|v]$ the projection of x on v. Theorem 3.2 thus implies that the projection of x on a two-dimensional subspace is the sum of the projections on an orthogonal basis in the subspace. The extension of these interpretations to higher dimensions is straightforward.

Notice that the geometric interpretations depend only on the properties of the second moments of the stochastic variables. This implies that it is not necessary to assume that the distributions are Gaussian. However, if the distributions are not Gaussian, the projections cannot be interpreted as conditional expectations. To give a probabilistic interpretation of the projections, we will use the following property.

THEOREM 3.4 (Projection Theorem)

Let $y_1, y_2, \ldots, y_l$ and x be elements in Euclidean space, and let Y be the linear subspace spanned by $y_1, y_2, \ldots, y_l$. Then there exists a unique element $\hat{x} \in Y$ such that

$$\|x - \hat{x}\| = \inf_{z \in Y} \|x - z\|$$

Proof

Let $\hat{x}$ be the orthogonal projection of x on Y, then

$$(x - \hat{x}, y_i) = 0 \quad \text{for} \quad i = 1, 2, \ldots, l$$

Take any $z \in Y$, that is,

$$z = \sum_{i=1}^{l} \alpha_i y_i$$

then

$$(x - \hat{x}, z) = (x - \hat{x}, \sum \alpha_i y_i) = \sum \alpha_i (x - \hat{x}, y_i) = 0$$

Hence

$$\|x - z\|^2 = (x - z, x - z) = (x - \hat{x} - (z - \hat{x}), x - \hat{x} - (z - \hat{x}))$$
$$= (x - \hat{x}, x - \hat{x}) - 2(x - \hat{x}, z - \hat{x}) + (z - \hat{x}, z - \hat{x})$$

where the second term vanishes because $z - \hat{x} \in Y$. We thus find

$$\|x - z\|^2 = \|x - \hat{x}\|^2 + \|z - \hat{x}\|^2$$

Hence

$$\|x - z\|^2 \geqslant \|x - \hat{x}\|^2$$

where equality is obtained for $z = \hat{x}$.

Thus if we have stochastic variables which are not Gaussian, the projection of x on Y can be interpreted as the linear estimate of x based on observations of $y_1, y_2, \ldots, y_l$ which minimizes the criterion

$$\|x - \hat{x}\|^2 = E(x - \hat{x})^2$$

Exercises

1. A stochastic variable x has a normal distribution $N(a, \sigma_0)$. To determine x it is measured with a measuring instrument. The measuring error can be modeled as a normal stochastic variable $N(0, \sigma)$ which is independent of x. Determine the estimate of x, which is best in the sense that it minimizes $E|x - \hat{x}|$ from the a priori knowledge and the measurement.

2. Consider the problem of Exercise 1, but assume that two measurements are used. The measurement errors are assumed to be independent and normal, $N(0, \sigma_1)$ and $N(0, \sigma_2)$ respectively. Determine the estimate and its variance. Also determine the limit of the estimate as $\sigma_0 \to \infty$.

3. A stochastic vector x is normal with mean m_x and covariance R_x. With a measuring instrument we can measure the linear combination

$$y = Cx + e$$

where the measuring error e is independent of x and normal with zero mean and covariance R_e. What is the best mean square estimate of x obtained by combining the measured value with the a priori information? What is the variance of the estimate?

4. Let x and y be normal stochastic vectors. Show that it is always possible to find a matrix A of appropriate dimensions and a normal stochastic vector e such that

$$x = Ay + e$$

where e is independent of y. Give a probabilistic interpretation of the quantity Ay.

5. Consider the scalar stochastic processes

State Estimation for Discrete Time Systems

$$x(t) = \sum_{s=t_0}^{t} g(t, s) v(s)$$

$$y(t) = \sum_{s=t_0}^{t} h(t, s) e(s)$$

where $\{v(t)\}$ and $\{e(t)\}$ are correlated sequences of independent normal $N(0, 1)$ variables and $h(t, t) \neq 0$. Determine the best mean square estimate of $x(t)$ based on $y(t), y(t-1), \ldots, y(t_0)$.

Hint: Use result of Exercise 4. As $\{v(t)\}$ and $\{e(t)\}$ are sequences of independent variables we must have

$$Ev(t)e(s) = \begin{cases} 0 & t-s \neq \tau \\ \alpha & t-s = \tau \end{cases}$$

$\tau > 0$ gives a prediction problem and $\tau < 0$ a smoothing problem.

6. Consider the scalar stochastic processes

$$x(t) = \int_{s=t_0}^{t} g(t, s)\, dv(s)$$

$$y(t) = \int_{s=t_0}^{t} h(t-s)\, de(s) \qquad (*)$$

where $\{e(t)\}$ and $\{v(s)\}$ are Wiener processes with unit variance parameters and the operator defined by (*) is assumed to have an inverse

$$e(t) = \int_{s=t_0}^{t} k(t, s)\, dy(s)$$

Determine the estimate of $x(t)$ based on $y(s)$, $t_0 \leqslant s \leqslant t$ which minimizes the mean square estimation error. Also determine the variance of the estimation error.

4. STATE ESTIMATION FOR DISCRETE TIME SYSTEMS

Formulation

It was shown in Section 2 that the solution of the estimation problem for Gaussian processes and a large class of criteria is given by the conditional mean. We will now consider the state estimation problem for a discrete time system which is described by the state equation

$$\begin{aligned} x(t+1) &= \Phi x(t) + v(t) \\ y(t) &= \theta x(t) + e(t) \end{aligned} \qquad (4.1)$$

where x is an n-dimensional state vector, y a p-dimensional vector of observed outputs, $\{v(t), t \in T\}$ and $\{e(t), t \in T\}$ are sequences of independent Gaussian vectors with zero mean values and the covariances

$$Ev(t)v^T(t) = R_1$$
$$Ev(t)e^T(t) = 0$$
$$Ee(t)e^T(t) = R_2 \qquad (4.2)$$

It is assumed that the initial state $x(t_0)$ of (4.1) is independent of v and e and normal with mean m and covariance R_0. The matrices Φ, θ, R_1 and R_2 may depend on time. It is assumed that R_2 is positive definite. We will now consider the problem of estimating $x(t + 1)$ based on observations of the output $y(t)$, $y(t - 1)$, ..., $y(t_0)$ in such a way that the criterion

$$Eg(a^T(x(t+1) - \hat{x})) \qquad (4.3)$$

is minimal. The function g is assumed to be symmetric and nondecreasing for positive arguments. Notice that a parametric version of this problem was solved in Section 4 of Chapter 5.

The estimate which minimizes (4.3) is denoted by $\hat{x}(t+1|t)$ and the estimation error $x - \hat{x}$ is denoted by $\tilde{x}(t+1|t)$. The double arguments are used to emphasize that it is the estimate of $x(t+1)$ based on measurements up to time t. The shorter notations $\hat{x}(t+1) = \hat{x}(t+1|t)$ and $\tilde{x}(t+1) = \tilde{x}(t+1|t)$ are also used.

Preliminaries

It follows from Theorem 2.1 that the solution to the problem is given by the conditional mean. Hence

$$\hat{x}(t+1|t) = E[x(t+1)|y(t_0) = \eta(t_0), \ldots, y(t) = \eta(t)] \qquad (4.4)$$

To simplify the writing we introduce the vector $\mathscr{Y}_t$ defined by

$$\mathscr{Y}_t^T = (y^T(t_0), y^T(t_0 + 1), \ldots, y^T(t)) \qquad (4.5)$$

Equation (4.4) can then be written as

$$\hat{x}(t+1|t) = E[x(t+1)|\mathscr{Y}_t] \qquad (4.4)$$

As the stochastic processes $\{x(t), t \in T\}$ and $\{y(t), t \in T\}$ are Gaussian, the conditional mean is given by Theorem 3.2. A direct evaluation using this theorem is, however, cumbersome. In practice we also very often have the situation that the measurements are obtained sequentially in time. We will therefore give a recursive formula for the estimate. Thus assuming that $\hat{x}(t|t-1)$ is known, we will give a formula for $\hat{x}(t+1|t)$.

Main Result

We have

State Estimation for Discrete Time Systems

$$\hat{x}(t+1|t) = E[x(t+1) \mid \mathcal{Y}_t] = E[x(t+1) \mid \mathcal{Y}_{t-1}, y(t)] \quad (4.6)$$

To evaluate the conditional expectation for given $\mathcal{Y}_{t-1}$ and $y(t)$, we will first change the variables so that we get independent variables. It follows from Theorem 3.2 that $\mathcal{Y}_{t-1}$ and

$$\begin{aligned}\tilde{y}(t) &= y(t) - E[y(t) \mid \mathcal{Y}_{t-1}] = y(t) - E[\theta x(t) + e(t) \mid \mathcal{Y}_{t-1}]\\ &= y(t) - \theta \hat{x}(t) = \theta \tilde{x}(t) + e(t)\end{aligned} \quad (4.7)$$

are independent. The quantity $\tilde{y}(t)$ is sometimes referred to as the *innovation* at time t because it is the part of the measured output signal which contains some information which was not previously available.

Thus instead of evaluating the conditional expectation of $x(t+1)$ given $\mathcal{Y}_{t-1}$ and $y(t)$, we will evaluate the conditional expectation given the transformed variables $\mathcal{Y}_{t-1}$ and $\tilde{y}(t)$. We thus have

$$\begin{aligned}\hat{x}(t+1) &= E[x(t+1) \mid \mathcal{Y}_{t-1}, y(t)] = E[x(t+1) \mid \mathcal{Y}_{t-1}, \tilde{y}(t)]\\ &= E[x(t+1) \mid \mathcal{Y}_{t-1}] + E[x(t+1) \mid \tilde{y}(t)] - Ex(t+1)\end{aligned} \quad (4.8)$$

where the last equality is obtained from Theorem 3.3. We will now evaluate the different terms of the right member of (4.8). We have

$$\begin{aligned}E[x(t+1) \mid \mathcal{Y}_{t-1}] &= E[\Phi x(t) + v(t) \mid \mathcal{Y}_{t-1}] = E[\Phi x(t) \mid \mathcal{Y}_{t-1}]\\ &= \Phi E[x(t) \mid \mathcal{Y}_{t-1}] = \Phi \hat{x}(t|t-1)\end{aligned} \quad (4.9)$$

where the first equality follows from (4.1), the second from the fact that $v(t)$ is independent of $x(s)$ and $e(s)$ for $s \leq t$. To evaluate $E[x(t+1) \mid \tilde{y}(t)]$, we will use Theorem 3.2. We have

$$\begin{aligned}R_{x\tilde{y}} &= \operatorname{cov}[x(t+1), \tilde{y}(t)] = \operatorname{cov}[\Phi x(t) + v(t), \theta \tilde{x}(t) + e(t)]\\ &= E[\Phi x(t) + v(t) - \Phi Ex(t)][\theta \tilde{x}(t) + e(t)]^T\\ &= E[\Phi(\hat{x}(t) + \tilde{x}(t))\tilde{x}^T(t)\theta^T] = \Phi(E\tilde{x}(t)\tilde{x}^T(t))\theta^T\end{aligned} \quad (4.10)$$

where the second equality follows from (4.1) and (4.7), the third from the definition of covariance, the fourth from $e(t)$, $v(t)$ and $x(t)$ being independent with zero mean. The fifth equality follows from Theorem 3.2 which implies that $\hat{x}(t)$ and $\tilde{x}(t)$ are independent. We also have

$$\begin{aligned}R_{\tilde{y}\tilde{y}} &= \operatorname{cov}[\tilde{y}(t), \tilde{y}(t)]\\ &= E[\theta \tilde{x}(t) + e(t)][\theta \tilde{x}(t) + e(t)]^T = \theta[E\tilde{x}(t)\tilde{x}^T(t)]\theta^T + R_2\end{aligned} \quad (4.11)$$

because $e(t)$ and $\tilde{x}(t)$ are independent. Introducing

$$P(t) = E\tilde{x}(t)\tilde{x}^T(t) \quad (4.12)$$

and using Theorem 3.2 we thus find

$$E[x(t+1) \mid \tilde{y}(t)] = Ex(t+1) + K(t)\tilde{y}(t) \quad (4.13)$$

where

$$K(t) = R_{x\tilde{y}}R_{\tilde{y}\tilde{y}}^{-1} = \Phi P(t)\theta^T[\theta P(t)\theta^T + R_2]^{-1} \tag{4.14}$$

Combining (4.8), (4.9), and (4.13), we find that the estimate is given by the recursive equations

$$\hat{x}(t+1|t) = \Phi\hat{x}(t|t-1) + K(t)\tilde{y}(t) \tag{4.15}$$
$$\tilde{y}(t) = y(t) - \theta\hat{x}(t|t-1) \tag{4.16}$$

To determine the initial condition of (4.15) we observe that

$$\hat{x}(t_0+1|t_0) = E[x(t_0+1)|y(t_0)]$$

Using Theorem 3.2 we find

$$\hat{x}(t_0+1|t_0) = E[x(t_0+1)|y(t_0)]$$
$$= \Phi m + \Phi R_0\theta^T[\theta R_0\theta^T + R_2]^{-1}[y(t_0) - \theta m] \tag{4.17}$$
$$P(t_0+1) = \Phi R_0\Phi^T + R_1 - \Phi R_0\theta^T[\theta R_0\theta^T + R_2]^{-1}\theta R_0\Phi^T \tag{4.18}$$

Noticing the similarity between (4.15) and (4.17), we find that the initial condition of (4.15) can be given as

$$\hat{x}(t_0|t_0-1) = m \tag{4.19}$$

It now remains to determine $P(t)$. Subtracting (4.15) from (4.1) we get

$$\tilde{x}(t+1|t) = \Phi\tilde{x}(t|t-1) + v(t) - K(t)\tilde{y}(t)$$
$$= [\Phi - K(t)\theta]\tilde{x}(t|t-1) + v(t) - K(t)e(t) \tag{4.20}$$

We thus find that the estimation error is governed by a stochastic difference equation. We find

$$E\tilde{x}(t+1|t) = 0$$

The quantity $P(t)$ defined by (4.12) is thus the covariance matrix of the estimation error. To obtain an equation for the covariance matrix $P(t)$, we multiply (4.20) with its transpose and take mathematical expectations. Compare with Theorem 3.1 of Chapter 3

$$P(t+1) = [\Phi - K(t)\theta]P(t)[\Phi - K(t)\theta]^T + R_1 + K(t)R_2K^T(t)$$
$$= \Phi P(t)\Phi^T + R_1 - \Phi P(t)\theta^T[\theta P(t)\theta^T + R_2]^{-1}\theta P(t)\Phi^T$$
$$= [\Phi - K(t)\theta]P(t)\Phi^T + R_1 \tag{4.21}$$

The last equalities are obtained by using (4.14). Observing the similarity between (4.18) and (4.21), we find that the initial condition of (4.21) can be taken as

$$P(t_0) = R_0 \tag{4.22}$$

Summing up we find Theorem 4.1.

THEOREM 4.1 (Kalman)

The estimate of the state at time $t+1$ of (4.1) based on $y(t_0), y(t_0+1)$,

..., $y(t)$, which minimizes the criterion (4.3), is the conditional mean $\hat{x}(t+1|t)$ which satisfies the recursive equation

$$\hat{x}(t+1|t) = \Phi\hat{x}(t|t-1) + K(t)[y(t) - \theta\hat{x}(t|t-1)]$$
$$\hat{x}(t_0|t_0-1) = m \tag{4.23}$$

The matrix $K(t)$ is given by

$$K(t) = \Phi P(t)\theta^T[\theta P(t)\theta^T + R_2]^{-1} \tag{4.14}$$

where $P(t)$ is the covariance of the estimation error

$$\begin{aligned}P(t+1) &= \Phi P(t)\Phi^T + R_1 - \Phi P(t)\theta^T[\theta P(t)\theta^T + R_2]^{-1}\theta P(t)\Phi^T\\ &= [\Phi - K(t)\theta]P(t)\Phi^T + R_1\\ &= [\Phi - K(t)\theta]P(t)[\Phi - K(t)\theta]^T + R_1 + K(t)R_2K^T(t)\end{aligned}$$
$$P(t_0) = R_0 \tag{4.21}$$

Remark 1

Notice that the covariance $P(t)$ does not depend on the observations. The values of $P(t)$ and thus also of $K(t)$ can be computed a priori. If the optimal filter is implemented on an on-line computer, we thus have a natural possibility of trading computing time against storage by precomputing $P(t)$ and $K(t)$, and storing $K(t)$.

Remark 2

Notice that in order to obtain the best estimate we must in fact compute the conditional distribution of $x(t+1)$ given $\mathcal{Y}_t$. As the distribution is Gaussian, it is completely characterized by the mean $\hat{x}$ and the covariance P. Equations (4.23) and (4.21) can thus be interpreted as an algorithm for computing the conditional probability distribution.

Remark 3

Notice that it is *not* necessary to store all observed outputs as the calculation progresses. Since P and K can be precomputed, the conditional distribution of $x(t+1)$ given $\mathcal{Y}_t$ is uniquely given by the conditional mean $\hat{x}(t+1) = E[x(t+1)|\mathcal{Y}_t]$. If p denotes the conditional density we thus have

$$p[x(t+1)|\mathcal{Y}_t] = p[x(t+1)|\hat{x}(t+1)] \tag{4.24}$$

The conditional mean is thus a *sufficient statistic* for the conditional distribution of $x(t+1)$ given $\mathcal{Y}_t$. This means, roughly speaking, that as far as the conditional distribution is concerned, knowledge of $\hat{x}(t+1)$ is equivalent to knowledge of $\mathcal{Y}_t$.

Also notice that this argument can be extended to the joint conditional distribution of future values of x

$$p[x(t+1), x(t+2), \ldots, x(t+k) | \mathcal{Y}_t]$$
$$= p[x(t+1), x(t+2), \ldots, x(t+k) | \hat{x}(t+1)] \quad (4.25)$$

Remark 4

Comparing Theorem 4.1 with the parametric approach to the state reconstruction problem given in Section 4 of Chapter 5, we find that the structure of the state reconstructor given by (4.4) of Chapter 5 is in fact optimal.

Remark 5

It follows immediately from the derivation that the theorem holds also if the matrices Φ, θ, R_1 and R_2 are time varying. Writing the time dependence explicitly the model (4.1) can be represented as

$$x(t+1) = \Phi(t+1; t)x(t) + v(t)$$
$$y(t) = \theta(t)x(t) + e(t)$$

where the covariances of $v(t)$ and $e(t)$ are $R_1(t)$ and $R_2(t)$. The optimal estimator is then given by (4.14), (4.23) and (4.21) where $\Phi = \Phi(t+1; t)$, $\theta = \theta(t)$, $R_1 = R_1(t)$ and $R_2 = R_2(t)$.

Figure 7.2 shows a block diagram of the optimal Kalman filter.

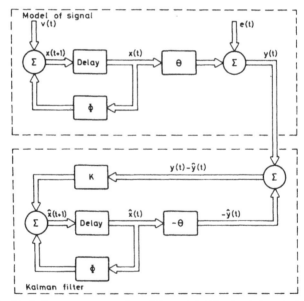

Fig. 7.2. Block diagram representation of the model of the signal given by (4.1) and the optimal Kalman filter (4.23).

State Estimation for Discrete Time Systems

The Innovations

We will now investigate some properties of the innovations $\tilde{y}(t)$ which were defined by (4.7). These quantities have physical interpretations as the difference between the output at time t and the estimate of $y(t)$ based on $y(t-1), y(t-2), \ldots, y(t_0)$.

THEOREM 4.2

The innovations $\tilde{y}(t)$ and $\tilde{y}(s)$ are independent if $t \neq s$ and

$$\text{cov}\,[\tilde{y}(t), \tilde{y}(t)] = \theta P(t)\theta^T + R_2 \tag{4.26}$$

Proof

We have

$$\tilde{y}(t) = y(t) - E[y(t) \mid \mathcal{Y}_{t-1}]$$

It then follows from Theorem 3.2 that $\tilde{y}(t)$ and $\mathcal{Y}_{t-1}$ are independent. Take $s < t$, then $\tilde{y}(s)$ is a linear function of $\mathcal{Y}_{s-1}$. As $\tilde{y}(t)$ is independent of $\mathcal{Y}_{t-1}$ it is then also independent of $\tilde{y}(s)$.

An Algebraic Proof of Theorem 4.2

We can also give a direct algebraic proof of Theorem 4.2. It follows from (4.7) and (4.20) that the innovations $\tilde{y}(t)$ are characterized by

$$\tilde{x}(t+1) = [\Phi - K(t)\theta]\tilde{x}(t) + v(t) - K(t)e(t)$$
$$\tilde{y}(t) = \theta\tilde{x}(t) + e(t) \tag{4.27}$$

In order to analyze these equations, we introduce the matrix Ψ defined by

$$\Psi(t+1;t) = \Phi - K(t)\theta$$

and

$$\Psi(t+k+1;t) = \Psi(t+k+1;t+k)\Psi(t+k;t)$$

For $s > t$ we get

$$\tilde{x}(s) = \Psi(s;t)\tilde{x}(t) + \sum_{k=t}^{s-1} \Psi(s;k+1)[v(k) - K(k)e(k)] \tag{4.27'}$$

We find

$$E[\tilde{y}(s)\tilde{y}^T(t)] = E[\theta\tilde{x}(s) + e(s)][\theta\tilde{x}(t) + e(t)]^T$$
$$= \theta E\tilde{x}(s)\tilde{x}^T(t)\theta^T + \theta E\tilde{x}(s)e^T(t)$$

But

$$E\tilde{x}(s)\tilde{x}^T(t) = \Psi(s;t)P(t)$$

and it follows from (4.27') that

$$E\tilde{x}(s)e^T(t) = -\Psi(s;t+1)K(t)R_2$$

Hence

$$E[\tilde{y}(s)\tilde{y}^T(t)] = \theta\Psi(s;t)P(t)\theta^T - \theta\Psi(s;t+1)K(t)R_2$$
$$= \theta\Psi(s;t+1)\{[\Phi - K(t)\theta]P(t)\theta^T + K(t)R_2\} = 0$$
$$\text{for } t \neq s$$

because it follows from (4.14) that the terms inside the curly brackets will cancel. As $\tilde{y}(s)$ and $\tilde{y}(t)$ are Gaussian, it now follows that they are independent. Equation (4.26) follows from (4.11).

Representation of the Process $\{y(t), t \in T\}$

Theorem 4.2 is of great importance for the representation of the stochastic process represented by the output signals. It follows from (4.1) that it can be represented by

$$x(t+1) = \Phi x(t) + v(t)$$
$$y(t) = \theta x(t) + e(t) \qquad (4.1)$$

where $\{v(t), t \in T\}$ and $\{e(t), t \in T\}$ are sequences of independent Gaussian vectors with zero mean values and the covariances

$$\text{cov}\,[v(t), v(t)] = R_1$$
$$\text{cov}\,[e(t), e(t)] = R_2$$

Due to Theorem 4.2, the stochastic vectors $\tilde{y}(s)$ and $\tilde{y}(t)$ are independent for $t \neq s$. Thus $\{\tilde{y}(t), t \in T\}$ is a sequence of independent Gaussian vectors. The process $\{y(t), t \in T\}$ can be represented by the stochastic difference equation

$$\hat{x}(t+1) = \Phi\hat{x}(t) + K\tilde{y}(t)$$
$$y(t) = \theta\hat{x}(t) + \tilde{y}(t) \qquad (4.28)$$

The representation of (4.28) is of interest because, for this particular representation, the estimation formula is trivial. We get, by eliminating $\tilde{y}$,

$$\hat{x}(t+1) = \Phi\hat{x}(t) + K(t)[y(t) - \theta\hat{x}(t)] \qquad (4.29)$$

Hence assuming that $x(t_0)$ is known, we find that the optimal state estimator is given by (4.29).

Also notice that the matrices Φ and θ of (4.1) and (4.28) are the same, and that the state variable of (4.28) has the physical interpretation as the best estimate of the state vector of (4.1). Compare with Theorem 4.1.

Geometric Interpretation of the State Estimation Problem

Using the geometric interpretation of multivariate Gaussian distributions introduced in Section 3, we can give a geometric interpretation of the state estimation problem. For simplicity we will assume that the initial state $x(t_0)$ has zero mean value. We then find that the stochastic variables $x(t)$ and $y(t)$ have zero mean values. Introducing an Euclidean space whose scalar product is defined by (3.7), it follows from the results of Section 3 that the best estimate of $x(t+1)$ is the projection of $x(t+1)$ on the linear space spanned by $y(t), y(t-1), \ldots, y(t_0)$. Let $Y(t)$ denote this space and denote the projection by Proj $[x(t+1) \mid Y(t)]$. The state estimation problem is then reduced to the problem of calculating a projection in an Euclidean space. To simplify the calculation of the projection, we will first introduce an orthogonal basis in $Y(t)$. This is done using the Gram-Schmidt procedure. We thus introduce

$$\tilde{y}(t_0) = y(t_0)$$
$$\tilde{y}(t_1) = y(t_1) - \text{Proj}\,[y(t_1) \mid \tilde{y}(t_0)] = y(t_1) - \text{Proj}\,[y(t_1) \mid Y(t_0)]$$
$$\tilde{y}(t_2) = y(t_2) - \text{Proj}\,[y(t_2) \mid \tilde{y}(t_0), \tilde{y}(t_1)] = y(t_2) - \text{Proj}\,[y(t_2) \mid Y(t_1)]$$
$$\vdots$$
$$\tilde{y}(t) = y(t) - \text{Proj}\,[y(t) \mid \tilde{y}(t_0), \ldots, \tilde{y}(t-1)] = y(t) - \text{Proj}\,[y(t) \mid Y(t-1)]$$

where $\tilde{y}(t_0), \tilde{y}(t_1), \ldots, \tilde{y}(t)$ is an orthogonal basis in $Y(t)$. Notice that $\tilde{y}(t)$ can be interpreted as the difference between the measured output and the best estimate of the output based on prior data. Compare with (4.7). The variable $\tilde{y}(t)$ thus corresponds to the innovations which were discussed in Section 4. Having obtained an orthogonal basis, it is now easy to calculate the projections. We get

$$\begin{aligned}\text{Proj}\,[x(t+1) \mid Y(t)] &= \text{Proj}\,[x(t+1) \mid Y(t-1), \tilde{y}(t)] \\ &= \text{Proj}\,[x(t+1) \mid Y(t-1)] \\ &\quad + \text{Proj}\,[x(t+1) \mid \tilde{y}(t)] \end{aligned} \quad (4.30)$$

because $\tilde{y}(t)$ is orthogonal to $Y(t-1)$. Compare with (4.8). Using (4.1) we thus get

$$\begin{aligned}\text{Proj}\,[x(t+1) \mid Y(t)] &= \text{Proj}\,[\Phi x(t) + e(t) \mid Y(t-1)] \\ &\quad + \text{Proj}\,[x(t+1) \mid \tilde{y}(t)] \\ &= \Phi\,\text{Proj}\,[x(t) \mid Y(t-1)] + K(t)\tilde{y}(t) \quad (4.31)\end{aligned}$$

because $e(t)$ is orthogonal to $Y(t)$. Equation (4.31) is identical to (4.15). The determination of $K(t)$ can now be done in a straightforward manner using the fact that $x(t+1) - \text{Proj}\,[x(t+1) \mid Y(t)]$ is orthogonal to $Y(t)$.

Using the geometric interpretation we also find that Theorem 4.2 is almost trivial because it follows directly from the construction of $\tilde{y}(t)$.

Exercises

1. Consider the dynamical system

$$x(t+1) = ax(t) + v(t)$$
$$y(t) = x(t) + e(t)$$

where $\{v(t), t \in T\}$ and $\{e(t), t \in T\}$ are sequences of independent normal $(0, 1)$ and $(0, \sigma)$ stochastic variables. The initial state is normal $(1, \sigma_0)$. Determine the optimal state estimator for the system and the steady state estimator when $a = 1$.

2. Consider a system according to (4.1) and (4.2) with

$$\Phi = \begin{pmatrix} 1 & 1 \\ 0 & 1 \end{pmatrix}, \quad \theta = (1 \quad 0)$$

$$R_1 = \begin{pmatrix} r_1 & 0 \\ 0 & r_2 \end{pmatrix}, \quad R_2 = 1$$

Determine the steady state filter gain K and the covariance P of the steady state estimate.

3. Consider the system

$$x(t+1) = \Phi x(t) + \Gamma u(t) + v(t)$$

whose output is given by

$$y(t) = \theta x(t) + e(t)$$

where $\{e(t)\}$ and $\{v(t)\}$ are discrete time, normal white noise with zero mean values and the covariances

$$E v(t) \, v^T(s) = \delta_{s,t} R_1$$
$$E v(t) \, e^T(s) = \delta_{s,t} R_{12}$$
$$E e(t) \, e^T(s) = \delta_{s,t} R_2$$

Show that the optimal estimates $\hat{x}(t) = \hat{x}(t \mid t - 1)$ are given by

$$\hat{x}(t+1) = \Phi \hat{x}(t) + \Gamma u(t) + K[y(t) - \theta \hat{x}(t)]$$

where the best value of K is given by

$$K = K(t) = [\Phi P(t) \theta^T + R_{12}][\theta P(t) \theta^T + R_2]^{-1}$$

where

$$P(t+1) = \Phi P(t) \Phi^T + R_1 - K(t)[R_2 + \theta P(t) \theta^T] K^T(t)$$

Compare Exercise 2 of Chapter 5, Section 4.

4. Consider the system

$$x(t+1) = \Phi x(t) + \Gamma e(t)$$
$$y(t) = \theta x(t) + e(t)$$

where $\{e(t)\}$ is a sequence of independent normal random variables with zero mean values and the covariance R_2. The initial state is normal with mean a and covariance R_0. It is assumed that the matrices Φ, Γ, and θ have constant elements and that the matrix $\Phi - \Gamma\theta$ has all eigenvalues inside the unit circle. Determine the steady state estimator $x(t + 1 \mid t)$ which is optimal in the sense that the criterion (4.3) is minimal. Also determine the covariance of the steady state estimation error.

Hint: Show that

$$P(t + 1) = [\Phi - K\theta]P(t)[\Phi - K\theta]^T$$
$$+ [K - R_{12}R_2^{-1}]R_2[K - R_{12}R_2^{-1}]^T + R_1 - R_{12}R_2^{-1}R_{21}$$

for arbitrary K.

5. Let x and y be scalars. Consider the system

$$x(t + 1) = x(t) + be(t)$$
$$y(t) = x(t) + e(t)$$

where $\{e(t)\}$ is a sequence of independent normal $(0, 1)$ random variables. Let the initial state $x(t_0)$ be normal $(0, \sigma_0)$ and independent of $e(t)$ for all t. Determine the best mean square predictors $\hat{x}(t + 1 \mid t)$ and $\hat{y}(t + 1 \mid t)$, and the prediction errors. Analyze in particular the case $t_0 \to -\infty$. Hint: Consider the two cases $|b - 1| < 1$ and $|b - 1| > 1$ separately.

6. Consider the problem of Exercise 5. Determine the predictor $\hat{y}(t + 1 \mid t)$ using Theorem 3.1 of Chapter 6. Compare with the results of Exercise 5.

7. Consider the system given by (4.1) and (4.2). Show that the best minimum mean square estimate $\hat{x}(t \mid t)$ is given by the equation

$$\hat{x}(t + 1) = \Phi\hat{x}(t) + K(t + 1)[y(t + 1) - \theta\Phi\hat{x}(t)]$$

where

$$K(t) = P(t)\theta^T[R_2 + \theta P(t)\theta^T]^{-1}$$
$$P(t) = \Phi S(t - 1)\Phi^T + R_1$$
$$S(t) = P(t) - K(t)\theta P(t)$$
$$P(t_0) = R_0$$

Compare Exercise 3 of Chapter 5, Section 4.

Hint: $P(t) = E\tilde{y}(t \mid t - 1)\tilde{y}^T(t \mid t - 1)$ and $S(t) = E\tilde{y}(t \mid t)\tilde{y}^T(t \mid t)$

8. Let $\hat{x}(t + k \mid t)$ denote the prediction of $x(t + k)$ based on $\mathcal{Y}_t$, and let $P(t + k \mid t)$ be the covariance of the corresponding prediction error. Show that

$$\hat{x}(t+k+1\,|\,t) = \Phi\hat{x}(t+k\,|\,t)$$
$$P(t+k+1\,|\,t) = \Phi P(t+k\,|\,t)\Phi^T + R_1.$$

and derive a recursive formula for $\hat{x}(t+k\,|\,t)$ for fixed k.

9. Consider the system of Exercise 4. Determine a recursive equation for the prediction $\hat{x}(t+k\,|\,t)$ in the steady state. Also determine the covariance of the steady state prediction error.

10. Prove Theorem 4.1 by showing that it follows from (4.8) and (4.9) that the reconstructor has the structure

$$\hat{x}(t+1) = \Phi\hat{x}(t) + K\tilde{y}(t)$$

Then use the results of Theorem 4.1 of Chapter 5 to find the optimal K.

11. The Kalman filter Theorem 4.1 can be applied to parameter identification. Consider the system

$$y(t) + a(t-1)y(t-1) = b(t-1)u(t-1) + e(t)$$

where

$$a(t+1) = a(t) + v_1(t)$$
$$b(t+1) = b(t) + v_2(t)$$

and $\{e(t)\}$, $\{v_1(t)\}$, and $\{v_2(t)\}$ are discrete time white Gaussian noise with variances r_2, r_{11} and r_{22}. Show that Theorem 4.1 can be applied in order to obtain estimates of a and b. Determine the recursive equations and give the least amount of information that has to be carried along in the computation.

Hint: Consider the pararmeters a and b as states of a dynamical system. Notice that in this case the matrices P and K cannot be precomputed.

12. Show that Theorem 4.1 can be generalized to the situation when the disturbances $\{e(t)\}$ and $\{v(t)\}$ have constant but unknown mean values.

Hint: Introduce the unknown mean values as auxiliary state variables.

13. Equation (4.1) can be written as

$$x(t+1) = \Phi(t+1;t_0)x(t_0) + \sum_{s=t_0}^{t} \Phi(t+1;s)v(s)$$
$$y(t) = \theta x(t) + e(t)$$

Derive the Kalman filtering theorem using the results of Exercise 3 of Section 2 in the special case $x(t_0) = 0$.
Hint: First write the output as

State Estimation for Discrete Time Systems

$$y(t) = \sum_{s=t_0}^{t} \varphi(t;s)\varepsilon(s)$$

14. Consider the system

$$x(t+1) = \begin{pmatrix} 1 & 2 \\ 0 & 1 \end{pmatrix} x(t) + \begin{pmatrix} 0 \\ 1 \end{pmatrix} e(t)$$

$$y(t) = (1 \quad 0)x(t) + e(t)$$

where $\{e(t)\}$ is a sequence of independent normal $(0, 1)$ random variables. Determine the steady state Kalman filter and the steady state covariance matrix of the estimates. Generalize the results.

15. Determine the covariance function of the estimation error of the Kalman filter.

16. Prove Theorem 4.1 by deriving the relation between the representations (4.1) and (4.27) directly, and then solve the filtering problem for (4.27).

17. Consider the system

$$x(t+1) = \Phi x(t) + \Gamma u(t) + v(t)$$
$$y(t) = \theta x(t) + e(t)$$

which is identical to (4.1) apart from the term $\Gamma u(t)$ of the right member of the first equation. Assume that the function $\{u(t), t \in T\}$ is known. Show that the conditional mean of $x(t)$ given $\mathcal{Y}_{t-1}$ is given by

$$\hat{x}(t+1) = \Phi \hat{x}(t) + \Gamma u(t) + K(t)[y(t) - \theta \hat{x}(t)]$$

18. Solve the prediction problem discussed in Section 3 of Chapter 6 using the Kalman filtering theorem.

19. Consider the stochastic process described by (4.1). Let $K(t; t)$ be the optimal gain of the state estimator

$$\hat{x}(t|t) = \Phi(t; t-1)\hat{x}(t-1|t-1)$$
$$+ K(t; t)[y(t) - \theta(t)\Phi(t; t-1)\hat{x}(t-1|t-1)]$$

and let $K(t+l; t)$ be the optimal gain of the l-step predictor

$$\hat{x}(t+l|t) = \Phi(t+l; t+l-1)\hat{x}(t+l-1|t-1)$$
$$+ K(t+l; t)[y(t) - \theta(t)\Phi(t; t+l-1)$$
$$\times \hat{x}(t+l-1|t-1)].$$

Show that

$$K(t+l; t) = \Phi(t+l; t)K(t; t)$$
$$= \Phi(t+l; t+l-1)\Phi(t+l-1,$$
$$t+l-2) \cdots \Phi(t+1; t)K(t).$$

5. DUALITY

This section will show that the state estimation problem is the dual of an optimal control problem. We will consider a system described by the equations

$$x(t+1) = \Phi x(t) + v(t) \quad (5.1)$$
$$y(t) = \theta x(t) + e(t) \quad (5.2)$$

where T is the set of integers, the initial state $x(t_0)$ has mean m, and covariance R_0 and $\{e(t), t \in T\}$ and $\{v(t), t \in T\}$ are sequences of uncorrelated random vectors with zero mean values and the covariances

$$\begin{aligned} Ev(t)v^T(t) &= R_1 \\ Ev(t)e^T(t) &= 0 \\ Ee(t)e^T(t) &= R_2 \end{aligned} \quad (5.3)$$

As usual, the matrices Φ, θ, R_1, and R_2 may depend on t. Assume that we want to estimate $a^T x(t_1)$ linearly in $y(t_1 - 1), y(t_1 - 2) \ldots, y(t_0)$, and m in such a way that the criterion

$$E[a^T x(t_1) - a^T \hat{x}(t_1)]^2 \quad (5.4)$$

is minimal.

As the estimate is linear we have

$$a^T \hat{x}(t_1) = -\sum_{t=t_0}^{t_1-1} u^T(t) y(t) + b^T m \quad (5.5)$$

The minus sign is introduced in order to obtain the final result in a nice form. The estimation problem is thus a problem of determining the vectors $b, u(t_1 - 1), u(t_1 - 2), \ldots, u(t_0)$. We will now determine the $u:s$ in such a way that the criterion (5.4) is minimal. To do so, we introduce the vectors $z(t)$ defined recursively from

$$z(t) = \Phi^T z(t+1) + \theta^T u(t+1) \quad (5.6)$$

with the initial condition

$$z(t_1 - 1) = a$$

Hence

$$\begin{aligned} a^T x(t_1) &= z^T(t_1 - 1) x(t_1) \\ &= z^T(t_0 - 1) x(t_0) + \sum_{t=t_0}^{t_1-1} [z^T(t) x(t+1) - z^T(t-1) x(t)] \quad (5.7) \end{aligned}$$

It now follows from (5.1) and (5.6) that

Duality

$$z^T(t)x(t+1) = z^T(t)\Phi x(t) + z^T(t)v(t)$$
$$z^T(t-1)x(t) = z^T(t)\Phi x(t) + u^T(t)\theta x(t)$$

Introducing this in (5.7), we find

$$a^T x(t_1) = z^T(t_0 - 1)x(t_0) + \sum_{t=t_0}^{t_1-1} [z^T(t)v(t) - u^T(t)\theta x(t)] \qquad (5.8)$$

Equations (5.2) and (5.5) give

$$a^T \hat{x}(t_1) = -\sum_{t=t_0}^{t_1-1} u^T(t)y(t) + b^T m = -\sum_{t=t_0}^{t_1-1} [u^T(t)\theta x(t) + u^T(t)e(t)] + b^T m \qquad (5.9)$$

Combining (5.8) and (5.9) we find

$$a^T x(t_1) - a^T \hat{x}(t_1) = z^T(t_0 - 1)x(t_0) - b^T m$$
$$+ \sum_{t=t_0}^{t_1-1} [z^T(t)v(t) - u^T(t)e(t)]$$

Squaring and taking mathematical expectations, we find that the criterion (5.4) can be expressed as follows

$$E[a^T x(t_1) - a^T \hat{x}(t_1)]^2 = [(z(t_0 - 1) - b)^T m]^2 + z^T(t_0 - 1)R_0 z(t_0 - 1)$$
$$+ \sum_{t=t_0}^{t_1-1} [z^T(t)R_1 z(t) + u^T(t)R_2 u(t)] \qquad (5.10)$$

To minimize the criterion, we thus have to choose the parameter $b = z(t_0 - 1)$ and the $u:s$ should be determined in such a way that the function

$$z^T(t_0 - 1)R_0 z(t_0 - 1) + \sum_{t=t_0}^{t_1-1} [z^T(t)R_1 z(t) + u^T(t)R_2 u(t)] \qquad (5.11)$$

is as small as possible. Summing up we find Theorem 5.1.

THEOREM 5.1 (Duality theorem)

The state estimation problem for the system described by (5.1) and (5.2) is equivalent to the problem of finding the control signal u for the system (5.6) which minimizes the criterion (5.11).

Exercises

1. Show that the problem of estimating $a^T x(t_1)$ of the systems (5.1) and (5.2) linearly in $y(t_1), y(t_1 - 1), \ldots, y(t_0)$ is the dual of the following control problem

$$z(t) = \Phi^T z(t+1) + \theta^T u(t+1)$$

$$z(t_1) = \Phi^{-1}a$$
$$z^T(t_0 - 1)R_0 z(t_0 - 1) + \sum_{t=t_0}^{t_1-1} [z^T(t)R_1 z(t) + u^T(t)R_2 u(t)]$$

2. It is possible to derive other duality theorems by applying the method of maximum likelihood to the state estimation problem. Consider the system described by (4.1) and (4.2). Apart from a normalizing factor, the likelihood function L equals the joint density function of $x(t_0), x(t_0 + 1), \ldots, x(t_1)$ and $y(t_0), y(t_0 + 1), \ldots, y(t_1)$. The logarithm of L then becomes

$$-2 \log L = \sum_{k=t_0}^{t_1} [y(k) - \theta x(k)]^T R_2^{-1} [y(k) - \theta x(k)]$$
$$+ \sum_{k=t_0}^{t_1-1} v^T(k) R_1^{-1} v(k)$$
$$+ [x(t_0) - m]^T R_0^{-1} [x(t_0) - m] + \text{const} \qquad (*)$$

where x and e are related through

$$x(t + 1) = \Phi x(t) + e(t) \qquad (**)$$

The state estimation problem can then be reduced to the problem of finding $\{v(t)\}$ such that the system $(**)$ is optimal with respect to the criterion $(*)$. Show that this approach gives the following equation for the best estimate

$$\hat{x}(t + 1 | t_1) = \Phi \hat{x}(t | t_1) + R_1 \lambda(t + 1)$$
$$\lambda(t) = \Phi^T \lambda(t + 1) + \theta^T R_2^{-1} [y(t) - \theta \hat{x}(t)] \qquad (***)$$

The boundary conditions are

$$\lambda(t_0) = R_0^{-1}[\hat{x}(t_0) - m]$$
$$\lambda(t_1 + 1) = 0$$

where

$$\hat{x}(t_0 | t_1) = \hat{x}(t_0)$$

3. Show that the solution to the estimation problem given in Exercise 2 is equivalent to the solution given by Theorem 4.1 by deriving the recursive equations for $\hat{x}(t | t)$ and $P(t)$ from $(***)$.

4. Show that the method used in Exercise 2 also provides a solution to the filtering problem.

6. STATE ESTIMATION FOR CONTINUOUS TIME PROCESSES

This section will discuss the state estimation problem for continuous time processes. The purpose is to derive the Kalman-Bucy equations for the state estimator. The solution of the continuous time problem is considerably more difficult than the discrete time problem. In the discrete time case, it was possible to carry out most of the analysis in a finite dimensional Euclidean space. When dealing with continuous time processes we need, however, infinite dimensional spaces. Hilbert space theory is the natural tool to handle the continuous time case. We will also encounter another difficulty—namely the fact that, even using Hilbert space theory, there is no natural way to introduce the concept of white noise which was found to be extremely useful in the discrete time case.

This section will derive the result indirectly via the concept of duality. We will first show that the state estimation problem is the dual of a deterministic control problem and we will then derive the formulas required using results from the theory of optimal control of deterministic systems.

Problem Statement

We will now formulate the state estimation problem for continuous time processes. Consider a continuous time stochastic process described by

$$dx = Ax\,dt + dv \tag{6.1}$$
$$dy = Cx\,dt + de \tag{6.2}$$

where the initial state $x(t_0)$ has the mean value m and the covariance matrix R_0. The stochastic processes $\{v(t), t \in T\}$ and $\{e(t), t \in T\}$ are assumed to be stochastic processes with uncorrelated increments. The incremental covariances are assumed to be $R_1\,dt$ and $R_2\,dt$ respectively. The processes $\{e(t), t \in T\}$ and $\{v(t), t \in T\}$ are also assumed to be mutually uncorrelated and also uncorrelated with $x(t_0)$.

The state estimation problem can now be formulated as follows. Assume that a realization of the output y has been observed over the interval (t_0, t). Determine the best estimate of the value of the state vector at time t. To complete the problem statement, we must also specify the admissible estimators and what is meant by the best estimate. It is assumed that the admissible estimates are linear functions of the observed output and that the criterion is to minimize the mean square estimation error.

Assuming that we want to estimate the linear function $a^T x(t_1)$, the admissible estimators are thus of the form

$$a^T \hat{x}(t_1) = -\int_{t_0}^{t_1} u^T(t)\,dy(t) + b^T m \tag{6.3}$$

and the criterion is to minimize

$$E[a^T x(t_1) - a^T \hat{x}(t_1)]^2 \qquad (6.4)$$

The minus sign in (6.3) is introduced in order to get the final result on a nice form. With this formulation, the state estimation problem is thus reduced to the problem of finding the function u and the vector b. It is assumed that u is a continuous function of time.

Duality

We will now show that the estimation problem is the dual of a deterministic control poroblem. The analysis follows the analysis of the discrete time problem in Section 5 closely. To do so we will first rewrite the criterion.

Equations (6.2) and (6.3) give

$$a^T \hat{x}(t_1) = -\int_{t_0}^{t_1} u^T(t)\, dy(t) + b^T m$$
$$= -\int_{t_0}^{t_1} [u^T(t)\, Cx(t)\, dt + u^T(t)\, de(t)] + b^T m \qquad (6.5)$$

Introduce the vector z defined as the solution of the differential equation

$$\frac{dz}{dt} = -A^T z - C^T u \qquad (6.6)$$

with the initial condition

$$z(t_1) = a \qquad (6.7)$$

Then

$$a^T x(t_1) = z^T(t_1)x(t_1) = z^T(t_0)x(t_0) + \int_{t_0}^{t_1} d[z^T(t)x(t)] \qquad (6.8)$$

But

$$d(z^T x) = dz^T x + z^T dx = -z^T Ax\, dt - u^T Cx\, dt + z^T Ax\, dt + z^T dv$$
$$= -u^T Cx\, dt + z^T dv$$

Hence

$$a^T x(t_1) = z^T(t_0)x(t_0) + \int_{t_0}^{t_1} [-u^T(t)Cx(t)\, dt + z^T(t)\, dv(t)] \qquad (6.9)$$

Equations (6.5) and (6.9) now give

$$a^T[x(t_1) - \hat{x}(t_1)] = z^T(t_0)x(t_0) - b^T m + \int_{t_0}^{t_1} [z^T(t)\, dv(t) + u^T(t)\, de(t)] \qquad (6.10)$$

Taking mathematical expectation, we get

$$E a^T[x(t_1) - \hat{x}(t_1)] = [z(t_0) - b]^T m$$

State Estimation For Continuous Time Processes

We thus find that the estimate given by (6.5) is unbiased for all a and all choices of u if the vector b is chosen as $b = z(t_0)$. Squaring (6.10) and taking mathematical expectation we get

$$E[a^T x(t_1) - a^T \hat{x}(t_1)]^2 = [(z(t_0) - b)^T m]^2 + z^T(t_0) R_0 z(t_0)$$
$$+ \int_{t_0}^{t_1} [z^T(t) R_1 z(t) + u^T(t) R_2 u(t)] dt \quad (6.11)$$

To find a function u, such that the linear estimate given by (6.3) is optimal in the sense of mean squares, is thus equivalent to the problem of finding a control signal for the dynamical system (6.6) with the initial condition (6.7) and the criterion

$$z^T(t_0) R_0 z(t_0) + \int_{t_0}^{t_1} [z^T(t) R_1 z(t) + u^T(t) R_2 u(t)] dt \quad (6.12)$$

Summing up we find

THEOREM 6.1 (Duality theorem)

The state estimation problem for the system described by (6.1) and (6.2) is equivalent to the problem of finding the best control law for the linear deterministic system (6.6) with the criterion (6.12).

The Deterministic Control Problem

The problem which we have just arrived at differs somewhat in notation from the standard formulation of the linear optimal control theory. To facilitate a comparison we will therefore state the results in standard form. Consider the system

$$\frac{dx}{dt} = Ax + Bu \quad (6.13)$$

with given initial condition $x(t_0)$. Find a control law such that the criterion

$$x^T(t_1) Q_0 x(t_1) + \int_{t_0}^{t_1} [x^T(t) Q_1 x(t) + u^T(t) Q_2 u(t)] dt \quad (6.14)$$

is minimal. It is assumed that the matrices Q_0 and Q_1 are positive semi-definite and that Q_2 is positive definite. The elements of all matrices may be piecewise continuous functions of time.

The solution to this problem is given by the linear control law

$$u = -Lx \quad (6.15)$$

where

$$L = Q_2^{-1} B^T S \quad (6.16)$$

and S is the solution to the Riccati equation

$$-\frac{dS}{dt} = A^T S + SA + Q_1 - SBQ_2^{-1}B^T S \quad (6.17)$$

with the initial condition
$$S(t_1) = Q_0 \qquad (6.18)$$

The solution exists and is unique if the Riccati equation has a solution. Proofs of those facts are found in standard texts on deterministic control theory. A proof is also found in Chapter 8, Section 7 of this book. Comparing with the standard formulation, we find that the problem given by (6.6) and (6.12) has the solution

$$u(t) = -K^T z(t) \qquad (6.19)$$

where
$$K = PC^T R_2^{-1} \qquad (6.20)$$

and

$$\frac{dP}{dt} = AP + PA^T + R_1 - PC^T R_2^{-1} CP$$
$$P(t_0) = R_0 \qquad (6.21)$$

The equivalence between the problem given by (6.6) and (6.12) and the standard optimal control problem given by (6.13) and (6.14) is further illustrated in the table below.

Standard optimal control problem	State estimation problem
t	$-t$
t_0	t_1
t_1	t_0
A	A^T
B	C^T
Q_0	R_0
Q_1	R_1
Q_2	R_2
S	P
L	K^T

Main Result

Using results from deterministic control theory, we have thus determined the function u which gives the best estimate. We will now rewrite the result so that a stochastic differential equation is obtained for the estimate. The estimate is given by

$$a^T \hat{x}(t_1) = -\int_{t_0}^{t_1} u^T(t) \, dy(t) + b^T m \qquad (6.5)$$

where u is given by (6.19).

To obtain a stochastic differential equation we will simply differentiate this equation. Notice, however, that both u and b depend implicitly

State Estimation For Continuous Time Processes

on t_1. We will thus rewrite the equation in such a way that the dependence is explicit. Equations (6.6) and (6.19) give

$$\frac{dz}{dt} = -A^T z - C^T u = -(A - KC)^T z \qquad (6.22)$$

Let $\Psi(t; t_1)$ be the solution of the differential equation

$$\frac{d\Psi}{dt} = (A - KC)\Psi \qquad (6.23)$$

with the initial condition

$$\Psi(t_1; t_1) = I \qquad (6.24)$$

The solution of (6.22) with the initial condition $z(t_1) = a$ then is

$$z(t) = \Psi^T(t_1; t)a \qquad (6.25)$$

Hence

$$u(t) = -K^T \Psi^T(t_1; t)a \qquad (6.26)$$
$$b = \Psi^T(t_1; t_0)a \qquad (6.28)$$

Equation (6.5) for the estimate now becomes

$$a^T \hat{x}(t_1) = a^T \int_{t_0}^{t_1} \Psi(t_1; t) K \, dy(t) + a^T \Psi(t_1; t_0) m \qquad (6.29)$$

Hence if we choose

$$\hat{x}(t_1) = \int_{t_0}^{t_1} \Psi(t_1; t) K \, dy(t) + \Psi(t_1; t_0) m \qquad (6.30)$$

we obtain an estimate $\hat{x}$ such that the mean square estimation error is minimal for all a.

Differentiating (6.30) we get

$$d\hat{x}(t_1) = \left[\int_{t_0}^{t_1} \frac{\partial \Psi(t_1; t)}{\partial t_1} K \, dy(t) + \frac{\partial \Psi(t_1; t_0)}{\partial t_1} m \right] dt_1 + K \, dy(t_1)$$
$$= (A - KC)\hat{x}(t_1) dt_1 + K \, dy(t_1)$$
$$= A\hat{x}(t_1) \, dt_1 + K[dy(t_1) - C\hat{x}(t_1) \, dt_1] \qquad (6.31)$$

The linear estimate which minimizes the mean square estimation error thus satisfies a linear stochastic differential equation. The initial value is obtained from (6.30). We have

$$\hat{x}(t_0) = m \qquad (6.32)$$

Subtracting (6.31) from (6.1) we find that the estimation error is governed by the stochastic differential equation

$$d\tilde{x} = (A - KC)\tilde{x} \, dt + dv - K \, de \qquad (6.33)$$

Using the results of Chapter 3 we thus find that the covariance of the estimation error is governed by the differential equation

$$\frac{dQ}{dt} = AQ + QA^T + R_1 - KCQ - QC^TK^T + KR_2K^T$$
$$= AQ + QA^T + R_1 - PC^TR_2^{-1}CQ - QC^TR_2^{-1}CP + PC^TR_2^{-1}CP$$
(6.34)

with the initial condition

$$Q(t_0) = R_0 \tag{6.35}$$

The second equality of (6.34) is obtained from (6.20). Subtracting (6.34) from (6.21) we get

$$\frac{d}{dt}(Q - P) = A(Q - P) + (Q - P)A^T$$
$$- (Q - P)CR_2^{-1}C^TP - PC^TR_2^{-1}C(Q - P)$$

As $Q(t_0) = P(t_0) = R_0$ we find that $Q(t) = P(t)$. Compare Lemma 5.1 of Chapter 5. The covariance of the estimation error is thus governed by Eq. (6.20). Summing up we find Theorem 6.2.

THEOREM 6.2 (Kalman and Bucy)

The linear estimate of the state vector of the system (6.1), (6.2) is governed by the stochastic differential equation

$$d\hat{x} = A\hat{x}\,dt + K[dy - C\hat{x}\,dt] \tag{6.31}$$
$$\hat{x}(t_0) = m \tag{6.32}$$

where

$$K = PC^TR_2^{-1} \tag{6.20}$$

and P is the covariance of the estimation error which is governed by the equation

$$\frac{dP}{dt} = AP + PA^T + R_1 - PC^TR_2^{-1}CP$$
$$P(t_0) = R_0 \tag{6.21}$$

Remark 1

Notice that since (6.31) is a stochastic differential equation, its solution can *not* be interpreted using ordinary integrals. Instead the solution of (6.31) must be defined using stochastic integrals as was discussed in Chapter 3. Strictly speaking this means that (6.31) can *not* be thought of as a linear filter which operates on the sample functions of the observation process.

Remark 2

Since the stochastic processes $\{x(t), t \in T\}$ and $\{y(t), t \in T\}$ are Gaus-

State Estimation For Continuous Time Processes

sian, the conditional distribution of $x(t)$ given $y(s)$, $t_0 \leq s \leq t$ is also Gaussian. The conditional mean is $\hat{x}$ and the conditional covariance P.

The Innovations

We will now investigate the properties of the innovations. The results given are analogous to those for discrete time processes discussed in Section 4. We have

THEOREM 6.3

The stochastic process $\{\tilde{y}(t), t \in T\}$ defined by

$$\tilde{y}(t) = y(t) - \hat{y}(t) \tag{6.36}$$

has increments which are independent and have zero mean values and the incremental covariances $R_2 dt$.

Proof

We have

$$d\tilde{y}(t) = dy(t) - d\hat{y}(t) = Cx(t)\,dt + de - C\hat{x}(t)\,dt = C\tilde{x}(t)\,dt + de(t) \tag{6.37}$$

Since $\tilde{x}$ and e have zero mean values we find that $d\tilde{y}$ also has zero mean value.

Let $t_1 < t_2 \leq t_3 < t_4$ and consider

$$E[\tilde{y}(t_4) - \tilde{y}(t_3)][\tilde{y}(t_2) - \tilde{y}(t_1)]^T$$

$$= E \int_{s=t_3}^{t_4} \int_{t=t_1}^{t_2} [C(s)\tilde{x}(s)\,ds + de(s)][C(t)\tilde{x}(t)\,dt + de(t)]^T$$

$$= \int_{s=t_3}^{t_4} \int_{t=t_1}^{t_2} C(s) R_{\tilde{x}}(s, t) C^T(t)\,ds\,dt + \int_{s=t_3}^{t_4} \int_{t=t_1}^{t_2} C(s) E[\tilde{x}(s) de^T(t)]\,ds \tag{6.38}$$

The first equality follows from (6.37) and the second from the fact that the process $\{e(t)\}$ has independent increments, and that $de(s)$ is independent of $\tilde{x}(t)$ for $s \geq t$.

It follows from (6.33) that

$$\tilde{x}(s) = \Psi(s; t)\tilde{x}(t) + \int_t^s \Psi(s; \tau)[dv(\tau) - K(\tau)de(\tau)]$$

where Ψ is defined by (6.23) and (6.24). Hence

$$E\tilde{x}(s)de^T(t) = -\Psi(s; t) K(t) R_2(t)\,dt \tag{6.39}$$

It follows from (6.33) and Theorem 6.1 of Chapter 3 that

$$R_{\tilde{x}}(s, t) = \Psi(s; t) P(t) \tag{6.40}$$

where P is given by (6.21). Equations (6.38), (6.39), and (6.40) now give

$$E[\tilde{y}(t_4) - \tilde{y}(t_3)][\tilde{y}(t_2) - \tilde{y}(t_1)]^T$$
$$= \int_{t_3}^{t_4} \int_{t_1}^{t_2} C(s)\Psi(s;t)[P(t)C^T(t) - K(t)R_2(t)]ds\,dt = 0 \quad (6.41)$$

The integral vanishes because it follows from (6.20) that the integrand is zero. Since $\{y(t)', t \in T\}$ is a Wiener process, it now follows that $[\tilde{y}(t_4) - \tilde{y}(t_3)]$ and $[\tilde{y}(t_2) - \tilde{y}(t_1)]$ are normal. The condition (6.41) then implies that the increments are independent. To find the incremental covariance of the process $\{\tilde{y}(t), t \in T\}$ we use arguments analogous to those used in the derivation of (6.38). Observing that the first term of (6.38) is of order $(dt)^2$ we find

$$E[\tilde{y}(t_2) - \tilde{y}(t_1)][\tilde{y}(t_2) - \tilde{y}(t_1)]^T$$
$$= E\int_{s=t_1}^{t_2} \int_{t=t_1}^{t_2} de(s)\,de^T(t) + o(dt) = \int_{t_1}^{t_2} R_2(t)\,dt + o(dt)$$

The incremental covariance is thus $R_2\,dt$, and the proof of the theorem is complete.

Representation of the Process $\{y(t), t \in T\}$

Theorem 6.3 makes it possible to find an interesting representation of the stochastic process $\{y(t), t \in T\}$ defined by (6.1) and (6.2). It follows from Theorem 6.3 that the process can be represented as

$$d\hat{x} = A\hat{x}\,dt + K\,d\tilde{y}$$
$$dy = C\hat{x}\,dt + d\tilde{y} \quad (6.42)$$

where

$$\hat{x}(t_0) = m \quad (6.43)$$

and $\{\tilde{y}(t), t \in T\}$ is a Wiener process with incremental covariance $R_2\,dt$. The representation (6.42) is of interest because it is invertible in the sense that y can be solved directly in terms of $\tilde{y}$ and vice versa. This means that the manipulations required to solve filtering and prediction problems are very easy to do in the representation (6.42). Compare with the analysis of Chapter 6 Section 3, and Exercise 6, Section 3 of this chapter. Also notice that the state $\hat{x}$ of (6.42) has physical interpretation as the conditional mean of the state of the system (6.1).

Exercises

1. Consider the stochastic process

$$dx = \alpha x\,dt + dv$$
$$dy = x\,dt + de$$

where $\{v(t)\}$ and $\{e(t)\}$ are independent Wiener processes with incre-

State Estimation For Continuous Time Processes

mental covariances $r_1 dt$ and $r_2 dt$ respectively. Assume that the initial state is normal $N(m, \sqrt{r_0})$. Show that the gain of the optimal filter is

$$K(t) = P(t)/r_2$$

where

$$P(t) = \frac{(r_1/\beta) \sinh \beta t + r_0[\cosh \beta t + (\alpha/\beta) \sinh \beta t]}{\cosh \beta t - (\alpha/\beta) \sinh \beta t + [r_0/(r_2\beta)] \sinh \beta t}$$

$$\beta = \sqrt{\alpha^2 + r_1/r_2}$$

2. Show that the solution of the Riccati equation

$$\frac{dP}{dt} = AP + PA^T + R_1 - PC^T R_2^{-1} CP$$

with the initial condition

$$P(t_0) = R_0$$

can be represented as

$$P(t) = [\Lambda_{21}(t;t_0) + \Lambda_{22}(t;t_0)R_0][\Lambda_{11}(t;t_0) + \Lambda_{12}(t;t_0)R_0]^{-1}$$

where

$$\Lambda(t;t_0) = \begin{bmatrix} \Lambda_{11}(t;t_0) & \Lambda_{12}(t;t_0) \\ \Lambda_{21}(t;t_0) & \Lambda_{22}(t;t_0) \end{bmatrix}$$

is the solution of the linear equation

$$\frac{d\Lambda}{dt} = \begin{bmatrix} -A^T & C^T R_2^{-1} C \\ R_1 & A \end{bmatrix} \Lambda$$

with initial condion

$$\Lambda(t_0;t_0) = I \quad \text{(the } 2n \times 2n \text{ identity matrix)}$$

3. Consider the stochastic differential equation

$$dy + ay\, dt = bu\, dt + de$$

where $\{e(t)\}$ is a Wiener process with variance parameter r and the parameters a and b are given by

$$da = -\alpha a\, dt + dv$$
$$db = -\beta b\, dt + dw.$$

where $\{v(t)\}$ and $\{w(t)\}$ are independent Wiener processes with variance parameters r_{11} and r_{22} respectively. Use Theorem 6.2 to derive recursive equations for tracking the parameters a and b as well as possible in the sense of minimum means square.

4. It is possible to derive a duality theorem which differs from Theorem 6.1 by applying the method of maximum likelihood to the state esti-

mation problem. It can be shown that the likelihood function for the problem given by (6.1) and (6.2) is given by

$$-2 \log L = [x(t_0) - m]^T R_0^{-1}[x(t_0) - m]$$
$$+ \int_{t_0}^{t} [y(t) - Cx(t)]^T R_2^{-1}[y(t) - Cx(t)] \, dt$$
$$+ \int_{t_0}^{t} v^T(t) R_1^{-1} v(t) \, dt + \text{const} \qquad (*)$$

The estimation problem can then be formulated to find the control signal v for the system

$$\frac{dx}{dt} = Ax + v$$

such that the criterion (*) is minimal. Show that this approach to the problem gives the same result as Theorem 6.2.

5. Show that Theorem 6.2 can be obtained formally by taking the formal limit of the corresponding discrete time problem discussed in Section 4 (Theorem 4.1).

6. Show that the Theorem 6.2 can be generalized to the situation when the processes $\{v(t)\}$ and $\{e(t)\}$ are sums of Wiener processes and deterministic functions $v_1 t$ and $e_1 t$ respectively where v_1 and e_1 are unknown constants.

   Hint: Introduce v_1 and e_1 as auxiliary state variables.

7. Consider the system with

$$A = \begin{bmatrix} 0 & 1 \\ 0 & 0 \end{bmatrix} \qquad C = [1 \quad 0]$$
$$R_1 = \begin{bmatrix} 1 & 0 \\ 0 & \sigma^2 \end{bmatrix} \qquad R_2 = r_2$$

   Determine the steady state Kalman filter for estimating the state of the system. Also determine the covariance matrix P_0 of the steady state estimator. Determine the limit of the transfer function of the optimal filter as $r_2 \to 0$.

8. Consider the problem of Exercise 7. Determine the optimal Kalman filter when $r_2 = 0$.

   Hint: When $r_2 = 0$, the state variable $x_1(t)$ can be measured exactly. The dimensionality of the problem can then be reduced.

9. Consider the system

$$dx = Ax \, dt + B \, de$$
$$dy = Cx \, dt + de$$

State Estimation for Discrete Time Systems

where $\{e(t), t \in T\}$ is a Wiener process with incremental covariance $R_2 dt$. Let $A, B, C,$ and R_2 be constant matrices. Assume that $A - BC$ has all eigenvalues in left half plane. Determine the gain of the steady state Kalman filter and the steady state covariance.

10. Consider the state estimation problems for the systems

$$\begin{cases} dx = Ax\, dt + B\, du \\ dy = Cx\, dt + de \end{cases}$$

$$\begin{cases} dz = -A^T z\, dt + C^T\, dv \\ dy = B^T z\, dt + dn \end{cases}$$

where $\{u(t), t \in T\}$, $\{e(t), t \in T\}$, $\{v(t), t \in T\}$, and $\{n(t), t \in T\}$ are processes with incremental covariances $R_1 dt$, $R_2 dt$, $R_2^{-1} dt$, and $R_1^{-1} dt$ respectively. The initial states are normal with covariances R_0 and R_0^{-1} respectively. Show that the state estimation problems are duals and find the relation-ships between the Riccati equations of the problems.

11. Consider the system given by (6.1) and (6.2); show that the minimum mean square predictor of $x(s)$ given $\mathcal{Y}_t$ for $s \geq t$ is given by

$$\hat{x}(s|t) = \Phi(s;t)\hat{x}(t)$$

and that the variance of the predictor $P(s|t)$ is given by

$$\frac{dP(s|t)}{ds} = A(s)P(s|t) + P(s|t)A^T(s) + R_1(s)$$

$$P(t|t) = P(t)$$

Derive a recursive formula for the predictor.

12. Let x and y be scalars. Consider the system

$$\begin{cases} dx = x\, dt + b\, de \\ dy = x\, dt + de \end{cases}$$

where $\{e(t), t \in T\}$ is a Wiener process with unit variance parameter. Let the initial state be normal $(0, \sqrt{r_0})$. Determine the best mean square estimate of $x(t)$ given $\mathcal{Y}_t$. Also determine the steady state filtergain and the variance of the steady state estimation error.

13. The random drift of a gyroscope has been described by the following simplified mathematical model

$$dx_1 = (x_2 + x_3)\, dt$$
$$dx_2 = 0$$
$$dx_3 = -\frac{1}{\tau} \cdot x_3\, dt + \sigma_s \sqrt{\frac{2}{\tau}}\, dv$$

where x_1 is the drift angle, x_2 day-to-day drift, x_3 "random drift", and $\{v(t),\ t \in T\}$ a Wiener process with unit variance parameter. Assume that $x_1(0) = 0$, $Ex_2(0) = 0$, $Ex_3(0) = 0$, var $x_2(0) = \sigma_d^2$, var $x_3(0) = \sigma_s^2$, and that x_1 can be measured without error. Determine the minimum variance estimates of the components x_2 and x_3.

14. Consider the Riccati equation

$$\frac{dP}{dt} = AP + PA^T + R_1, \qquad P(t_0) = P_0$$

Introduce $P = QQ^T$, and $P_0 = Q_0 Q_0^T$. Show that the matrix Q satisfies the differential equation

$$\frac{dQ}{dt} = AQ + \frac{1}{2} R_1; \qquad Q(t_0) = Q_0.$$

15. Assume that the Riccati equation (6.21) has a solution which is positive definite for all t. Show that the inverse P^{-1} satisfies the equation

$$\frac{dP^{-1}}{dt} = -P^{-1}A - A^T P^{-1} + P^{-1} R_1 P^{-1} + C^T R_2^{-1} C$$

$$P^{-1}(t_0) = P_0^{-1}.$$

16. Combine the results of Exercises 14 and 15 to obtain a simplified algorithm for solving the Riccati equation (6.21) in the special case when $R_1 = 0$.

7. BIBLIOGRAPHY AND COMMENTS

Filtering and prediction problems were first discussed and solved by Kolmogorov and Wiener in

Kolmogorov, A. N., "Interpolation and Extrapolation of Stationary Random Sequences," *Bull. Moscow Univ.*, USSR, Ser. Math. **5**, (1941).

Wiener, N., *The Extrapolation, Interpolation, and Smoothing of Stationary Time Series with Engineering Applications*, Wiley, New York, 1949. Originally issued as a classified MIT Rad. Lab. Report in February, 1942.

Kolmogorov studied the discrete time case and Wiener the continuous time case. Kolmogorov exploited a representation of the random processes suggested by Wold in

Wold, H., *A Study in the Analysis of Stationary Time Series*, Almqvist and Wiksell, Uppsala, 1938.

An account of Kolmogorovs approach which is easy accessible is given in

Whittle, P., *Prediction and Regulation*, Van Nostrand, Princeton, New Jersey, 1963.

Wiener reduced the problem to a certain integral equation, the so-called Wiener Hopf equation. This is required in order to represent the processes by a stationary dynamical system driven by white noise. Wiener's original work is no easy reading. A simplified derivation of Wiener's theory is given in

Bode, H. W., and Shannon, C. E., "A Simplified Derivation of Linear Least-Square Smoothing and Prediction Theory," *Proc. IRE* **38**, 417 (April 1950).

A similar approach was used in

Zadeh, L. A. and Ragazzini, J. R., "An Extension of Wiener's Theory of Prediction," *J. Appl. Phys.* **21**, 645–655 (1950).

This paper also generalizes Wiener's results to the case when the signal is a sum of a stochastic process and a polynomial with unknown coefficients.

The Wiener-Kolmogorov theory is now well established in textbooks. See for example,

Davenport, W. B. Jr. and Root, W. L., *An Introduction to the Theory of Random Signals and Noise*, McGraw-Hill, New York, 1958.
Lee, Y. W., *Statistical Theory of Communication*, Wiley, New York, 1960.
Yaglom, A.M., *Theory of Stationary Random Functions*, transl. from Russian by R.A. Silverman, Prentice-Hall, Englewood Cliffs, New Jersey, 1966.

Kalman and Bucy made an essential contribution to the theory. They extended the theory to finite observation intervals and nonstationary processes in

Kalman, R. E., "A New Approach to Linear Filtering and Prediction Problems," *J. Basic Eng.* **82**, 34–45 (March 1960).
Kalman, R. E., "New Methods in Wiener Filtering Theory," *Proceedings of First Symp. on Eng. Appl. of Random Function Theory and Probability*, J. L. Bogdanoff and F. Kozin (eds.), Wiley, New York, 1963.
Bucy, R. S., "Optimum Finite-Time Filters for a Special Nonstationary Class of Inputs," Internal Memorandum, BBD-600, Johns Hopkins Univ. Applied Physics Lab., 1959.
Kalman, R. E. and Bucy, R. S., "New Results in Linear Filtering and Prediction Theory," *Trans. ASME*, Ser. D., *J. Basic Eng.* **83**, 95–107 (December 1961).

An advantage of the Kalman-Bucy theory is that it is very well suited for digital computations.

An interesting viewpoint as well as a new proof of the Kalman-Bucy theorem is given in

Kailath, T., "An Innovations Approach to Least-Squares Estimation Part I: Linear Filtering in Additive White Noise," *IEEE Trans. Autom. Control* **AC-13**, 646–655 (1968).

Kailath's approach also admits a very elegant solution to the smoothing problem see

Kailath, T. and Frost, P., "An Innovations Approach to Least-Squares Estimation Part II: Linear Smoothing in Additive White Noise," *IEEE Trans. Autom. Control* **AC-13**, 655–660 (1968).

The relationships between the Wiener-Kolmogorov theory and the Kalman-Bucy theory are now clearly understood. It has been shown that the Wiener-Hopf integral equation associated with the Kalman-Bucy problem can be reduced to an initial value problem for a Riccati equation.

Schumitzky, A., "On the Equivalence between Matrix Riccati Equations and Fredholm Resolvents," *J. Computer and Sys. Sci.* **2**, 76-87 (1968).

The idea of reducing the solution of an integral equation to an initial value problem for a differential equation was proposed in

Bellman, R. E., "Functional Equations in the Theory of Dynamic Programming-VII: A Partial Differential Equation for the Fredholm Resolvent," *Proc. Am. Math. Soc.* **8**, 435-440 (1957).

The problem is also persued in

Kailath, T., "Fredholm Resolvents, Wiener-Hopf Equations and Riccati Differential Equations," Report, Stanford University.

The observation that the least squares estimates for normal processes also is optimal for many different criteria is due to

Sherman, S., "Non Mean-Square Error Criteria," *Trans. IRE* **IT-4**, 125-126 (1958).

This paper also gives a more general version of Theorem 2.1.

The proofs of the Kalman filtering theorem follows Kalman's original papers. Kalman's equations have also been derived by Bryson using the method of maximum likelihood. See

Bryson, A. E. and Ho, Y. C., *Optimal Programming, Estimation and Control*, Blaisdell New York, 1968.

Bryson has also extended Kalman's results to the smoothing problem. Compare Exercise 2 of Section 5 and Exercise 4 of Section 7. A very neat solution to the smoothing problem has been given in the previously mentioned paper by Kailath and Frost. The smoothing problem has also been solved by

Zachrisson, L. E., "An Optimal Smoothing of Continuous Time Kalman Processes," Rep. no. R 24 (1968), The Royal Institute of Technology, Sweden.

The duality theorem was first formulated in Kalman and Bucy's 1961 paper quoted above. Another duality theorem is found in

Pearson, J. D., "On the Duality between Estimation and Control," *SIAM J. Control* **4**, 594-600 (1966).

It may happen that an estimator for an n:th order system is of order lower than n. Examples of this were found in Chapter 6 when the C-polynomial is of lower degree than the A-polynomial. In the state-space formulation this situation can occur when the matrix R_2 is singular. This case is usually referred to as the "Colored measurement noise" case. An intuitive method to solve this problem is given in

Bryson, A. E. and Johansen, D. E., "Linear Filtering for Time-Varying Systems Using Measurements Containing Coloured Noise," *Trans. IEEE on Autom. Control* **10**, No. 1, 4-10 (January 1965).

The problem is also discussed in

Bucy, R. S. "Optimum Finite Time Filters for a Special Nonstationary Class of Inputs," Rep. BBD 600, Johns Hopkins Applied Physics Lab., 1959.

Bucy, R. S. "Optimal Filtering for Correlated Noise," *J. Math. Anal. and Appl.* **20**, 1-8 (1967).

Bucy, R. S., Rappaport, D., and Silverman, L. M., "Correlated Noise Filtering and Invariant Directions for the Riccati Equation," Report, University of Southern California, 1969.

A direct proof of Theorem 6.2 based on a direct computation of the conditional probability density functional is found in

Bucy, R. S. and Joseph, P. D., *Filtering for Stochastic Processes with Applications to Guidance*, Wiley (Interscience), New York, 1968.

The results of Chapter 7 can be generalized to the nonlinear problem of estimating the state of the system

$$dx = f(x, t)dt + \sigma(x, t)\, dv$$
$$dy = g(x, t)dt + \lambda(x, t)\, de$$

where $\{e(t), t \in T\}$ and $\{v(t), t \in T\}$ are Wiener processes in the sense that it is possible to derive functional equations for the conditional distribution of $x(t)$ given $\mathscr{Y}_t$. In the scalar case we get the following equation for the density of the conditional distribution of $x(t)$ given $\mathscr{Y}_t$

$$dp = \left[-\frac{\partial}{\partial x}(fp) + \frac{1}{2}\frac{\partial^2}{\partial x^2}(\sigma^2 p) \right] dt$$
$$+ \frac{1}{\lambda^2}\left[g - \int gp\, dx \right]\left[dy - \left(\int gp\, dx\right) dt \right]$$

These functional equations were first derived by Stratonovich. See the book

Stratonovich, R. L., *Conditional Markov Processes and Their Applications to the Theory of Optimal Control*, Elsevier, New York, 1968.

which contains an exposé of the theory as well as many references. Stratonovich interpreted the stochastic differentials using the Stratonovich integral. Derivations of the functional equations for conditional distributions using the Ito integrals are given in

Kushner, H. J., "On the Dynamical Equations of Conditional Probability Density Functions with Applications to Optimal Stochastic Control," *J. Math. Anal. and Appl.* **8**, 332-344 (1964).

CHAPTER 8

LINEAR STOCHASTIC CONTROL THEORY

1. INTRODUCTION

In Chapter 6 we discussed a simple regulation problem for a system with one input and one output. The disturbances acting on the system were described as stochastic processes. The control problem was formulated as a variational problem by choosing the variance of the output signal as the criterion. We found in the simple example that there were strong relations between stochastic control theory and prediction theory. The essential result obtained in Chapter 6 said that the behavior of the optimal regulator can be explained as follows:

The control variable should be chosen in such a way that the predicton of the output over the time delay in the process equals the desired output. The control error then equals the prediction error.

The present chapter will discuss a much more general control problem. We will still assume a linear process model, however, we will allow the model to be time varying and we will also allow several input signals and several output signals. The criterion is taken as to minimize the mathematical expectation of a quadratic form in the state variables and the control variables. The central result which will be derived is the *separation theorem* or the so-called *certainty equivalence principle*. This result implies that the optimal control strategy can be separated in two parts, one state estimator which produces the best estimate of the state vector of the system from the observed outputs, and one linear feedback law which gives the control signal as a linear function of the estimated state. The linear control law is the same as would be used if there were no disturbances and if the state vector was known exactly. This explains the names "certainty equivalence principle" and "separation theorem."

The result is very important for the development of control theory. It is well known that in the theory of optimal control of deterministic systems there is no difference between a control strategy and a control program, i.e., between the performances of a system with feedback and an open loop system. To obtain a difference it is necessary to introduce disturbances. The linear stochastic control theory is a framework which provides this.

The formulation of the problem for discrete time systems is discussed in Section 2. Section 3 is devoted to the derivation of some preliminary results which are needed in the following. The central result is Lemma 3.1 which can be interpreted as the solution to an optimization problem with incomplete state information which does not involve any dynamics.

The case of complete state information is solved in Section 4 using Dynamic Programming. In Section 5 we solve the problem for the case of incomplete state information, again using Dynamic Programming. The proof of the separation theorem is straight-forward but it involves many different steps. With this approach it is also necessary to make explicit use of the solution of the Kalman filtering problem. An indirect proof of the separation theorem is given in Section 6. The proof which is based on an identity from the calculus of variations does not require the explicit solution of the filtering problems, and we will thus obtain a very general version of the separation theorem which also covers the case of delayed, as well as advanced, measurements. The proof will also make it possible to give a physical interpretation of the different terms of the minimal expected loss, as well as to compare the deterministic case with the cases of complete and incomplete state information.

The continuous time problem is discussed in Section 7, which covers problem statement as well as the proof of the separation theorem. The proof is the continuous time analog of the proof given in Section 6.

2. FORMULATION

The formulation of a stochastic control problem will now be discussed. We will consider a system governed by the stochastic difference equation

$$x(t+1) = \Phi x(t) + \Gamma u(t) + v(t) \quad (2.1)$$
$$y(t) = \theta x(t) + e(t) \quad (2.2)$$

where $t \in T = \{\ldots, -1, 0, 1, \ldots\}$, x is an $n \times 1$ state vector, u a $p \times 1$

vector of control variables, y an $r \times 1$ vector of outputs, and $\{v(t), t \in T\}$ and $\{e(t), t \in T\}$ are sequences of independent normal random variables with zero mean values and the covariances

$$\operatorname{cov}[v(t), v(t)] = R_1$$
$$\operatorname{cov}[v(t), e(t)] = 0$$
$$\operatorname{cov}[e(t), e(t)] = R_2 \qquad (2.3)$$

The matrices Φ, Γ, θ, R_1, and R_2 may depend on time. It is assumed that $e(t)$ and $v(t)$ are independent of $x(t)$, and that the initial state $x(t_0)$ is normal with

$$Ex(t_0) = m$$
$$\operatorname{cov}[x(t_0), x(t_0)] = R_0 \qquad (2.4)$$

It is also assumed that R_0 and R_1 are nonnegative definite and that R_2 is positive definite. The performance of the system is characterized by the scalar loss function

$$l = x^T(N)Q_0 x(N) + \sum_{t=t_0}^{N-1} [x^T(t)Q_1 x(t) + u^T(t)Q_2 u(t)] \qquad (2.5)$$

The matrices Q_0 and Q_1 are symmetric and nonnegative, the matrix Q_2 is assumed to be positive definite. The assumption on Q_2 can be relaxed. All matrices may depend on time.

As the loss l is a stochastic variable there is no straightforward way to define what we mean by the smallest value of l. We can, for example, say that l_1 is smaller than l_2 if $l_1 < l_2$ with probability one, if $\max_\omega l_1 < \min_\omega l_2$ or if $El_1 < El_2$, to give a few examples. In the following we will choose the expected loss as a criterion.

$$El = E\{x^T(N)Q_0 x(N) + \sum_{t=t_0}^{N-1} x^T(t)Q_1 x(t) + u^T(t)Q_2 u(t)\} \qquad (2.6)$$

The stochastic control problem can thus be formulated as follows.

PROBLEM 2.1

Find an admissible control strategy for the system described by (2.1) and (2.2) such that the criterion (2.6) is minimal.

To complete the problem statement, we must also specify what we mean by *admissible control strategy*. For a stochastic control problem it is very important to specify the data which is available for determining the control action. This is in sharp contrast with the deterministic case. If the matrix θ of (2.2) equals the unit matrix, and $e(t) = 0$, (2.2) reduces to

$$y(t) = x(t)$$

Preliminaries

This implies that the output signal at time t will give the exact value of the state vector. This situation is referred to as *complete state information*. In this case the control law or the control strategy is a function which maps the state space R^n into the control space R^p. Notice that since (2.1) is a stochastic state model, no additional information about the future development of the system is obtained if past measurements are included.

In most cases the state variables are not known exactly. We will call this situation the case of *incomplete state information*. In case of incomplete state information, the value of the control signal at time t is a function of all observed outputs up to time t. In analogy with the analysis of the filtering problem in Chapter 7, we introduce the quantity $\mathcal{Y}_t$ to denote the observed outputs or the available information. In the discrete time case we thus have

$$\mathcal{Y}_t^T = [y^T(t_0), y^T(t_0 + 1), \ldots, y^T(t)]$$

where $\mathcal{Y}_t \in Y_t$. The control strategy is thus a function which maps the space of observed outputs into the space of possible control actions.

It is clear from this discussion that the case of incomplete state information is much more difficult, particularly because the dimensions of the space Y_t will increase as t increases. Ways to overcome this are the central problems of stochastic control theory.

3. PRELIMINARIES

The derivation of the main result involves several steps. To make the derivation easier to follow, this section will discuss some of the steps involved in a simple setting.

A Static Optimization Problem

We will first study a stochastic optimization problem which does not involve any dynamics. A result of this will be a very important result—the fundamental lemma of stochastic control theory. This lemma clearly shows the difference between the cases of complete state information and incomplete state information. It will also illustrate the importance of specifying the information pattern or the data available for the decision in a stochastic optimization problem.

Let $x \in X$ and $y \in Y$ be two scalar stochastic variables defined on a probability space and let the control or decision variable be $u \in U$. Let the loss function l be a function which maps $X \times Y \times U$ into the real numbers. The expected loss is then

$$El(x, y, u) \tag{3.1}$$

where E denotes mathematical expectation with respect to x and y.

The minimization of (3.1) will now be discussed. It is first assumed that the admissible control strategies are all functions which map $X \times Y$ into U (complete state information). We will then consider the case when the admissible control strategies are functions which map Y into U (incomplete state information).

Complete State Information

Let $\min_{u(x,y)} El(x, y, u)$ denote the minimum of $El(x, y, u)$ with respect to all control strategies which map $X \times Y$ into U. We then have Lemma 3.1.

LEMMA 3.1

Assume that the function $l(x, y, u)$ has a unique minimum with respect to $u \in U$ for all $x \in X$ and all $y \in Y$. Let $u^o(x, y)$ denote the value of u for which the minimum is achieved. Then

$$\min_{u(x,y)} El(x, y, u) = El(x, y, u^o(x, y)) = E \min_u l(x, y, u) \quad (3.2)$$

Proof

For all admissible strategies we have

$$l(x, y, u) \geqslant l(x, y, u^o(x, y)) = \min_u l(x, y, u)$$

Hence

$$El(x, y, u) \geqslant El(x, y, u^o(x, y)) = E \min_u l(x, y, u)$$

Minimizing the left hand side with respect to all admissible strategies we get

$$\min_{u(x,y)} El(x, y, u) \geqslant El(x, y, u^o(x, y)) = E \min_u l(x, y, u) \quad (3.3)$$

Since $u^o(x, y)$ is an admissible strategy we also have

$$El((x, y, u^o(x, y)) \geqslant \min_{u(x,y)} El(x, y, u) \quad (3.4)$$

Combining the inequalities (3.3) and (3.4) we now find (3.2) and the lemma is proven.

Remark

Notice that $\min_u l(x, y, u)$ defines u as a function of x and y and that the Lemma implies that the operations of minimizing, with respect to the admissible control strategies, and taking mathematical expectation, with respect to the joint distribution of x and y, commute.

Incomplete State Information

The admissible control strategies will now be restricted to all functions which map Y into U. The control action thus has to be based on information of one of the variables only. Let $\min_{u(y)} El(x, y, u)$ denote the minimum of $El(x, y, u)$ with respect to all admissible control strategies. We have the following result.

LEMMA 3.2

Let $E[\cdot\,|y]$ denote the conditional mean given y. Assume that the function $f(y, u) = E[l(x, y, u)\,|y]$ has a unique minimum with respect to $u \in U$ for all $y \in Y$. Let $u^o(y)$ denote the value of u for which the minimum is achieved. Then

$$\min_{u(y)} El(x, y, u) = El(x, y, u^o(y)) = \underset{y}{E}\{\min_u E[l(x, y, u)\,|y]\} \quad (3.5)$$

where $\underset{y}{E}$ denotes the mean value with respect to the distribution of y.

Proof

For all admissible strategies we have

$$f(y, u) \geqslant f(y, u^o(y)) = \min_u f(y, u)$$

Hence

$$El(x, y, u) = \underset{y}{E}f(y, u) \geqslant \underset{y}{E}f(y, u^o(y)) = El(x, y, u^o(y))$$
$$= \underset{y}{E}\{\min_u E[l(x, y, u)\,|y]\}$$

Minimizing the left hand side with respect to all admissible strategies we find

$$\min_{u(y)} El(x, y, u) \geqslant El(x, y, u^o(y)) = \underset{y}{E}\{\min_u E[l(x, y, u)\,|y]\} \quad (3.6)$$

Since $u^o(y)$ is an admissible strategy we also have

$$El(x, y, u^o(y)) \geqslant \min_{u(y)} El(x, y, u) \quad (3.7)$$

Combining the inequalities (3.6) and (3.7), we now find (3.5) and the Lemma is proven.

Remark 1

Notice that $E[\cdot\,|y]$ is a function of y and that the operation $\min_u f(x, u)$ defines a function $X \to U$. The operation $\min_u E[\cdot\,|y]$ thus defines a function $Y \to U$.

Remark 2

Notice that the lemma can be interpreted that the operation of minimizing, with respect to admissible strategies $u: Y \to U$ and taking conditional expectations given y, commute.

Remark 3

Notice that in a case like the one discussed, it is very important to specify the admissible control strategies. We have

$$\min_{u(y)} El(x, y, u) \geqslant \min_{u(x,y)} El(x, y, u) \tag{3.8}$$

and we can thus conclude that the loss for the case of complete state information will not be larger than in the case of incomplete state information.

Mean Value of a Quadratic Form of Normal Stochastic Variables

In the following we will also frequently have to calculate quantities of the type

$$Ex^T Sx$$

where x is a normal random variable. We have Lemma 3.3.

LEMMA 3.3

Let x be normal with mean m and covariance R. Then

$$Ex^T Sx = m^T Sm + \operatorname{tr} SR \tag{3.9}$$

Proof

We have

$$\begin{aligned} Ex^T Sx &= E(x - m)^T S(x - m) + Em^T Sx + Ex^T Sm - Em^T Sm \\ &= E(x - m)^T S(x - m) + m^T Sm \end{aligned} \tag{3.10}$$

because

$$Ex = m$$

Further

$$(x - m)^T S(x - m) = \operatorname{tr}\,(x - m)^T S(x - m) = \operatorname{tr} S(x - m)(x - m)^T$$

Taking mathematical expectation we get

$$\begin{aligned} E(x - m)^T S(x - m) &= E \operatorname{tr} S(x - m)(x - m)^T \\ &= \operatorname{tr} SE(x - m)(x - m)^T = \operatorname{tr} SR \end{aligned} \tag{3.11}$$

Introduce (3.11) into (3.10) and the proof is completed.

Preliminaries

Exercises

1. Let u denote the control vector, y the output, and x the state which is assumed normal with mean value m and covariance R. The vectors x, y, and u are related by

$$y = Au + x$$

Consider an optimization with the loss function

$$l = q_0 + q_1^T y + q_2^T u + \frac{1}{2} y^T Q_1 y + y^T Q_{12} u + \frac{1}{2} u^T Q_2 u$$

Determine a control u which minimizes the expected loss when there is no state information (that is, u should be a function of the a priori information only), or when there is complete state information (that is, u is permitted to be a function of x). Also determine the minimal expected loss in the two cases.

2. The model of Exercise 1 has been applied in economics. See H. Theil, *Optimal Decision Rules for Government and Industry*, North Holland, Amsterdam, 1964. In this book, the American economy during the depression has been described by the model of Exercise 1. The variables have the following meaning:

   u_1 = government wage bill
   u_2 = indirect taxes
   u_3 = government expenditure on goods and services
   y_1 = total consumption
   y_2 = net investment
   y_3 = "distribution variable" = private wage bill$-2 \times$ (total real profit)

   All units are billions of dollars. The values are deviations from "desired" values. The parameters were chosen as follows

$$A = \begin{bmatrix} 0.666 & -0.188 & 0.671 \\ -0.052 & -0.296 & 0.259 \\ 0.285 & 2.358 & -1.427 \end{bmatrix}$$

$$m = \begin{bmatrix} -5.39 \\ -3.704 \\ -0.729 \end{bmatrix}$$

$$q_0 = 0; \; q_1 = 0, \; q_2 = 0, \; Q_1 = Q_2 = I, \; Q_{12} = 0$$

   Determine the optimal control for incomplete state information. Hint: Roosevelt's decisions were

$$u_1 = 0.54, \; u_2 = -2.00, \; \text{and} \; u_3 = -1.14$$

4. COMPLETE STATE INFORMATION

We will now solve the stochastic control problem formulated in Section 2 for the particular case of complete state information. It is thus assumed that the system is described by (2.1), the criterion by (2.6), and that the admissible control strategies are functions which map the state space X into the space of control variables U. To solve problem 2.1 we will use dynamic programming. We will first derive a functional equation, and we will then give an explicit solution to the functional equation.

A Functional Equation

Consider the situation at time t. The state variables $x(t_0)$, $x(t_0 + 1)$, ..., $x(t)$ have been observed and the control signal $u(t)$ should be determined. As the system is governed by a stochastic difference equation, the conditional probability distributions of future states, given past values of the states, are functions of $x(t)$ only. It is thus sufficient to choose $u(t)$ as a function of $x(t)$. The expected loss can be written as a sum of two terms

$$E\left[x^T(N)Q_0 x(N) + \sum_{s=t_0}^{N-1} x^T(s)Q_1 x(s) + u^T(s)Q_2 u(s) \right]$$
$$= E\left[\sum_{s=t_0}^{t-1} x^T(s)Q_1 x(s) + u^T(s)Q_2 u(s) \right]$$
$$+ E\left[x^T(N)Q_0 x(N) + \sum_{s=t}^{N-1} x^T(s)Q_1 x(s) + u^T(s)Q_2 u(s) \right] \quad (4.1)$$

The first term does not depend on $u(t), u(t+1), \ldots, u(N-1)$. To minimize the loss function with respect to these control variables is thus equivalent to minimize the second term of (4.1). Assuming that the minimum exists, we apply Lemma 3.1 and find

$$\min E\left[x^T(N)Q_0 x(N) + \sum_{s=t}^{N-1} x^T(s)Q_1 x(s) + u^T(s)Q_2 u(s) \right] = E\{V(x,t)\} \quad (4.2)$$

where

$$V(x,t) = \min_{u(t),\ldots,u(N-1)} E\left[x^T(N)Q_0 x(N) + \sum_{s=t}^{N-1} x^T(s)Q_1 x(s) + u^T(s)Q_2 u(s) \mid x \right] \quad (4.3)$$

and the minimum is taken with respect to $u(t), u(t+1), \ldots, u(N-1)$. It follows from Lemma 3.1 that

Complete State Information

$$V(x(t), t) = \min_{u(t)} E\{x^T(t)Q_1x(t) + u^T(t)Q_2u(t)$$
$$+ \min_{u(t+1)} E[x^T(t+1)Q_1x(t+1)$$
$$+ u^T(t+1)Q_2u(t+1) + \min_{u(t+2)} E \cdots | x(t+1)]|x(t)\}$$

Notice, however, that the definition of $V(x, t)$ (4.3) gives

$$V(x(t+1), t+1) = \min_{u(t+1)} E\{x^T(t+1)Q_1x(t+1)$$
$$+ u^T(t+1)Q_2u(t+1) + \min_{u(t+2)} E \cdots | x(t+1)\}$$
$$= \min E\left\{x^T(N)Q_0x(N) + \sum_{s=t+1}^{N-1} x^T(s)Q_1x(s)\right.$$
$$\left. + u^T(s)Q_2u(s) | x(t+1)\right\} \tag{4.4}$$

Equations (4.3) and (4.4) give the following functional equation for V

$$V(x, t) = \min_{u(t)} E[x^T(t)Q_1x(t) + u^T(t)Q_2u(t) + V(x(t+1), t+1)|x]$$
$$= \min_{u(t)} [x^T(t)Q_1x(t) + u^T(t)Q_2u(t) + E\{V(x(t+1), t+1)|x\}] \tag{4.5}$$

For $t = N$ we find

$$V(x, N) = \min_u E[x^T(N)Q_0x(N)|x] = x^TQ_0x \tag{4.6}$$

This gives the initial condition for (4.5) which is called the *Bellman equation*.

Solution of the Bellman Equation

We will now show that the solution of the functional equation (4.5) with the initial condition (4.6) is a quadratic function

$$V(x, t) = x^TS(t)x + s(t) \tag{4.7}$$

where S is a nonnegative definite matrix.

This is apparently true for $t = N$. Proceeding by induction we assume that (4.7) holds for $t + 1$ and we will then show that it holds also for t.
We have thus

$$V(x(t+1), t+1) = x^T(t+1)S(t+1)x(t+1) + s(t+1) \tag{4.7'}$$

To evaluate the functional (4.5) we must first determine the conditional distribution of $x(t + 1)$ given $x(t) = x$. Equation (2.1) gives

$$x(t+1) = \Phi x(t) + \Gamma u(t) + v(t)$$

The conditional distribution of $x(t + 1)$ given $x(t) = x$ is thus normal with mean $\Phi x + \Gamma u$ and covariance R_1. Using Lemma 3.3 we get

$$E[V(x(t+1), t+1)|x] = [\Phi x + \Gamma u]^T S(t+1)[\Phi x + \Gamma u] \\ + \text{tr } R_1 S(t+1) + s(t+1) \quad (4.8)$$

Introducing (4.7') and (4.8) into the functional (4.5) we get

$$\begin{aligned} V(x, t) &= \min_u \{x^T Q_1 x + u^T Q_2 u + (\Phi x + \Gamma u)^T S(t+1)(\Phi x + \Gamma u) \\ &\quad + \text{tr } R_1 S(t+1) + s(t+1)\} \\ &= \min_u \{x^T [\Phi^T S(t+1)\Phi + Q_1 - L^T(Q_2 + \Gamma^T S(t+1)\Gamma)L]x \\ &\quad + (u + Lx)^T (Q_2 + \Gamma^T S(t+1)\Gamma)(u + Lx) \\ &\quad + \text{tr } R_1 S(t+1) + s(t+1)\} \\ &= x^T [\Phi^T S(t+1)\Phi + Q_1 - L^T(Q_2 + \Gamma^T S(t+1)\Gamma)L]x \\ &\quad + \text{tr } R_1 S(t+1) + s(t+1) \quad (4.9) \end{aligned}$$

where

$$L = L(t) = [Q_2 + \Gamma^T S(t+1)\Gamma]^{-1} \Gamma^T S(t+1)\Phi \quad (4.10)$$

and the second equality is obtained by completing the squares. The inverse in (4.10) exists because Q_2 is positive definite and $S(t+1)$ is nonnegative definite.

The minimum is thus obtained for

$$u = -L(t)x \quad (4.11)$$

and the optimal strategy is a linear function represented by the matrix $L(t)$ or a linear feedback from all state variables. We thus find that the functional (4.5) has a solution of the form (4.7) with

$$S(t) = \Phi^T S(t+1)\Phi + Q_1 - L^T[Q_2 + \Gamma^T S(t+1)\Gamma]L \quad (4.12)$$
$$s(t) = s(t+1) + \text{tr } R_1 S(t+1) \quad (4.13)$$

To complete the induction it now remains to show that $S(t)$ is nonnegative definite. For this purpose we will rewrite (4.12). Equations (4.10) and (4.12) give

$$S(t) = \Phi^T S(t+1)[\Phi - \Gamma L] + Q_1 = [\Phi - \Gamma L]^T S(t+1)\Phi + Q_1 \quad (4.12')$$
$$Q_2 L = \Gamma^T S(t+1)[\Phi - \Gamma L] \quad (4.14)$$

Premultiplying (4.14) by L^T and subtracting we find

$$S(t) = [\Phi - \Gamma L]^T S(t+1)[\Phi - \Gamma L] + L^T Q_2 L + Q_1 \quad (4.15)$$

Hence if $S(t+1)$ is nonnegative definite, then $S(t)$ is also nonnegative definite.

The minimum of $V(x, N)$ given by (4.6) will always exist. Backtracking step by step we can then show that the minimum of $V(x, t)$ also exists and we have thus shown that the minima exist. Summing up we find Theorem 4.1.

THEOREM 4.1

Let the admissible control strategies be such that $u(t)$ is a function of $x(t)$ and t. Assume that Q_2 is positive definite and Q_1 nonnegative definite. The solution to Problem 2.1 is then given by

$$u(t) = -L(t)x(t) \tag{4.11}$$

where

$$L(t) = [Q_2 + \Gamma^T S(t+1)\Gamma]^{-1}\Gamma^T S(t+1)\Phi \tag{4.10}$$

and $S(t)$ is given by the recursive equation

$$\begin{aligned}S(t) &= \Phi^T S(t+1)\Phi + Q_1 - \Phi^T S(t+1)\Gamma \\ &\quad \times [Q_2 + \Gamma^T S(t+1)\Gamma]^{-1}\Gamma^T S(t+1)\Phi \\ &= [\Phi - \Gamma L(t)]^T S(t+1)\Phi + Q_1 \\ &= [\Phi - \Gamma L(t)]^T S(t+1)[\Phi - \Gamma L(t)] + L^T(t)Q_2 L(t) + Q_1\end{aligned} \tag{4.12}$$

with the initial condition

$$S(N) = Q_0$$

The minimum value of the loss function is given by

$$\min E\left[x^T(N)Q_0 x(N) + \sum_{s=t_0}^{N-1} x^T(s)Q_1 x(s) + u^T(s)Q_2 u(s)\right] = E[V(x, t_0)|x]$$

$$= m^T S(t_0)m + \operatorname{tr} S(t_0)R_0 + \sum_{s=t_0}^{N-1} \operatorname{tr} R_1 S(s+1) \tag{4.16}$$

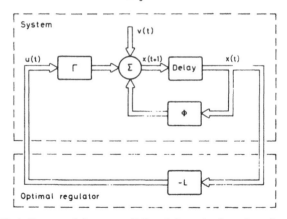

Fig. 8.1. Block diagram of the system (2.1) and the optimal regulator in the case of complete state information.

Figure 8.1 gives a block diagram of the system (2.1) and the optimal regulator (4.11).

Remark 1

Notice that the optimal strategy is the same as in the deterministic case when there are no disturbances.

Remark 2

The assumption that Q_2 is positive definite is not necessary. In order to get a unique control law it is sufficient that the quantity $Q_2 + \Gamma^T S(t+1)\Gamma$ is positive definite for all t. This can of course be true even if Q_2 is not positive definite. If $Q_2 + \Gamma^T S(t+1)\Gamma$ is not positive definite, there may exist many control laws which give the minimal value of the expected loss.

Remark 3

Notice that the terms $m^T S(t_0)m + \text{tr } S(t_0)R_0$ in (4.16) for the minimal value of the loss function depend on the distribution of the initial state, and that the terms $\sum \text{tr } R_1 S(t+1)$ are due to the disturbance v which is forcing the system (2.1).

Exercises

1. Consider the system
$$x(t+1) = x(t) + u(t) + v(t)$$
where x and u are scalars and $\{v(t)\}$ is a sequence of independent normal random variables with zero mean values and covariances r. Assume that the initial state is normal (m, σ) and let the loss function be
$$l = \sum_{k=1}^{N} x^2(k) + u^2(k)$$
Determine the control strategy which minimizes the expected loss and the minimum value of the loss function when the admissible strategies are such that $u(t)$ is a function of $x(t)$. Also determine the limiting control law as $N \to \infty$.

2. Consider the system
$$x(t+1) = ax(t) + bu(t) + v(t)$$
with the loss function
$$l = \sum_{t=1}^{N} x^2(t)$$

Let the admissible control strategies be such that $u(t)$ is a function of $x(t)$. Determine the strategy which minimizes the expected loss.
3. Consider the system and the loss function given in Exercise 2. Determine the control strategy which minimizes the expected loss when the admissible control strategies are such that $u(t)$ is a function of $x(t-1)$.
4. Show that Theorem 4.1 still holds if the random variables occuring are assumed to be uncorrelated instead of independent and if the admissible control strategies are restricted in such a way that $u(t)$ is a linear function of $x(t)$.
5. Show that (4.11) is also the optimal strategy among all admissible strategies with the property that $u(t)$ is a function of $x(t), x(t-1), \ldots$

5. INCOMPLETE STATE INFORMATION 1

We will now generalize the analysis of Section 4 to the case of incomplete state information. Thus consider the linear system (2.1), (2.2) with the criterion (2.6). It is assumed that the admissible control strategies are such that the value of the control signal at time t is a function of observed outputs up to time $t-1$. The only difference compared to the problem of Section 4 is thus in the data available for determining the control variable.

Introduce $\mathscr{Y}_t$ as the vector of observed outputs up to time t

$$\mathscr{Y}_t = [y^T(t_0), y^T(t_0+1), \ldots, y^T(t)]^T \tag{5.1}$$

The vector $\mathscr{Y}_t$ is apparently a vector in an $r \times (t - t_0 + 1)$ dimensional space Y_t and the admissible control strategies are functions which map $Y_t \times T$ into U. Notice that the dimension of Y_t increases with t.

To find the control strategy which minimizes the expected loss (2.6) we will use an argument similar to those of Section 4. We will thus first derive a functional equation using dynamic programming, and we will then procede to solve the functional equation.

A Functional Equation

Consider the situation at time t. The output signals $\mathscr{Y}_{t-1}$ or $y(t_0)$, $y(t_0+1), \ldots, y(t-1)$ have been observed and the problem is to determine a control strategy u such that the criterion is minimal. The criterion can be written as follows

$$E\left[\sum_{s=t_0}^{t-1} x^T(s)Q_1 x(s) + u^T(s)Q_2 u(s)\right]$$
$$+ E\left[x^T(N)Q_0 x(N) + \sum_{s=t}^{N-1} x^T(s)Q_1 x(s) + u^T(s)Q_2 u(s)\right] \tag{5.2}$$

Only the second term of the expression (5.2) depends on $u(t)$. Assuming that a unique minimum exists, it follows from Lemma 3.2 that

$$\min_{u(t)} E\left[x^T(N)Q_0 x(N) + \sum_{s=t}^{N-1} x^T(s)Q_1 x(s) + u^T(s)Q_2 u(s) \right]$$
$$= E \min_{u(t)} E\left[x^T(N)Q_0 x(N) + \sum_{s=t}^{N-1} x^T(s)Q_1 x(s) + u^T(s)Q_2 u(s) \mid \mathcal{Y}_{t-1} \right]$$
(5.3)

where $E[\cdot \mid \mathcal{Y}_{t-1}]$ denotes conditional expectation given $\mathcal{Y}_{t-1}$, the first E of the right member denotes expectation with respect to the distribution of $\mathcal{Y}_{t-1}$, and the minimum is taken with respect to all strategies which express $u(t)$ as a function of $\mathcal{Y}_{t-1}$.

Repeating the arguments given above for $t = N-1, N-2, \ldots$, under the assumption that all minima exist and are unique, we find

$$\min_{u(t),\ldots,u(N-1)} E\left[x^T(N)Q_0 x(N) + \sum_{s=t}^{N-1} x^T(s)Q_1 x(s) + u^T(s)Q_2 u(s) \right] = EV(\mathcal{Y}_{t-1}, t)$$
(5.4)

where the minima are taken with respect to all admissible control strategies which give $u(t)$ as a function of $\mathcal{Y}_{t-1}$ and where the function V satisfies the following functional equation

$$V(\mathcal{Y}_{t-1}, t) = \min_u E[x^T(t)Q_1 x(t) + u^T(t)Q_2 u(t) + V(\mathcal{Y}_t, t+1) \mid \mathcal{Y}_{t-1}] \quad (5.5)$$

which is the Bellman equation for the case of incomplete state information. This functional equation is quite complicated for the reason that the dimension of $\mathcal{Y}_t$ increases with t. To simplify the functional equation, we will exploit the particular structure of the system given by (2.1) and (2.2). For this purpose we will investigate the conditional distribution of $x(t)$ and $\mathcal{Y}_t$ given $\mathcal{Y}_{t-1}$.

It follows from (5.1) that

$$\mathcal{Y}_t^T = [\mathcal{Y}_{t-1}^T, y^T(t)]$$

The first components of $\mathcal{Y}_t$ are thus identical to those of $\mathcal{Y}_{t-1}$. To determine the conditional distribution of $\mathcal{Y}_t$ given $\mathcal{Y}_{t-1}$ it is thus sufficient to know the distribution of $y(t)$ given $\mathcal{Y}_{t-1}$. We have (2.2)

$$y(t) = \theta x(t) + e(t)$$

The conditional distribution of $y(t)$ given $\mathcal{Y}_{t-1}$ is thus uniquely determined by the conditional distribution of $x(t)$ given $\mathcal{Y}_{t-1}$.

To proceed we will use the results on filtering theory developed in Chapter 7. It follows from Remark 3 of Theorem 4.1 of Chapter 7 that the conditional mean

Incomplete State Information 1

$$\hat{x}(t) = E[x(t)) \mid \mathcal{Y}_{t-1}] \tag{5.6}$$

is a sufficient statistic for the conditional distribution of $x(t)$ given $\mathcal{Y}_{t-1}$. The left member of (5.5) is thus a function of $\hat{x}(t)$. Compare (4.24) of Chapter 7. We can thus introduce

$$W(\hat{x}(t), t) = V(\mathcal{Y}_{t-1}, t)$$

$$= \min_{u(t),\ldots,u(N-1)} E\left[x^T(N)Q_0 x(N) + \sum_{s=t}^{N-1} x^T(s)Q_1 x(s) + u^T(s)Q_2 u(s) \mid \mathcal{Y}_{t-1} \right]$$

$$= \min_{u(t),\ldots,u(N-1)} E\left[x^T(N)Q_0 x(N) + \sum_{s=t}^{N-1} x^T(s)Q_1 x(s) + u^T(s)Q_2 u(s) \mid \hat{x}(t) \right] \tag{5.7}$$

where the last equality follows from (4.24) of Chapter 7 and the minimum is taken with respect to all strategies which give $u(t)$ as a function of $\mathcal{Y}_{t-1}$. Using the function W, we now find that the functional (5.5) can be written as

$$W(\hat{x}(t), t) = \min_u E[x^T(t)Q_1 x(t) + u^T Q_2 u + W(\hat{x}(t+1), t+1) \mid \mathcal{Y}_{t-1}]$$

$$= \min_u E[x^T(t)Q_1 x(t) + u^T Q_2 u + W(\hat{x}(t+1), t+1) \mid \hat{x}(t)] \tag{5.8}$$

This is a considerable simplification because the argument $\hat{x}$ of W is of constant dimension. The dimension of $\hat{x}$ is also frequently much lower than the dimension of $\mathcal{Y}_t$. The initial condition for the functional (5.8) is

$$W(\hat{x}, N) = E[x^T(N)Q_0 x(N) \mid \hat{x}] \tag{5.9}$$

Solution of the Functional Equation

We will now solve the functional (5.8) with the initial condition (5.9). To do this we will use the result of Theorem 4.1 of Chapter 7 which says that the conditional distribution of $x(N)$ given $\mathcal{Y}_{N-1}$ is normal with mean $\hat{x}$ and covariance $P(N)$. It then follows from Lemma 3.3 that

$$W(\hat{x}, N) = \hat{x}^T Q_0 \hat{x} + \operatorname{tr} Q_0 P(N) \tag{5.10}$$

We will now show that the functional (5.8) has a solution which is a quadratic function

$$W(\hat{x}, t) = \hat{x}^T S(t) \hat{x} + s(t) \tag{5.11}$$

This is apparently true for $t = N$. Proceeding by induction we assume that (5.11) holds for $t + 1$ and we will then show that it holds also for t.

To evaluate the right member of (5.8) we must know the conditional distributions of $x(t)$ and $\hat{x}(t+1)$ given $\mathcal{Y}_{t-1}$. It follows from Theorem 4.1 of Chapter 7 that the conditional distribution of $x(t)$ given $\mathcal{Y}_{t-1}$ is normal with mean $\hat{x}(t)$ and covariance matrix $P(t)$. Lemma 3.3 then gives

$$E[x^T(t)Q_1x(t) \mid \mathscr{Y}_{t-1}] = \hat{x}^T(t)Q_1\hat{x}(t) + \operatorname{tr} Q_1P(t) \quad (5.12)$$

It also follows from Theorem 4.1 of Chapter 7 that

$$\hat{x}(t+1) = \Phi\hat{x}(t) + \Gamma u(t) + K(t)[y(t) - \theta\hat{x}(t)] \quad (5.13)$$

Compare with (4.23) of Chapter 7. We have further

$$y(t) - \theta\hat{x}(t) = \theta(x(t) - \hat{x}(t)) + e(t) = \tilde{y}(t) \quad (5.14)$$

Since the conditional distribution of $y(t) - \theta\hat{x}(t)$ given $\mathscr{Y}_{t-1}$ is normal with zero mean and covariance matrix $\theta P(t)\theta^T + R_2$ we get

$$E[\hat{x}(t+1) \mid \mathscr{Y}_{t-1}] = \Phi\hat{x}(t) + \Gamma u(t) \quad (5.15)$$

and

$$\operatorname{cov}[\hat{x}(t+1) \mid \mathscr{Y}_{t-1}] = K(t)[\theta P(t)\theta^T + R_2]K^T(t) \quad (5.16)$$

Summing up we find

$$\begin{aligned}
W(\hat{x}(t), t) &= \min_u \{\hat{x}(t)^T Q_1 \hat{x}(t) + \operatorname{tr} Q_1 P(t) + u^T Q_2 u \\
&\quad + [\Phi\hat{x}(t) + \Gamma u]^T S(t+1)[\Phi\hat{x}(t) + \Gamma u] \\
&\quad + \operatorname{tr} S(t+1)K(t)[\theta P(t)\theta^T + R_2]K^T(t) + s(t+1)\} \\
&= \min_u \{\hat{x}^T[\Phi^T S(t+1)\Phi + Q_1 - L^T(Q_2 + \Gamma^T S(t+1)\Gamma)L]\hat{x} \\
&\quad + (u + L\hat{x})^T[Q_2 + \Gamma^T S(t+1)\Gamma](u + L\hat{x}) + \operatorname{tr} Q_1 P(t) \\
&\quad + \operatorname{tr} S(t+1)K(t)[\theta P(t)\theta^T + R_2]K^T(t) + s(t+1)\} \\
&= \hat{x}^T[\Phi^T S(t+1)\Phi + Q_1 - L^T(Q_2 + L^T S(t+1)L)L]\hat{x} \\
&\quad + \operatorname{tr} Q_1 P(t) + \operatorname{tr} S(t+1)K(t) \\
&\quad \times [\theta P(t)\theta^T + R_2]K^T(t) + s(t+1)
\end{aligned} \quad (5.17)$$

where

$$L(t) = [Q_2 + \Gamma^T S(t+1)\Gamma]^{-1}\Gamma^T S(t+1)\Phi \quad (5.18)$$

and the minimum is assumed for

$$u(t) = -L(t)\hat{x}(t) \quad (5.19)$$

The details of the analysis presented above are completely analogous to the corresponding analysis for the case of complete state information given in Section 4. We thus find that the functional (5.8) has a solution of the form (5.11) with

$$\begin{aligned}
S(t) &= \Phi^T S(t+1)\Phi + Q_1 - L^T(t)[Q_2 + \Gamma^T S(t+1)\Gamma]L(t) \\
&= [\Phi - \Gamma L(t)]^T S(t+1)\Phi + Q_1 \\
&= [\Phi - \Gamma L(t)]^T S(t+1)[\Phi - \Gamma L(t)] + L(t)Q_2 L^T(t) + Q_1
\end{aligned} \quad (5.20)$$

$$s(t) = s(t+1) + \operatorname{tr} Q_1 P(t) + \operatorname{tr} S(t+1)K(t)[\theta P(t)\theta^T + R_2]K^T(t) \quad (5.21)$$

The minimum value of the loss function is given by

Incomplete State Information 1

$$\min_{u(t_0),\ldots,u(N-1)} E\left[x^T(N)Q_0 x(N) + \sum_{s=t_0}^{N-1} x^T(s)Q_1 x(s) + u^T(s)Q_0 u(s) \right]$$
$$= E[W(\hat{x}(t_0), t_0)] = E[\hat{x}^T(t_0)S(t_0)\hat{x}(t_0) + s(t_0)]$$
$$= m^T S(t_0)m + \sum_{s=t_0}^{N-1} \text{tr } Q_1 P(s)$$
$$+ \sum_{s=t_0}^{N-1} \text{tr } S(s+1)K(s)[\theta P(s)\theta^T + R_2]K^T(s) + \text{tr } Q_0 P(N) \quad (5.22)$$

The minimum value of the loss function can be written in a slightly different form which admits physical interpretations. To obtain this, consider (5.20) and (4.21) of Chapter 7. We get

$$P(t+1) = \Phi P(t)\Phi^T + R_1 - K(t)[\theta P(t)\theta^T + R_2]K^T(t)$$
$$S(t) = \Phi^T S(t+1)\Phi + Q_1 - L^T(t)[\Gamma^T S(t+1)\Gamma + Q_2]L(t)$$

Premultiply the first equation by $S(t+1)$, the second by $P(t)$, take the trace of the difference, and we find

$$\text{tr } P(t+1)S(t+1) - \text{tr } P(t)S(t)$$
$$= \text{tr } S(t+1)R_1 - \text{tr } S(t+1)K(t)[\theta P(t)\theta^T + R_2]K^T(t)$$
$$- \text{tr } P(t)Q_1 + \text{tr } P(t)L^T(t)[\Gamma^T S(t+1)]\Gamma + Q_2]L(t)$$

Summing in t from t_0 to $N-1$ and we find

$$\text{tr } P(N)S(N) + \sum_{t=t_0}^{N-1}\{\text{tr } P(t)Q_1 + \text{tr } S(t+1)K(t)[\theta P(t)\theta^T + R_2]K^T(t)\}$$
$$= \text{tr } P(t_0)S(t_0) + \sum_{t=t_0}^{N-1}\{\text{tr } R_1 S(t+1) + \text{tr } P(t)L^T(t)$$
$$\times [\Gamma^T S(t+1)\Gamma + Q_2]L(t)\} \quad (5.23)$$

Combining (5.22) with (5.23), and using (4.10), we thus find that the minimum value of the risk function is

$$\min_{u(t_0),\ldots,u(N-1)} E\left[x^T(N)Q_0 x(N) + \sum_{t=t_0}^{N-1} x^T(t)Q_1 x(t) + u^T(t)Q_2 u(t) \right]$$
$$= m^T S(t_0)m + \text{tr } R_0 S(t_0) + \sum_{t=t_0}^{N-1} \text{tr } R_1 S(t+1)$$
$$+ \sum_{t=t_0}^{N-1} \text{tr } P(t)L^T(t)\Gamma^T S(t+1)\Phi \quad (5.24)$$

It now remains to show that the minimum exists. We find trivially that the minimum of $W(\hat{x}, N)$ exists. Backtracking we find that, as all functions involved are quadratic, the minima of all $W(\hat{x}, t)$ will also exist. Summing up we find Theorem 5.1.

THEOREM 5.1

The solution of the optimal control problem in the case of incomplete state information is given by the control strategy

$$u(t) = - L(t)\hat{x}(t) \qquad (5.19)$$

where

$$L(t) = [Q_2 + \Gamma^T S(t + 1)\Gamma]^{-1}\Gamma^T S(t + 1)\Phi \qquad (5.18)$$
$$S(t) = \Phi^T S(t + 1)\Phi + Q_1 - L^T(t)[Q_2 + \Gamma^T S(t + 1)\Gamma]L(t)$$
$$= [\Phi - \Gamma L(t)]^T S(t + 1)\Phi + Q_1$$
$$= [\Phi - \Gamma L(t)]^T S(t + 1)[\Phi - \Gamma L(t)] + L^T(t)Q_2 L(t) + Q_1 \qquad (5.20)$$
$$S(N) = Q_0$$

and $\hat{x}(t)$ is the conditional mean given by

$$\hat{x}(t + 1) = \Phi\hat{x}(t) + \Gamma u(t) + K(t)[y(t) - \theta\hat{x}(t)] \qquad (5.25)$$

The minimal expected loss is

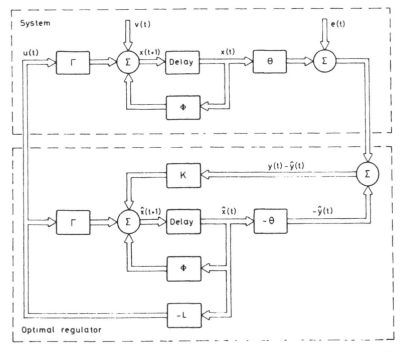

Fig. 8.2. Block diagram of the system (2.1), (2.2), and the optimal regulator (5.19), (5.25) in the case of incomplete state information.

$$\min El = m^T S(t_0) m + \operatorname{tr} S(t_0) R_0 + \sum_{t=t_0}^{N-1} \operatorname{tr} S(t+1) R_1(t)$$
$$+ \sum_{t=t_0}^{N-1} \operatorname{tr} P(t) L^T(t) \Gamma^T S(t+1) \Phi$$

A block diagram of the optimal system is shown in Fig. 8.2.

Remark 1

Notice that the optimal strategy given by (5.19) and (5.25) can be separated into two parts. Equation (5.25) is an algorithm for computing the minimum mean square estimate of the state variables from the observed data, and (5.19) is a linear feedback from the estimated state. The dynamics of the optimal feedback law arise from the state estimation. The feedback matrix L depends only on the system dynamics Φ and Γ and the parameters of the loss function Q_0, Q_1, and Q_2, but does not depend on the characteristics of the disturbances.

Remark 2

The feedback matrix L is the same matrix as is obtained when solving the optimal control problem for the deterministic system

$$x(t+1) = \Phi x(t) + \Gamma u(t)$$

with the criterion

$$x^T(N) Q_0 x(N) + \sum_{t=t_0}^{N-1} x^T(t) Q_1 x(t) + u^T(t) Q_2 u(t)$$

minimal.

Remark 3

It follows from the derivation of Theorem 5.1 that the matrices Φ, Γ, θ, Q_1, Q_2, R_1 and R_2 may depend on time.

Properties of the Closed Loop System

Theorem 5.1 is a powerful tool for the design of linear control systems. We will now discuss the properties of the closed loop system obtained when the system described by (2.1) and (2.2) is controlled using the optimal strategy given by (5.19) and (5.25).

The closed loop system is thus described by

$$x(t+1) = \Phi x(t) + \Gamma u(t) + v(t)$$
$$y(t) = \theta x(t) + e(t)$$
$$u(t) = -L\hat{x}(t)$$
$$\hat{x}(t+1) = \Phi \hat{x}(t) + \Gamma u(t) + K[y(t) - \theta \hat{x}(t)]$$

The closed loop system is thus of order $2n$. Changing the coordinates in the state space from x and $\hat{x}$ to x and $\tilde{x}$ where

$$\tilde{x}(t) = \hat{x}(t) - x(t)$$

we find

$$x(t+1) = [\Phi - \Gamma L]x(t) + \Gamma L \tilde{x}(t) + v(t)$$
$$\tilde{x}(t+1) = [\Phi - K\theta]\tilde{x}(t) + v(t) - Ke(t)$$

The dynamics of the closed loop system are thus determined by the matrices $\Phi - \Gamma L$ and $\Phi - K\theta$, that is, the dynamics of the corresponding deterministic system and the dynamics of the optimal filter. If the matrices $[\Phi - \Gamma L]$ and $[\Phi - K\theta]$ are constant, the eigenvalues of the closed loop system will be the eigenvalues of $[\Phi - \Gamma L]$ and $[\Phi - K\theta]$.

Exercises

1. Consider the system

$$x(t+1) = x(t) + u(t) + v(t)$$
$$y(t) = x(t) + e(t)$$

where $\{v(t)\}$ and $\{e(t)\}$ are sequences of independent Gaussian random variables with zero mean values and variances r_1 and r_2. Assume that the initial state is normal (m, σ) and let the loss function be

$$l = \sum_{k=1}^{N} x^2(k) + qu^2(k)$$

Determine the control strategy which minimizes the expected loss when

(a) $r_1 = r_2 = 0$ (b) $r_1 \neq 0, r_2 = 0$
(c) $r_1 \neq 0, r_2 \neq 0$

In case (c) it is assumed that $u(t)$ is a function of $y(t-1), y(t-2), \ldots$.

2. Consider the problem of Exercise 1. Determine the pulse transfer function representing the optimal control law when $N \to \infty$. Give an asymptotic expression for the expected loss and interpret the different terms physically.

3. Consider the system

$$x(t+1) = \Phi x(t) + e(t)$$

where $\{e(t)\}$ is a sequence of normal random variables with zero mean values. The covariance of $e(t)$ is $R(t)$. The initial state is normal with mean value m and covariance R_0. Show that

Incomplete State Information 1

$$E\left[x^T(t_1)Q_0x(t_1) + \sum_{s=t_0}^{t_1-1} x^T(s)Q_1x(s)\right]$$

$$= m^T S(t_0)m + \operatorname{tr} S(t_0)R_0 + \sum_{k=t_0}^{t_1-1} \operatorname{tr} S(k+1)R(k)$$

where

$$S(t) = \Phi^T S(t+1)\Phi + Q_1$$
$$S(t_1) = Q_0$$

4. Consider the system (2.1) with the criterion (4.1). Let the admissible control strategies be such that the value of the control signal at time t is a function of the state x at time $t - 1$. Show that the optimal strategy is given by

$$u(t) = -L(t)\Phi x(t-1) - L(t)\Gamma u(t-1)$$
$$= -L(t)\Phi(t; t-1)x(t-1) - L(t)\Gamma(t-1)u(t-1)$$

where L is given by (5.18). Also determine the minimal value of the loss function.

Hint: Introduce

$$V(x(t-1), t) = \min E\left[x^T(N)Q_0x(N) + \sum_{k=t}^{N-1} x^T(k)Q_1x(k) + u^T(k)Q_2u(k) \,\Big|\, x(t-1)\right]$$

and use dynamic programming.

5. Consider the system (2.1) with the criterion (4.1). Let the admissible controls be such that u is a function of apriori data only. (Open loop control.) Show that the optimal control signal is given by

$$u(t) = -L(t)m(t)$$

where

$$\begin{cases} m(t+1) = \Phi m(t) + \Gamma u(t) = (\Phi - \Gamma L)m(t) \\ m(t_0) = Ex(t_0) = m \end{cases}$$

Also determine the minimal value of the loss function.

Hint: Introduce

$$V(m(t), t) = \min E[x^T(N)Q_0x(N) + \sum_{k=t}^{N-1} x^T(k)Q_1x(k) + u^T(k)Q_2u(k)]$$

and use dynamic programming.

6. Compare the results of Theorem 4.1 with those of Exercises 4 and 5 and discuss the importance of specifying the information available to determine the value of the control signal at time t. Compare the results of open loop control and closed loop control. Analyze the effect on the loss function of a delay of one time unit in the measurement of the state vector.

6. INCOMPLETE STATE INFORMATION 2

This section will prove the separation theorem by methods which differ from the straightforward proof of Section 5. The approach, which is based on an identity from the calculus of variations, will provide additional insight into the problem and make it possible to generalize Theorem 5.1.

We will first prove the identity and we will then show how it can be applied to the linear quadratic control problem both in the deterministic case and in the cases of complete and incomplete state information.

An Identity

Consider a system described by the stochastic difference equation

$$x(t+1) = \Phi x(t) + \Gamma u(t) + v(t) \qquad (6.1)$$

We have Lemma 6.1.

LEMMA 6.1

Assume that the difference equation

$$S(t) = \Phi^T S(t+1)\Phi + Q_1 \\ - \Phi^T S(t+1)\Gamma[Q_2 + \Gamma^T S(t+1)\Gamma]^{-1}\Gamma^T S(t+1)\Phi \qquad (6.2)$$

with the initial condition

$$S(N) = Q_0 \qquad (6.3)$$

has a solution which is positive semidefinite for $t_0 \leqslant t \leqslant N$. Let the matrix L be defined by

$$L(t) = [Q_2 + \Gamma^T S(t+1)\Gamma]^{-1}\Gamma^T S(t+1)\Phi \qquad (6.4)$$

Then

$$x^T(N)Q_0 x(N) + \sum_{t=t_0}^{N-1} x^T(t)Q_1 x(t) + u^T(t)Q_2 u(t) = x^T(t_0)S(t_0)x(t_0) \\ + \sum_{t=t_0}^{N-1} [u(t) + L(t)x(t)]^T [\Gamma^T S(t+1)\Gamma + Q_2][u(t) + L(t)x(t)]$$

Incomplete State Information 2   279

$$+ \sum_{t=t_0}^{N-1} \{v^T(t)S(t+1)[\Phi x(t) + \Gamma u(t)] \\ + [\Phi x(t) + \Gamma u(t)]^T S(t+1)v(t) + v^T(t)S(t+1)v(t)\} \quad (6.5)$$

Proof

The proof is straightforward. We have the indentity

$$x^T(N)Q_0 x(N) = x^T(N)S(N)x(N) = x^T(t_0)S(t_0)x(t_0) \\ + \sum_{t=t_0}^{N-1} [x^T(t+1)S(t+1)x(t+1) - x^T(t)S(t)x(t)] \quad (6.6)$$

Consider the different terms of the sum. We have

$$x^T(t+1)S(t+1)x(t+1) \\ = [\Phi x(t) + \Gamma u(t) + v(t)]^T S(t+1)[\Phi x(t) + \Gamma u(t) + v(t)] \quad (6.7)$$

and

$$x^T(t)S(t)x(t) = x^T(t)\{\Phi^T S(t+1)\Phi + Q_1 \\ - L^T(t)[\Gamma^T S(t+1)\Gamma + Q_2]L(t)\}x(t) \quad (6.8)$$

Introducing (6.7) and (6.8) into (6.6) we find

$$x^T(N)Q_0 x(N) = x^T(t_0)S(t_0)x(t_0) + \sum_{t=t_0}^{N-1}\{[\Phi x(t) + \Gamma u(t)]^T S(t+1)v(t) \\ + v^T(t)S(t+1)[\Phi x(t) + \Gamma u(t)] + v^T(t)S(t+1)v(t)\} \\ + \sum_{t=t_0}^{N-1} \{u^T(t)[\Gamma^T S(t+1)\Gamma + Q_2]u(t) \\ + u^T(t)\Gamma^T S(t+1)\Phi x(t) + x^T(t)\Phi^T S(t+1)\Gamma u(t) \\ + x^T(t)L^T(t)[\Gamma^T S(t+1)\Gamma + Q_2]L(t)x(t) \\ - x^T(t)Q_1 x(t) - u^T(t)Q_2 u(t)\}$$

where the term $u^T Q_2 u$ has been added and subtracted in the last sum. Rearrangement of the terms now completes the proof of the lemma.

Having obtained Lemma 6.1, we have the tool to construct sufficient conditions for optimality.

The Deterministic Case

In order to make comparisons we will first consider the deterministic case, i.e., $v(t) = 0$. Lemma 6.1 then gives

$$x^T(N)Q_0 x(N) + \sum_{t=t_0}^{N-1} x^T(t)Q_1 x(t) + u^T(t)Q_2 u(t) = x^T(t_0)S(t_0)x(t_0)$$

$$+ \sum_{t=t_0}^{N-1} [u(t) + L(t)x(t)]^T$$
$$\times [\Gamma^T S(t+1)\Gamma + Q_2][u(t) + L(t)x(t)] \tag{6.9}$$

Since $S(t)$ was asssumed nonnegative definite, the second term of (6.9) is thus nonnegative. As S does not depend on u, we get

$$x^T(N)Q_0 x(N) + \sum_{t=t_0}^{N-1} x^T(t)Q_1 x(t) + u^T(t)Q_2 u(t) \geq x^T(t_0)S(t_0)x(t_0) \tag{6.10}$$

where equality is obtained for the control law

$$u(t) = -L(t)x(t) \tag{6.11}$$

Summing up we get Theorem 6.1.

THEOREM 6.1

Consider the system (6.1) with $v(t) \equiv 0$

$$x(t+1) = \Phi x(t) + \Gamma u(t) \tag{6.12}$$

where the initial state is normal with mean value m and covariance R_0. Let the criterion be defined by the loss function

$$l = x^T(N)Q_0 x(N) + \sum_{t=t_0}^{N-1} [x^T(t)Q_1 x(t) + u^T(t)Q_2 u(t)] \tag{6.13}$$

and let the admissible control strategies be such that $u(t)$ is a function of $x(t)$. Assume that the equation (6.2) with initial condition (6.3) has a solution S which is nonnegative definite, such that $Q_2 + \Gamma^T S \Gamma$ is positive definite for all t. Then there exists a unique control strategy (6.11) which minimizes the expected loss. The minimal value of the expected loss is

$$\min El = m^T S(t_0) m + \text{tr } S(t_0) R_0 \tag{6.14}$$

Equation (6.14) follows from Eq. (6.10) and Lemma 3.3.

Complete State Information

We will now consider the system (6.1). Control strategies which minimize the expected loss when the admissible strategies are such that $u(t)$ is a function of $x(t)$ will be derived from Lemma 6.1.

Taking the mathematical expectation of (6.5) we find

$$E\left[x^T(N)Q_0 x(N) + \sum_{t=t_0}^{N-1} x^T(t)Q_1 x(t) + u^T(t)Q_2 u(t) \right]$$
$$= E\left\{ x^T(t_0) S(t_0) x(t_0) + \sum_{t=t_0}^{N-1} v^T(t)S(t+1)v(t) + \sum_{t=t_0}^{N-1} [u(t) + L(t)x(t)]^T \right.$$
$$\left. \times [\Gamma^T S(t+1)\Gamma + Q_2][u(t) + L(t)x(t)] \right\} \tag{6.15}$$

Incomplete State Information 2

because $v(t)$ is independent of $x(t)$ and $u(t)$. Lemma 3.3 gives

$$Ex^T(t_0)S(t_0)x(t_0) = m^TS(t_0)m + \text{tr } S(t_0)R_0$$
$$Ev^T(t)S(t+1)v(t) = \text{tr } S(t+1)R_1(t)$$

Hence

$$E\left[x^T(N)Q_0x(N) + \sum_{t=t_0}^{N-1} x^T(t)Q_1x(t) + u^T(t)Q_2u(t)\right]$$
$$\geq m^TS(t_0)m + \text{tr } S(t_0)R_0 + \sum_{t=t_0}^{N-1} \text{tr } S(t+1)R_1(t) \quad (6.16)$$

because the last term in (6.15) is the mean value of nonnegative terms. Equality in (6.16) is obtained for the control law

$$u(t) = -L(t)x(t) \quad (6.17)$$

which is an admissible strategy in the case of complete state information.

THEOREM 6.2

Consider the system (6.1). Let the admissible control strategies be such that $u(t)$ is a function of $x(t)$. Assume that (6.2) with initial condition (6.3) has a solution S such that S is nonnegative definite and $Q_2 + \Gamma^TS\Gamma$ is positive definite. Then there exists a unique admissible control strategy (6.17) which minimizes the expected loss. The minimal value of the expected loss is

$$\min El = m^TS(t_0)m + \text{tr } S(t_0)R_0 + \sum_{t=t_0}^{N-1} \text{tr } S(t+1)R_1(t) \quad (6.18)$$

Remark

Notice that the control law (6.17) is identical to (6.11). Also compare the minimal values of the expected loss in the deterministic case and in the case of complete state information. Notice that $\sum \text{tr } S(t+1)R_1(t)$ is the contribution to the loss function due to the disturbance v.

Incomplete State Information

We will now turn to the case of incomplete state information. It is assumed that the admissible control strategies are such that $u(t)$ is a function of $\mathcal{Y}_{t-1}$. Equation (6.15) still holds in this case. Since (6.17) is not an admissible control law, we find, however, that it is not possible to make the term

$$E \sum_{t=t_0}^{N-1} [u(t) + L(t)x(t)]^T[\Gamma^TS(t+1)\Gamma + Q_2][u(t) + L(t)x(t)] \quad (6.19)$$

of (6.15) equal to zero.

Let $\hat{x}$ be the conditional mean
$$\hat{x}(t) = E[x(t) \mid \mathcal{Y}_{t-1}] \tag{6.20}$$
and P the conditional covariance
$$P(t) = \text{cov}\,[x(t) \mid \mathcal{Y}_{t-1}] \tag{6.21}$$
It follows from Lemma 3.2 that
$$\min_{u(t)} E\{[u(t) + L(t)x(t)]^T[\Gamma^T S(t+1)\Gamma + Q_2][u(t) + L(t)x(t)]\}$$
$$= E \min_{u(t)} E\{[u(t) + L(t)x(t)]^T[\Gamma^T S(t+1)\Gamma + Q_2]$$
$$\times [u(t) + L(t)x(t)] \mid \mathcal{Y}_{t-1}\} \tag{6.22}$$
where the minimum is taken with respect to all admissible control strategies, i.e., all strategies such that $u(t)$ is a function of $\mathcal{Y}_{t-1}$.

The conditions of Lemma 3.2 are fulfilled because the function is quadratic. Lemma 3.3 now gives
$$E\{[u(t) + L(t)x(t)]^T[\Gamma^T S(t+1)\Gamma + Q_2][u(t) + L(t)x(t)] \mid \mathcal{Y}_{t-1}\}$$
$$= \text{tr}\, P(t) L^T(t) [\Gamma^T S(t+1)\Gamma + Q_2] L(t)$$
$$+ [u(t) + L(t)\hat{x}(t)]^T[\Gamma^T S(t+1)\Gamma + Q_2][u(t) + L(t)\hat{x}(t)] \tag{6.23}$$
Since $P(t)$ does not depend on u (compare Exercise 4.3 of Chapter 7), we thus find that the optimal strategy is given by
$$u(t) = -L(t)\hat{x}(t) = -L(t) E[x(t) \mid \mathcal{Y}_{t-1}] \tag{6.24}$$
and the minimal value of the loss function is
$$\min E\left[x^T(N) Q_0 x(N) + \sum_{t=t_0}^{N-1} x^T(t) Q_1 x(t) + u^T(t) Q_2 u(t)\right]$$
$$= m^T S(t_0) m + \text{tr}\, S(t_0) R_0 + \sum_{t=t_0}^{N-1} \text{tr}\, S(t+1) R_1$$
$$+ \sum_{t=t_0}^{N-1} \text{tr}\, P(t) L^T(t) [\Gamma^T S(t+1)\Gamma + Q_2] L(t) \tag{6.25}$$
Summing up we find Theorem 6.3.

THEOREM 6.3

Consider the system (6.1). Let the admissible control strategies be such that $u(t)$ is a function of $\mathcal{Y}_{t-1}$. Assume that (6.2) with initial condition (6.3) has a solution S such that S is nonnegative definite and $Q_2 + \Gamma^T S \Gamma$ is positive definite for all t. Then there exists a unique admissible control strategy (6.24) which minimizes the expected loss. The minimal value of the expected loss is given by (6.25).

Remark

Combining this theorem with Theorem 4.1 of Chapter 7, we have an alternative proof of the separation Theorem 5.1. Notice, however, that the alternative proof does not require the recursive equations for the conditional mean explicitly. It is thus easy to modify Theorem 6.3 to other situations, that is, the case when $u(t)$ is a function of $\mathcal{Y}_t$.

Interpretation of the Minimal Loss Function

It is interesting to compare the minimal values of the expected loss in the different cases. In the deterministic case, the minimal value of the loss function is simply a quadratic form in the initial state, $m^T S(t_0) m$, where m denotes the initial state. If we have a stochastic initial condition characterized by a normal distribution with covariance matrix R_0 we get the additional term tr $R_0 S(t_0)$.

If we have the stochastic case with complete state information, we get the additional term $\sum \text{tr } R_1 S(t+1)$ due to the disturbances which are acting on the system. Finally in the case of incomplete state information, the minimal risk increases with the term

$$\sum \text{tr } P(t) L^T(t) [\Gamma^T S(t+1) \Gamma + Q_2] L(t)$$

which represents the additional risk due to the uncertainty in the state estimation.

Exercises

1. Consider the system

$$x(t+1) = ax(t) + u(t-k) + v(t)$$
$$y(t) = x(t) + e(t)$$

where $\{v(t)\}$ and $\{e(t)\}$ are sequences of independent Gaussian random variables with zero mean values and variances r_1 and r_2. Assume that the initial state is normal (m, σ) and let the loss function be

$$l = \sum_{k=1}^{N} x^2(k) + qu^2(k)$$

Determine the control strategy which minimizes the expected loss when the admissible control strategy is such that $u(t)$ is a function of $y(t)$, $y(t-1), \ldots$. Also determine the asymptotic expression for the minimal value of the expected loss as $N \to \infty$.

2. Prove the separation theorem when the admissible control strategies are such that $u(t)$ is a function of $\mathcal{Y}_t$.

3. Consider the system described by
$$y(t) + ay(t-1) = u(t-1) + bu(t-2) + e(t) + ce(t-1)$$
where $\{e(t)\}$ is a sequence of independent normal $(0, 1)$ random variables. Show that the system can be represented as
$$x(t+1) = \begin{bmatrix} -a & 1 \\ 0 & 0 \end{bmatrix} x(t) + \begin{bmatrix} 1 \\ b \end{bmatrix} u(t) + \begin{bmatrix} 1 \\ c \end{bmatrix} e(t+1)$$
$$y(t) = [1 \quad 0] x(t)$$
Determine a control strategy which minimizes the expected loss
$$Ex_1^2(t+1)$$
when the admissible control strategies are such that $u(t)$ is a function of $y(t), y(t-1), \ldots$.

4. Consider the system of Exercise 3. Let the criterion be to minimize the expected loss
$$E \frac{1}{N} \sum_{k=0}^{N} y^2(k)$$
when the admissible control strategies are such that $u(t)$ is a function of $y(t), y(t-1), \ldots$.

5. Consider the system
$$x(t+1) = \Phi x(t) + v(t)$$
where $\{v(t)\}$ is a sequence of independent normal random variables with zero mean. The covariance of $v(t)$ is $R_1(t)$ and $v(t)$ is assumed independent of $x(t)$. The initial state is normal with mean value m and covariance R_0. Show that
$$E[x^T(N)Q_0 x(N) + \sum_{s=t_0}^{N-1} x^T(s) Q_1 x(s)]$$
$$= m^T S(t_0) m + \text{tr } S(t_0) R_0 + \sum_{s=t_0}^{N-1} \text{tr } S(k+1) R_1(k)$$
$$S(t) = \Phi^T S(t+1) \Phi + Q_1$$
$$S(N) = Q_0.$$

6. Consider the system
$$x(t+1) = \Phi x(t) + \Gamma u(t) + v(t)$$
where $\{v(t)\}$ is a sequence of independent, normal random variables with zero mean value. It is assumed that $v(t)$ is independent of $x(t)$. The covariance of $v(t)$ is $R_1(t)$. The initial state is normal with mean m_0 and covariance R_0. Let the loss function be

$$l = x^T(N)Q_0 x(N) + \sum_{k=t_0}^{N-1} x^T(k)Q_1 x(k) + u^T(k)Q_2 u(k).$$

Show that the optimal open loop control is

$$u(t) = -L(t)m(t)$$

where L is given by (6.4) and $m(t)$ is given by

$$m(t+1) = (\Phi - \Gamma L)m(t)$$
$$m(t_0) = m_0$$

and that the minimal value of the expected loss is

$$El = m_0^T S(t_0) m_0 + \operatorname{tr} S(t_0) R_0 + \sum_{t=t_0}^{N-1} \operatorname{tr} L^T(t)[\Gamma S(t+1)\Gamma + Q_2]L(t)R(t)$$
$$= m_0^T S(t_0) m_0 + \operatorname{tr} Q_0 R(N) + \sum_{t=t_0}^{N-1} \operatorname{tr} Q_1(t) R(t)$$

where

$$R(t+1) = \Phi R(t)\Phi^T + R_1(t)$$
$$R(t_0) = R_0.$$

Compare the result with that of the optimal closed loop system (Theorem 6.2). Also compare the technique of solving the problem with the technique used in Exercise 5.

Hint: Use Lemma 6.1.

7. Consider the system given by (6.1) with the criterion (6.13). Let the admissible control strategies be such that $u(t)$ is a function of $x(t-1)$. Find the optimal strategy and show that the minimal loss is given by

$$El = m^T S(t_0) m + \operatorname{tr} S(t_0) R_0 + \sum_{t=t_0}^{N-1} \operatorname{tr} S(t+1) R_1(t)$$
$$+ \sum_{t=t_0+1}^{N-1} \operatorname{tr} \Phi^T S(t+1)\Gamma L(t) R_1(t) + \operatorname{tr} \Phi^T S(t_0+1)\Gamma L(t_0) R_0$$

Use the result to find the proper way of assigning a value to a delay in obtaining the information about the value of the state.

Hint: Use Lemma 6.1.

8. Consider the problem with incomplete state information. Assume that the admissible strategies are such that $u(t)$ is a function of $\mathscr{Y}_{t-k}$ where k is a fixed number. Find the optimal strategy and the minimal expected loss.

7. CONTINUOUS TIME PROBLEMS

This section will consider the continuous time version of the linear quadratic control problem. To prove the separation theorem we will use the continuous time analog of the method used in Section 6.

Formulation

We will consider a system described by the stochastic differential equations

$$dx = Ax\,dt + Bu\,dt + dv \qquad (7.1)$$
$$dy = Cx\,dt + de \qquad (7.2)$$

where x is an $n \times 1$ state vector, u a $p \times 1$ control vector, and y an $r \times 1$ vector of outputs. The stochastic processes $\{v(t), t \in T\}$ and $\{e(t), t \in T\}$ are independent Wiener processes with zero mean values and the incremental covariances $R_1\,dt$ and $R_2\,dt$ respectively. A, B, C, R_1, and R_2 are matrices of appropriate dimensions whose elements may be piecewise continuous functions of time.

The initial state $x(t_0)$ is assumed normal with mean m and covariance R_0. The stochastic processes $\{v(t), t \in T\}$ and $\{e(t), t \in T\}$ are independent of $x(t_0)$. The matrices R_0 and R_1 are assumed to be symmetric and nonnegative definite, and R_2 symmetric and positive definite. In analogy with the discrete time case the criterion is chosen as the expected loss

$$E\left\{x^T(t_1)Q_0 x(t_1) + \int_{t_0}^{t_1} [x^T(t)Q_1 x(t) + u^T(t)Q_2 u(t)]\,dt\right\} \qquad (7.3)$$

The matrices Q_0 and Q_1 are symmetric and nonnegative definite and Q_2 is symmetric and positive definite. The admissible control strategies are such that the value of the control signal at time t is a function of the output signals which have been observed up to time t.

In analogy with the discrete time case we will consider two separate cases, complete state information and incomplete state information. Complete state information means that the state vector can be measured without error. As the system is governed by a stochastic differential equation, the state vector is a Markov process and the conditional distributions of future states given $x(t)$ are the same as the conditional distributions given $x(t)$ and other past values, $x(s)$, $s < t$. In the case of complete state information the admissible control strategy is such that $u(t)$ is a function of $x(t)$ and t. In the case of incomplete state information, the value of the control signal at time t, $u(t)$ is a function of $\mathcal{Y}_t = \{y(s), t_0 \leqslant s \leqslant t\}$. The stochastic control problem can now be stated as follows.

Continuous Time Problems

PROBLEM 7.1

Consider the system described by the stochastic differential Eqs. (7.1) and (7.2). Find an admissible control strategy such that the criterion (7.3) is minimal.

This problem is considerably more difficult than the corresponding discrete time problem. The reason for this is that the space spanned by the observed outputs is infinite dimensional. To solve the problem we will use an indirect method which closely resembles the one used in section 6. We will thus first derive an identity which can be used to give inequalities for the expected loss functions.

An Identity

The following result is the continuous time analogue of Lemma 6.1.

LEMMA 7.1

Assume that the Riccati equation

$$-\frac{dS}{dt} = A^T S + SA + Q_1 - SBQ_2^{-1}B^T S \qquad (7.4)$$

with the initial condition

$$S(t_1) = Q_0 \qquad (7.5)$$

has a solution which is nonnegative definite on the interval $t_0 \leqslant t \leqslant t_1$. Let x be the solution of the stochastic differential (7.1) then

$$x^T(t_1)Q_0 x(t_1) + \int_{t_0}^{t_1} [x^T(t)Q_1 x(t) + u^T(t)Q_2 u(t)] \, dt$$
$$= x^T(t_0) S(t_0) x(t_0) + \int_{t_0}^{t_1} (u + Q_2^{-1}B^T Sx)^T Q_2 (u + Q_2^{-1}B^T Sx) \, dt$$
$$+ \int_{t_0}^{t_1} tr \, R_1 S \, dt + \int_{t_0}^{t_1} dv^T Sx + \int_{t_0}^{t_1} x^T S \, dv \qquad (7.6)$$

Proof

We have

$$x^T(t_1) Q_0 x(t_1) = x^T(t_1) S(t_1) x(t_1) = x^T(t_0) S(t_0) x(t_0) + \int_{t_0}^{t_1} d(x^T Sx) \qquad (7.7)$$

As x is a solution of the stochastic differential (7.1), it does not have a time derivative. The differential $d(x^T Sx)$ thus does not obey the rules of ordinary calculus. Using the Ito differentiation rule, (Theorem 8.1 of Chapter 3) we get

$$d(x^T Sx) = dx^T Sx + x^T S \, dx + x^T \frac{dS}{dt} x \, dt + (tr \, SR_1) \, dt \qquad (7.8)$$

because the covariance of the increment dx equals $R_1 dt$. It follows from (7.1) that

$$x^T S\, dx = [x^T SAx + x^T SBu]dt + x^T S\, dv \qquad (7.9)$$
$$dx^T\, Sx = [x^T A^T Sx + u^T B^T Sx]dt + dv^T\, Sx \qquad (7.10)$$

Equation (7.4) gives

$$x^T \frac{dS}{dt} x\, dt = [-x^T A^T Sx - x^T SAx - x^T Q_1 x + x^T SBQ_2^{-1} B^T Sx]\, dt \qquad (7.11)$$

Equations (7.7), (7.8), (7.9), (7.10), and (7.11) give

$$\begin{aligned} d(x^T Sx) &= [u^T B^T Sx + x^T SBu - x^T Q_1 x + x^T SBQ_2^{-1} B^T Sx]\, dt \\ &\quad + \mathrm{tr}(R_1 S)\, dt + dv^T\, Sx + x^T S\, dv \\ &= [-u^T Q_2 u - x^T Q_1 x + (u + Q_2^{-1} B^T Sx)^T Q_2 (u + Q_2^{-1} B^T Sx)]\, dt \\ &\quad + \mathrm{tr}\, R_1 S\, dt + dv^T\, Sx + x^T S\, dv \end{aligned} \qquad (7.12)$$

where the last equality is obtained by adding and subtracting the quantity $u^T Q_2 u$. Reorganizing the terms of (7.12) we obtain (7.6) and the lemma is proven.

Using Lemma 7.1 we can now solve the problem of optimal control of the system (7.1) with the criterion (7.3) in different cases.

The Deterministic Case

For the purpose of making comparisons we will first consider the deterministic case when $v \equiv 0$. Lemma 7.1 give the following inequality for the loss function

$$\begin{aligned} x^T(t_1) Q_0 x(t_1) &+ \int_{t_0}^{t_1} [x^T(t) Q_1 x(t) + u^T(t) Q_2 u(t)]\, dt \\ &= x^T(t_0) S(t_0) x(t_0) + \int_{t_0}^{t_1} [u + Q_2^{-1} B^T Sx]^T Q_2^{-1} [u + Q_2^{-1} B^T Sx]\, dt \\ &\geqslant x^T(t_0) S(t_0) x(t_0) \end{aligned} \qquad (7.13)$$

where the equality is obtained for the control strategy

$$u = -Q_2^{-1} B^T Sx = -Lx \qquad (7.14)$$

The optimal strategy is unique because Q_2 is positive definite. It follows from Lemma 3.1 that the minimal value of the expected loss is

$$\min El = Ex^T(t_0) S(t_0) x(t_0) = m^T S(t_0) m + \mathrm{tr}\, S(t_0) R_0 \qquad (7.15)$$

Compare with (6.13)–(6.18) of Chapter 7.

Complete State Information

Taking mathematical expectation of (7.6) we get

$$E\left\{x^T(t_1)Q_0x(t_1) + \int_{t_0}^{t_1} [x^T(t)Q_1x(t) + u^T(t)Q_2u(t)]\,dt\right\}$$
$$= E\left\{x^T(t_0)S(t_0)x(t_0) + \int_{t_0}^{t_1} [u + Q_2^{-1}B^TSx]^TQ_2[u + Q_2^{-1}B^TSx]\,dt\right.$$
$$\left. + \int_{t_0}^{t_1} (\text{tr } R_1S)\,dt\right\} \geqslant m^TS(t_0)m + \text{tr } S(t_0)R_0 + \int_{t_0}^{t_1} (\text{tr } R_1S)\,dt$$
(7.16)

where equality is obtained for the control strategy

$$u = -Q_2^{-1}B^TSx = -Lx \tag{7.17}$$

In the case of complete state information, (7.17) is an admissible control strategy. The optimal strategy is thus given by (7.17) and the minimal value of the loss function is

$$\min El = m^TS(t_0)m + \text{tr } S(t_0)R_0 + \int_{t_0}^{t_1} (\text{tr } R_1S)\,dt \tag{7.18}$$

Incomplete State Information

Taking mathematical expectation of (7.6) we find

$$\min_u E\left\{x^T(t_1)Q_0x(t_1) + \int_{t_0}^{t_1} [x^T(t)Q_1x(t) + u^T(t)Q_2u(t)]\,dt\right\}$$
$$= m^TS(t_0)m + \text{tr } R_0S(t_0) + \int_{t_0}^{t_1} (\text{tr } R_1S)\,dt$$
$$+ \min_u E\left\{\int_{t_0}^{t_1} (u + Lx)^TQ_2(u + Lx)\,dt\right\} \tag{7.19}$$

We have further

$$E\left[\int_{t_0}^{t_1} (u + Lx)^TQ_2(u + Lx)\,dt\right]$$
$$= E_{y_t}\int_{t_0}^{t_1} E[(u + Lx)^TQ_2(u + Lx) \mid \mathscr{Y}_t]\,dt$$
$$= E_{y_t}\left[\int_{t_0}^{t_1} (u + L\hat{x})^TQ_2(u + L\hat{x})\,dt + \int_{t_0}^{t_1} (\text{tr } L^TQ_2LP)\,dt\right]$$
(7.20)

because the conditional distribution of $x(t)$ given $\mathscr{Y}_t$ is normal with mean $\hat{x}$ and covariance P. Since P does not depend on u we have

$$E\left[x^T(t_1)Q_0x(t_1) + \int_{t_0}^{t_1} (x^TQ_1x + uQ_2u)\,dt\right]$$
$$\geqslant m^TS(t_0)m + \text{tr } S(t_0)R_0 + \int_{t_0}^{t_1} (\text{tr } R_1S)\,dt$$
$$+ \int_{t_0}^{t_1} (\text{tr } L^TQ_2LP)\,dt \tag{7.21}$$

where equality is obtained for the control strategy

$$u = -L\hat{x} = -LE[x(t) \mid \mathcal{Y}_t] \qquad (7.22)$$

The optimal strategy is thus a linear function which expresses $u(t)$ as a function of the conditional mean $\hat{x}(t)$. Notice that the matrix L is the same matrix as is obtained in the deterministic case. Summing up we find Theorem 7.1.

THEOREM 7.1

Consider a system described by the stochastic differential (7.1) and (7.2). Let the admissible control strategy be such that the value of the control signal at time t is a function of the outputs observed up to time t. Assume that the Riccati equation (7.4) has a solution for $t_0 \leqslant t \leqslant t_1$. The control law

$$u = -L\hat{x}$$

where L is given by (7.14) and $\hat{x}$ is the conditional mean of $x(t)$ given $\mathcal{Y}_t$ then minimizes the criterion (7.3). The minimal value of the expected loss is

$$\min El = m^T S(t_0) m + \operatorname{tr} S(t_0) R_0 + \int_{t_0}^{t_1} (\operatorname{tr} SR_1) \, dt + \int_{t_0}^{t_1} (\operatorname{tr} SBQ_2^{-1} B^T SP) \, dt \qquad (7.23)$$

Remark

Notice that the different terms of the minimal value of the expected loss can be given interpretations which are analogous to those given in the discrete time case. The term $m^T S(t_0) m$ is thus the contribution due to the mean value of the initial state. The term $\operatorname{tr} S(t_0) R_0$ is the contribution due to the uncertainty of the initial state. The term $\int \operatorname{tr}(SR_1) \, dt$ is due to the disturbances which are forcing the system, and the last term of (7.23) is due to the uncertainty in the state estimate. Also compare (7.15) and (7.18).

Properties of the Closed Loop System

Combining the results of Theorem 7.1 of this Chapter and Theorem 6.2 of Chapter 7 we find that the optimal closed loop system is governed by the following equations

$$\begin{aligned} dx &= Ax \, dt + Bu \, dt + dv \\ dy &= Cx \, dt + de \\ d\hat{x} &= A\hat{x} \, dt + Bu \, dt + K[dy - C\hat{x} \, dt] \\ u &= -L\hat{x} \end{aligned}$$

Continuous Time Problems

Introducing x and $\hat{x}$ as state variables, we find that these equations reduce to

$$d\begin{bmatrix} x \\ \hat{x} \end{bmatrix} = \begin{bmatrix} A - BL & BL \\ 0 & A - KC \end{bmatrix} \begin{bmatrix} x \\ \hat{x} \end{bmatrix} dt + \begin{bmatrix} dv \\ dv - K\,de \end{bmatrix}$$

The dynamics of the closed loop system are thus determined by the dynamics of the optimal deterministic system $[A - BL]$ and the dynamics of the Kalman filter $[A - KC]$.

Exercises

1. Consider the system

   $$dx = u\,dt + dv$$
   $$dy = x\,dt + de$$

   where $\{v(t)\}$ and $\{e(t)\}$ are Wiener processes with the variance parameters r_1 and r_2. The initial state is normal with mean m and covariance r_0. Let the loss function be

   $$l = \int_0^T [x^2(t) + qu^2(t)]\,dt$$

   and let the purpose of the control be to minimize the average loss. Determine the optimal open loop control signal, the optimal control strategy when $e \equiv 0$, and the optimal strategy for incomplete state information.

2. Consider the problem in Exercise 1. Determine the transfer function of the steady state control law obtained when $T \to \infty$. Also determine the minimal values of the loss function in the different cases.

3. Consider the system

   $$dx = Ax\,dt + dv$$

   where $\{v(t),\ t \in T\}$ is a Wiener process with incremental covariance $R\,dt$ and the initial state $x(t_0)$ is normal with mean value m and the covariance matrix R_0. Show that

   $$E\left[x^T(t_1)Q_0 x(t_1) + \int_{t_0}^{t_1} x^T(s)Q_1 x(s)\,ds \right]$$
   $$= m^T S(t_0) m + \operatorname{tr} S(t_0)\,R_0 + \int_{t_0}^{t_1} \operatorname{tr} S(t) R(t)\,dt$$

   when

   $$-\frac{dS}{dt} = A^T S + SA + Q_1, \qquad S(t_1) = Q_0$$

4. Consider the system given by (7.1) and (7.2). Show that the loss function of the opimal open loop system is given by

$$m^T S(t_0) m + \text{tr } R(t_0) R_0 + \int_{t_0}^{t_1} \text{tr } R(s) R_1(s) \, ds$$

where

$$-\frac{dR}{dt} = A^T R + RA + Q_1$$

with the initial condition $R(t_1) = Q_0$.
Compare with the results for the optimal closed loop system.

5. Consider the system

$$dx = Ax \, dt + Bu \, dt + dv$$
$$dy = Cx \, dt + de$$

where $\{v(t)\}$ and $\{e(t)\}$ are Wiener processes with incremental variances $R_1 \, dt$ and $R_2 \, dt$ and the initial state is normal with mean m and covariance R_0. Find a control strategy which minimizes the expected loss

$$E\left[x^T(t_1) Q_0 x(t_1) + \int_{t_0}^{t_1} \sqrt{u^T(t) u(t)} \, dt \right]$$

The admissible control strategies are such that $u(t)$ is a function of the outputs observed up to time t.

6. Consider the system

$$dx = \begin{bmatrix} 0 & 1 \\ 0 & 1 \end{bmatrix} x \, dt + \begin{bmatrix} 0 \\ 1 \end{bmatrix} u \, dt + dv$$
$$dy = [1 \quad 0] x \, dt + de$$

where $\{v(t)\}$ and $\{e(t)\}$ are independent Wiener processes with incremental covariances $I \, dt$ and $r \, dt$. Let the loss function be

$$l = \int_{t_0}^{t_1} [x_1^2(t) + qu^2(t)] \, dt$$

Determine the control strategy which minimizes the expected loss El when the admissible strategies are such that $u(t)$ is permitted to be a functional of $\{y(s), t_0 \leqslant s \leqslant t\}$. Determine the limiting strategy as $t_0 \to -\infty$. Give a physical interpretation of the different terms of the minimal loss function.

8. BIBLIOGRAPHY AND COMMENTS

The concepts loss function, risk, and decision which are fundamental to stochastic control theory were first introduced in statistical decision theory. See

Wald, A., *Statistical Decision Functions*, Wiley, New York, 1950.
Lehman, E., *Testing Statistical Hypotheses*, Wiley, New York, 1952.
Blackwell, D. and Girshick, A., *Theory of Games and Statistical Decision*, Wiley, New York, 1954.

The problems analyzed in statistical decision theory are, however, usually confined to the static case discussed in Section 3.

Surprisingly enough the concept of randomized strategy has not yet been applied to stochastic control theory.

An early discussion of stochastic control problems is found in

Bellman, R., *Adaptive Control Processes*, Princeton Univ. Press, Princeton, New Jersey, 1961.

The certainty equivalence principle did first appear in economic literature. See

Simon, H. A., "Dynamic Programming under Uncertainty with a Quadratic Criterion Function," *Econometrica* **24**, 74 (1956).
Theil, H., "A Note on Certainty Equivalence in Dynamic Planning," *Econometrica* **25**, 346 (1959).

The discrete version of the separation theorem is proved in

Joseph, P. D. and Tou, J. T., "On Linear Control Theory," *Trans. AIEE* (Applications and Industry) **80**, 193-196 (1961).
Gunkel, T. L. III and Franklin, G. F., "A General Solution for Linear Sampled Data Control," *Trans. ASME J. Basic Eng.* **85-D**, 197-201 (1963).

Section 5 is based on

Åström, K. J., Koepcke, R. W., and Tung, F., "On the Control of Linear Discrete Dynamic Systems with Quadratic Loss," IBM Research Rep. RJ-222, September 1962.

The discrete time version of the separation theorem is also discussed in

Meier, L., "Combined Control and Estimation Theory," Report, Stanford Research Institute, California, 1965.
Gittelman, J. N., "Optimal Control of Discrete time Random Parameter Systems," Report 07303-1-T, Dept. EE, Systems Engineering Laboratory, University of Michigan, Ann Arbor, Michigan, July 1967.

The last mentioned report is of particular interest because it points out several obscurities in earlier proofs.

Continuous time versions of the separation theorem are found in

Potter, J. E., "A Guidance-Navigation Separation Theorem," MIT Exper. Astronom. Lab, Rep. RE-11, August 1964.
Striebel, C., "Sufficient Statistics in the Optimum Control of Stochastic Systems," *JMAA* **12**, 576-592 (1965).
Wonham, W. M., "On the Separation Theorem of Stochastic Control," *SIAM J. Control* **6**, (1968).

Wonham, W. M., "Random Differential Equations in Control Theory," in *Probabilistic Methods in Applied Mathematics*, A. T. Bharucha-Reid (editors), Academic Press, New York, 1969.

The identity (7.6) is a generalization of a result of Lagrange which he derived in connection with his work on the calculus of variations. See

Gelfand, I. M. and Fomin, S. V., *Calculus of Variations*, Prentice-Hall, Englewood Cliffs, New Jersey, 1963.

INDEX

A

admissible control strategy, 160, 162, 174, 258, 281
predictor, 162
Agniel, R.G., 157
analysis of dynamical systems, 91
of discrete time systems, 92
of continuous time systems, 104
Aoki, M., 12
Åström, K. J., 11, 89, 90, 157, 209, 293
Athans, M., 11
auto-correlation, 23
autocovariance function, 23
autoregressive process, 5, 15, 16, 22, 46

B

backward difference, 55, 56
difference operator, 198
band limited white noise, 31
Barnes, R. B., 87
basis weight control, 190, 191, 196, 198
basis weight loop, 194
Battin, R. H., 112
Bellman, R., 11, 12, 254, 293
Bellman equation, 265
Bernstein, S. N., 57, 87
Bertram, J. E., 158
Bharucha-Reid, A.T., 43, 89
Bode, H. W., 253
Bohlin, T., 209
Boltyanskii, V. G., 12
Boltzmann's constant, 70, 80
Box, G. E. P., 209
Brown, R., 20
Brownian motion, 19, 78
Bryson, A.E., 12, 254, 255

Bucy, R.S., 7, 12, 246, 253, 255

C

canonical form, 173
certainty equivalence principle, 8, 256, 257
characterization of disturbances, 5
characteristic equation, 182
Clark, J. M. C., 90
closed loop, 2
complete state information, 259, 260, 264, 286, 288
completely deterministic stochastic process, 5
completely uncorrelated process, 30
conditional distribution, 219
conditional expectation, 213
conditional mean, 8, 229
consistency, 14
continuity, 35
mean square, 35
Wiener process, 36
control law, 281
control error, 161
convergence concept, 33, 34
in probability one, 33
mean square, 33, 92
with probability one, 33
converter
A/D, 196
D/A, 196
correlation function, 24
cross-correlation function, 23
cross spectral density, 96
covariance, 15
covariance function, 18, 23, 24, 48, 64, 66
Cox, D, R., 42
Cramér, H., 43, 113
criterion, 173, 211

D

Davenport, W. B. Jr., 112, 253
dead-beat regulator, 159
  strategy, 180
decomposition of stationary processes, 28
determination of the transfer function, 97
deterministic case, 288
  control problem, 243
  control theory, 1
  output, 202
differentiability, 37
  mean square, 37
  Wiener process, 39
discrete parameter process, 14
Doob, J. L., 43, 87
dry basis weight, 198
dryer section, 191
duality, 238, 240, 242
duality theorem, 239, 243
dynamic programming, 257, 264, 269

E

Ekström, Å., 209
Einstein, A., 20, 87
environment, 172
equipartition law, 79
estimate, 211
estimator, 211
evaluation of performance, 115
  loss functions for continuous time systems, 128
  loss functions for discrete time systems, 116

F

Falb, P., 11
filtering problem, 210, 211
finite dimensional distribution, 14
Fisher's information matrix, 199
Fokker-Planck equation, 72
Fomin, S. V., 294
formal integration, 62
formal manipulations of stochastic d.e., 67
formulation of prediction and estimation problem, 211

forward difference, 56
shift operator q 164
Franklin, G. F., 293
Frost, P., 253
Fuller, A. T., 89
functional equation, 265, 269

G

Gamkrelidze, R. V., 12
Gaussian distribution, 17, 218
Gaussian process, 17
Gauss-Markov stochastic process, 151
Gelfand, I. M., 88, 89, 294
geometric interpretation of estimation problem, 221
Gikhman, I. I., 42, 57, 87, 88
Gittelman, J. N., 293
Gould, L. A., 112, 156
Gunkel, T. L., 293

H

Hall, A. C., 4
headbox, 191
Ho, Y. C., 12, 254

I

identification 196, 202
identification problem, 160, 188
identity, 168, 169, 278, 287
impulse response, 180
incomplete state information, 259, 261, 269, 274, 278, 286, 289
incremental covariance, 19
industrial application, 188
information matrix, 196
initial probability distribution, 18
innovations, 102, 170, 227, 231, 247
innovations representation, 102
input signal, 197
integral of a deterministic function, 58
integral of stochastic processes, 59
integrability, 40
integration by parts, 61
interpolation problem, 211
Ito, K., 43, 87, 89
Ito integral, 60
Ito differentiation rule, 74

J

James, H. M., 7, 12, 112, 156
Jenkins, G. M., 209
Johansen, D. E., 255
Joseph, P. D., 12, 255, 293
Jury, E. I., 157

K

Kailath, T., 114, 253, 254
Kalaba, R., 11
Kalman, R. E., 7, 12, 158, 228, 246, 253
Kalman-Bucy filtering algorithm, 158, 246
Kalman filter, 8, 116, 230, 250
Kalman filtering, 228, 257
Kaiser, J. F., 112, 156
Karlin, S., 42
Karhunen, K., 114
Karhunen-Loève expansion, 114
Koepcke, R. W., 293
Kolmogorov, A. N., 43, 252
Kolmogorov's theorem, 14
Kolmogorov backward operator, 72, 76
Kolmogorov forward operator, 72
Kolmogorov forward equation, 72
kraft paper, 189
k-step predictor, 166, 168, 169
Kushner, H. J., 89, 255

L

Laning, J. H., 112
Langevin equation, 70, 79
Lee, E. B., 12
Lee, Y. W., 253
Leadbetter, M. R., 43
least squares estimate, 199
Levy, P., 43, 57, 87
linear quadratic control problem, 286
linear stochastic control theory, 8
linear stochastic difference equation, 46, 47
linear stochastic differential equation, 54, 63
linearly singular process, 21
Loève, M., 43
loss function, 76, 115, 195, 199, 212
Luenberger, D. G., 158

M

Markov process, 18, 45
Markus, L., 12
maximum likelihood estimate, 195, 199
McKean, H. P., 43
mean value, 15, 66
mean value function, 18, 48, 64
mean square convergence, 35
Meier, L., 293
Miller, H. D., 42
minimal variance control strategy, 160, 172, 174, 175
minimum mean square prediction, 216
Mishchenko, E. F., 12
modeling stochastic d. e., 78
moisture control system, 194, 196
"most probable" prediction, 216
moving average, 15, 16, 22, 46, 169, 175

N

Nekolný, J., 157
Newton, G. C., 112, 156
Nichols, N. B., 7, 12, 112, 156
noise, 211
nonlinear stochastic d. e., 71
nonminimum phase, 182
normal, 17
normal processes, 17
null hypothesis, 200
numerical identification, 198

O

one step predictor, 163, 164, 165
open loop, 2
optimal control of deterministic processes, 2
optimal prediction, 162
 regulator, 159
 strategy, 160
optimization of performance, 115
orthogonal increments, 18, 113

P

Paley, R. E. A. C., 112
Paley-Wiener condition, 29

parameter optimization, 115
parameter set, 13
Parzen, E., 42
path, 14
Prabhu, N. U., 42
Pearson, J. D., 254
Peterka, V., 157
Phillips, R. S., 7, 12, 112, 156
PI-regulator, 180
Pontryagin, L. S., 12.
Potter, L. E., 293
predictor, 159, 202
prediction, 210
   error, 161, 164, 169, 195, 202
   problem, 211
   theory, 162, 210
process control, 188
process identification, 195
processes of second order, 18
processes with independent increments, 18
projection theorem, 223
properties of the closed loop system, 275, 290
pseudo-random binary signals, 198
purely deterministic process, 21
purely radom process, 30

Q

quality control, 188

R

Ragazzini, J. R., 253
random telegraph wave, 26
Rappaport, D., 255
rational spectrum, 99
   spectral density, 99, 128
realization, 14
reciprocal polynomial, 118, 166
reconstruction error, 143
reconstruction of state variables, 142, 150
refiner control, 196
regulator, 159
regulation error, 175
representation the processes, 232
representation theorem, 92, 101, 111
residuals, 195, 203
Riemann integrable, 41

Riesz-Fisher theorem, 35
Riccati equation, 7, 152, 153
Root, W. L., 112, 253
Routh-Hurwitz theorem, 157
Růžička, J., 157

S

sample function, 14
sample function space, 14
sampling interval, 194
sampling a stochastic d. e., 82
Schumitzky, A., 254
Schwartz inequality, 25
sensitivity of optimal system, 181
separation, theorem, 7, 159, 176, 256, 257
Shannon, C. E., 253
Sherman, S., 254
signal, 211
Silverman, S., 87
Silverman, L. M., 255
Simon, H. A., 12, 293
singular stochastic process, 5, 21, 26
singular Gaussian distribution, 218
Skorokhod, A. V., 42, 87, 88
smoothing problem, 211
Solodovnikov, V. V. 112
spectral density, 26, 27, 91, 96
   density function, 27
   distribution function, 27
   factorization, 92, 99
   factorization theorem, 92, 99, 108
   factorization of continuous time processes, 107
   factorization of discrete time processes, 98
stability properties, 3
state, 45
   models, 51, 91
   estimation, 212
   estimation for continuous time processes, 241
   estimation for discrete time systems, 225
static optimization problem, 259
stationary processes, 17, 95, 106
stochastic calculus, 74
   control problem, 210, 257
   control systems, 91
   control theory, 6, 188
   difference equations solutions, 47

Index

difference equation, 46
differential equation, 51, 54, 151
differential equation as a model for a physical process, 55
 integrals, 57
 optimal control, 159
 processes, 13, 35, 91
 state models, 92
Stratonovich, R. L., 88, 90, 255
Stratonovich integral, 60, 62
Strejc, V., 157
Striebel, C., 293
suboptimal strategies, 160, 182
sufficient statistic, 229

T

T (index set), 13
Theil, H., 12, 263, 293
time-series analysis, 193
Toma, M., 157
Tou, J. T., 12, 293
trajectory, 14
transfer function, 180
transition probability distribution, 18
Tung, F., 293
two step predictor, 165, 166

U

uncertainty of the initial state, 290
uncorrelate increments, 18

V

variance parameter, 20
vertical alignment, 154
Vidinčev, P., 157

W

Wax, N., 12, 87
weakly stationary processes, 17, 98
Wensmark, S., 209
wet basis weight, 198
Whittle, P., 209, 252
white noise, 29, 30, 92
Wilenkin, N. J., 88
Wiener, N., 12, 88, 112, 252
Wiener-Hopf equation, 7, 217
Wiener-Kolmogorov theory, 7
Wiener process, 19
Wold, H., 12, 87, 113, 252
Wonham, W. M., 293, 294
Wong, E., 90

Y

Yaglom, A. M., 43, 89, 253
Youla, D. C., 113
Yule, G. U., 12, 86, 87

Z

Zachrisson, L. E., 254
Zadeh, L. A., 253
Zakai, M., 90